Green Chemistry and Chemical Engineering

PROTON EXCHANGE MEMBRANE FUEL CELLS

CONTAMINATION AND MITIGATION STRATEGIES

T0179285

GREEN CHEMISTRY AND CHEMICAL ENGINEERING

Series Editor: Sunggyu Lee
Missouri University of Science and Technology, Rolla, USA

Proton Exchange Membrane Fuel Cells: Contamination and Mitigation Strategies
Hui Li, Shanna Knights, Zheng Shi, John W. Van Zee, and Jiujun Zhang

Proton Exchange Membrane Fuel Cells: Materials Properties and Performance
David P. Wilkinson, Jiujun Zhang, Rob Hui, Jeffrey Fergus, and Xianguo Li

Solid Oxide Fuel Cells: Materials Properties and Performance
Jeffrey Fergus, Rob Hui, Xianguo Li, David P. Wilkinson, and Jiujun Zhang

Efficiency and Sustainability in the Energy and Chemical Industries: Scientific Principles and Case Studies
Krishnan Sankaranarayanan, Jakob de Swaan Arons, and Hedzer van der Kooi

Green Chemistry and Chemical Engineering

PROTON EXCHANGE MEMBRANE FUEL CELLS

CONTAMINATION AND MITIGATION STRATEGIES

EDITED BY
HUI LI
SHANNA KNIGHTS
ZHENG SHI
JOHN W. VAN ZEE
JIUJUN ZHANG

CRC Press
Taylor & Francis Group
Boca Raton London New York

CRC Press is an imprint of the
Taylor & Francis Group, an **informa** business

CRC Press
Taylor & Francis Group
6000 Broken Sound Parkway NW, Suite 300
Boca Raton, FL 33487-2742

First issued in paperback 2019

ISBN-13: 978-1-4398-0678-4 (hbk)
ISBN-13: 978-0-367-38427-2 (pbk)

Library of Congress Cataloging-in-Publication Data

Proton exchange membrane fuel cells : contamination and mitigation strategies / editors, Hui Li ... [et al.].
    p. cm. -- (Green chemistry and chemistry engineering)
    Includes bibliographical references and index.
    ISBN 978-1-4398-0678-4 (alk. paper)
    1. Proton exchange membrane fuel cells. I. Li, Hui, 1964-

TK2931.P7846 2010
621.31'2429--dc22
                           2009045645

Visit the Taylor & Francis Web site at
http://www.taylorandfrancis.com

and the CRC Press Web site at
http://www.crcpress.com

Contents

Contents

Series Preface

The subjects and disciplines of chemistry and chemical engineering have encountered a new landmark in the way of thinking about, developing, and designing chemical products and processes. This revolutionary philosophy, termed *green chemistry and chemical engineering*, focuses on the designs of products and processes that are conducive to reducing or eliminating the use and/or generation of hazardous substances. In dealing with hazardous or potentially hazardous substances, there may be some overlaps and interrelationships between environmental chemistry and green chemistry. While environmental chemistry is the chemistry of the natural environment and the pollutant chemicals in nature, green chemistry proactively aims to reduce and prevent pollution at its very source. In essence, the philosophies of green chemistry and chemical engineering tend to focus more on industrial application and practice rather than academic principles and phenomenological science. However, as both a chemistry and chemical engineering philosophy, green chemistry and chemical engineering derives from and builds upon organic chemistry, inorganic chemistry, polymer chemistry, fuel chemistry, biochemistry, analytical chemistry, physical chemistry, environmental chemistry, thermodynamics, chemical reaction engineering, transport phenomena, chemical process design, separation technology, automatic process control, and more. In short, green chemistry and chemical engineering is the rigorous use of chemistry and chemical engineering for pollution prevention and environmental protection.

The Pollution Prevention Act of 1990 in the United States established a national policy to prevent or reduce pollution at its source whenever feasible. And adhering to the spirit of this policy, the Environmental Protection Agency (EPA) launched its Green Chemistry Program in order to promote innovative chemical technologies, which reduce or eliminate the use or generation of hazardous substances in the design, manufacture, and use of chemical products. The global efforts in green chemistry and chemical engineering have recently gained a substantial amount of support from the international community of science, engineering, academia, industry, and governments in all phases and aspects.

Some of the successful examples and key technological developments include the use of supercritical carbon dioxide as green solvent in separation technologies, application of supercritical water oxidation for destruction of harmful substances, process integration with carbon dioxide sequestration steps, solvent-free synthesis of chemicals and polymeric materials, exploitation of biologically degradable materials, use of aqueous hydrogen peroxide for efficient oxidation, development of hydrogen proton exchange membrane (PEM) fuel cells for a variety of power generation needs, advanced biofuel productions, devulcanization of spent tire rubber, avoidance of the use of chemicals and processes causing a generation of volatile organic compounds (VOCs), replacement of traditional petrochemical processes by microorganism-based bioengineering processes, replacement of chlorofluorocarbons (CFCs) with nonhazardous alternatives, advances in design of energy efficient processes, use of clean, alternative, and renewable energy sources in manufacturing, and much more. This list, even though it is only a partial compilation, is undoubtedly growing exponentially.

This book series on Green Chemistry and Chemical Engineering by CRC Press/Taylor & Francis is designed to meet the new challenges of the twenty-first century in the chemistry and chemical engineering disciplines by publishing books and monographs based on cutting-edge research and development to the effect of reducing adverse impacts on the environment by chemical enterprise. And, in achieving this, the series will detail the development of alternative sustainable technologies, which will minimize the hazard and maximize the efficiency of any chemical choice. The series aims at delivering the readers in academia and industry with an authoritative information source in the field of green chemistry and chemical engineering. The publisher and the series editor are fully aware of the rapidly evolving nature of the subject and its long-lasting impact on the quality of human life in both the present and future. As such, the team is committed to making this series the most comprehensive and accurate literary source in the field of green chemistry and chemical engineering.

Sunggyu Lee

Preface

Proton exchange membrane (PEM) fuel cells have drawn a great deal of attention in recent years due to their high efficiency, high energy density, and low or zero emissions. The rapid development of such clean energy-converting devices has been driven by several important applications including transportation, stationary and portable power, and micropower. Although great progress has been made in the research and development of PEM fuel cells in terms of stack power density increases and cost reduction, there are still two major technical challenges hindering commercialization: insufficient reliability/durability and high cost. Many failure modes of the fuel cell have been identified regarding the low reliability/durability, such as the degradation caused by contamination. This contamination effect is mainly induced by fuel and/or oxidant impurities as well as metal ions brought along with the fuel and air feed streams and cell components into the anode and cathode that can considerably degrade the fuel cell.

There are three major areas that have been addressed at the current stage of research for fuel cell contamination study: (1) theoretical and empirical modeling of contamination to provide a fundamental understanding of the mechanisms, (2) experimental observation and validation, and (3) contamination mitigation. In recent years, work toward identifying the potential impacts of contamination, understanding the contamination mechanisms, and developing mitigation strategies has drawn a great deal of attention.

In general, most of the published work has focused on the influence of individual and particular contaminants on PEM fuel cells. A general and comprehensive overview covering every aspect of fuel cell contamination from fundamentals to applications in a detailed level has not yet appeared in literature. Therefore, in order to identify the problems and gain knowledge and a fundamental understanding of contamination mechanisms, it is necessary to obtain as broad a range of updated and detailed information as possible. Then, based on validated mechanisms, effective control strategies of contamination can be developed to improve the reliability and durability of PEM fuel cells.

In order to facilitate PEM fuel cell research and development (R&D) and to accelerate sustainable commercialization, a comprehensive and in-depth book that focuses on both fundamental and application aspects of PEM fuel cell contamination is definitely needed. With high enthusiasm and great efforts from a group of fuel cell scientists and engineers with excellent academic records and strong industrial fuel cell expertise, this book is achieved.

In *Proton Exchange Membrane Fuel Cells: Contamination and Mitigation Strategies*, the nature, sources, and electrochemistry of contaminants, their impacts on fuel cell performance and lifetime, the poisoning mechanisms of contamination, and contamination mitigation strategies are reviewed in a broad scope. The major findings from both experimental and theoretical studies in contamination-related research, the methods or tools developed to diagnose various contamination phenomena, and the existing strategies to mitigate the adverse effects of contamination are summarized. Additionally, key issues in the future R&D of fuel cell contamination and control are also discussed. The comprehensive information on PEM fuel cell contamination is separated into nine chapters. Chapter 1 introduces the PEM fuel cell principles including fuel cell thermodynamics, kinetics, and operation, as well as a high-level review of the contamination issues. Chapter 2 describes the contamination sources and their corresponding electrochemical reactions in PEM fuel cell operation environments. Chapter 3 discusses the contamination issues at the cathode side of PEM fuel cells as well as some impurities from fuel cell components that degrade the cathode. In chapter 4, anode contamination is reviewed from both theoretical and experimental aspects. Chapter 5 mainly focuses on the membrane contamination issues. In terms of PEM fuel cell contamination modeling and model validation, chapters 6, 7, and 8 are designed to focus on the cathode side, anode side, and membrane, respectively. Finally, in chapter 9, contamination mitigation strategies are reviewed.

Because this book contains the latest research and development as well as important new directions in PEM fuel cell contamination, we believe that readers will be able to easily find numerous figures, photos, and data tables as well as comprehensive reference materials throughout. Each chapter is also designed to be relatively independent of the others. We hope this structure will help readers quickly find topics of interest without necessarily having to read through the entire book. However, there is unavoidably some overlap, reflecting the interconnectedness of the research and development in this dynamic field. We deeply hope that industry researchers, scientists, engineers, and postsecondary students working in the areas of energy, electrochemistry science/technology, fuel cells, and electrocatalysis will use this book.

We would like to acknowledge with appreciation the support from the PEM Fuel Cell Contamination Consortium led by the Institute for Fuel Cell Innovation, National Research Council of Canada (NRC-IFCI). This consortium currently has four members (NRC-IFCI, Ballard Power Systems Inc., Hydrogenics Corporation, and Angstrom Power Inc.) Special thanks go to Maja Veljkovic, Dr. Dave Ghosh, Dr. Yoga Yogendran, Dr. Simon Liu, Dr. Francois Girard, Dr. Haijiang Wang, Dr. Jun Shen, and Jeremy Rose from NRC-IFCI; Dr. Silvia Wessel and Nengyou Jia from Ballard; Dr. David Frank, Rami Abouatallah, and Nathan Joos from Hydrogenics; and Dr. Jeremy Schrooten from Angstrom Power for their strong support, encouragement, and suggestions.

If technical errors exist in this book, all co-editors and the authors would deeply appreciate the readers' constructive comments for correction and further improvement.

Hui Li
Shanna Knights
Zheng Shi
John W. Van Zee
Jiujun Zhang

Editors

Hui Li, PhD, is a research associate and PEMFC (proton exchange membrane fuel cells) Contamination Consortium project leader at the National Research Council of Canada Institute for Fuel Cell Innovation (NRC-IFCI). Dr. Li received her BS and MSc in chemical engineering from Tsinghua University (Beijing) in 1987 and 1990, respectively. After completing her MSc, she joined Kunming Metallurgical Institute as a research engineer for four years and then took a position as an associate professor at Sunwen University (then a branch of Zhongshan University/Guangzhou) for eight years. In 2002, she started her PhD program in electrochemical engineering at the University of British Columbia. After obtaining her PhD in 2006, she did one term of postdoctoral research at the Clean Energy Research Centre (CERC) at the University of British Columbia.

In 2007, she joined the low-temperature PEMFC Group at NRC-IFCI, under the supervision of Dr. Jiujun Zhang. Dr. Li has years of research and development experience in theoretical and applied electrochemistry and in electrochemical engineering. Her research is based on PEMFC contamination testing and contamination mitigation, preparation and development of electrochemical catalysts with long-term stability, catalyst layer/cathode structure, and catalyst layer characterization and electrochemical evaluation. Dr. Li has co-authored more than 20 research papers published in refereed journals and has one technology licensed to Mantra Energy Group. She has also produced many industrial technical reports.

Shanna Knights, P Eng, MASc, received her bachelor's and master's degrees in chemical engineering from the University of British Columbia (Vancouver). She started her career in 1988 at British Columbia Research Inc., Vancouver, where she worked as a research engineer and project manager for a variety of contract research projects related to waste management and chemical process and reaction engineering. Knights moved to Ballard Power Systems in 1995, where she is currently Program Manager, MEA Technology, in the R&D department. Her fuel cell research areas cover durability, reliability, performance, and operational behavior, including:

- Research conducted to elucidate the major failure mechanisms occurring under field trial operation including design, operational aspects, and degradation characterization, and provided guidance on operating limits for safety and degradation. Applications worked on included stationary power generation, cogeneration, bus, and light duty automotive.

- Conducted and led several research activities focused on failure modes including mechanistic understanding, accelerated stress test development, influence of operational conditions, control strategies, failure mitigation, and predictive degradation relationships.
- Failure modes studied included fuel starvation and cell reversal, partial fuel starvation, carbon corrosion, membrane thinning and hole formation; oxidant starvation; excessive drying and flooding, air and fuel contamination, freeze-start and freeze-thaw degradation.
- Conducted and led several research activities focused on performance including water management, performance effects of operational conditions and operating strategies, performance relationships as a function of design, novel cell designs for improved performance, and modeling and analysis of cell performance to target areas for performance improvement.

Zheng Shi, PhD, is a research officer at the National Research Council of Canada Institute for Fuel Cell Innovation. Dr. Shi received her BSc in chemistry from Shandong Normal University (Jinan City, China) in 1982 and her PhD in physical chemistry from Dalhousie University (Halifax, Nova Scotia) in 1990. Dr. Shi has been conducting theoretical modeling and simulation in different fields with first principle methods, molecular dynamic methods, and kinetic model. When working at Simon Fraser University (Burnaby, British Columbia), Dr. Shi focused on simulation of electron momentum distribution and conducted thermodynamic and kinetic studies of organic reactions, including enzyme catalyzed hydrolyses of esters, amides, and β-lactam. While working at the University of British Columbia (Vancouver), Dr. Shi concentrated on protein structure simulation and docking with molecular dynamic methods. She also worked on pharmacophore models and structure activity relationship development pertinent to drug discovery. Dr. Shi joined the Institute for Fuel Cell Innovation in 2004. Currently, she is working in the areas of fuel cell catalysis, fuel cell durability, and contamination. For fuel cell catalysis, Dr. Shi's research focuses on fundamental understanding of the electrocatalysis reaction mechanism with theoretical simulation and development of structure activity relationship. For durability and contamination research, Dr. Shi concentrates on mechanism understanding, and fuel cell performance simulation and prediction with transient kinetic model.

John W. Van Zee, PhD, is a professor in the Department of Chemical Engineering at the University of South Carolina (Columbia).

During 2001 and 2002, Dr. Van Zee created the basis for the National Science Foundation (NSF) Center for Fuel Cells and he has served as its director since it was officially established in 2003 as part of the Industry/University Cooperative Research Centers program. This is NSF's only Center for Fuel

Cells. In 2009, the center expanded to include the University of Connecticut (Storrs). As director, he has engaged 24 dues-paying companies to work with multiple professors and graduate students on precompetitive research.

In 2006 and 2007, Dr. Van Zee served as the founding director of the Future Fuels™ research initiative at USC during which he established international research relationships for professors and students with the Fraunhofer Institute for Solar Energy and the Korea Institute for Energy Research. Prior to working on fuel cells, he consulted for Milliken & Company and worked with researchers at the Naval Undersea Warfare Center in Newport, Rhode Island.

Dr. Van Zee enjoys teaching both undergraduate and graduate courses and maintaining a funded research program. He has graduated 19 PhD and 11 MS students, served the Electrochemical Society and America Institute of Chemical Engineers, co-edited 7 books, published over 120 papers in journals and conference proceedings, and produced 18 invention disclosures and 3 U.S. patents.

Jiujun Zhang, PhD, is a senior research officer and PEM (proton exchange membrane) Catalysis Core Competency leader at the National Research Council of Canada Institute for Fuel Cell Innovation (NRC-IFCI). Dr. Zhang received his BS and MSc in electrochemistry from Peking University in 1982 and 1985, respectively, and his PhD in electrochemistry from Wuhan University (Hubei Province, China) in 1988. Beginning in 1990, he carried out three terms of postdoctoral research at the California Institute of Technology (Caltech/Pasadena), York University (Toronto), and the University of British Columbia (Vancouver).

Dr. Zhang has over 27 years of research and development experience in theoretical and applied electrochemistry, including over 13 years of fuel cell R&D (among these two years at Caltech, six years at Ballard Power Systems, and five years at NRC-IFCI), and three years of electrochemical sensor experience. He holds several adjunct professorships, including one at the University of Waterloo (Ontario) and one at the University of British Columbia. His research is mainly based on fuel cell catalysis development. Up to now, Dr. Zhang has co-authored more than 200 publications including 150 refereed journal papers and 4 edited books. He also holds over 10 U.S. patents and patent publications.

Contributors

S. R. Dhanushkodi
Department of Chemical
   Engineering
University of Waterloo
Waterloo, Ontario, Canada

M. W. Fowler
Department of Chemical
   Engineering
University of Waterloo
Waterloo, Ontario, Canada

Cunping Huang
Florida Solar Energy Center
University of Central Florida
Cocoa, Florida

Xinyu Huang
Florida Solar Energy Center
University of Central Florida
Cocoa, Florida

Brian Kienitz
Los Alamos National Laboratory
Los Alamos, New Mexico

Shanna Knights
Ballard Power Systems
Burnaby, British Columbia, Canada

Hui Li
NRC Institute for Fuel Cell
   Innovation
Vancouver, British Columbia,
   Canada

Xianguo Li
Professor in Mechanical and
   Mechatronics Engineering
University of Waterloo
Waterloo, Ontario, Canada

Zhong-Sheng Liu
NRC Institute for Fuel Cell
   Innovation
Vancouver, British Columbia,
   Canada

Jianxin Ma
School of Automotive Studies
Clean Energy Automotive
   Engineering Center
Tongji University
Shanghai, China

A. G. Mazza
Faculty of Engineering and
   Applied Science
University of Ontario Institute
   of Technology
Oshawa, Ontario, Canada

M. D. Pritzker
Department of Chemical
   Engineering
University of Waterloo
Waterloo, Ontario, Canada

Jinli Qiao
School of Automotive Studies
Clean Energy Automotive
   Engineering Center
Tongji University
Shanghai, China

Marianne Rodgers
Florida Solar Energy Center
University of Central Florida
Cocoa, Florida

Zheng Shi
Institute for Fuel Cell Innovation
National Research Council
  of Canada
Vancouver, British Columbia,
  Canada

Datong Song
Institute for Fuel Cell Innovation
National Research Council
  of Canada
Vancouver, British Columbia,
  Canada

Thomas E. Springer
Los Alamos National Laboratory
Los Alamos, New Mexico

Daijun Yang
School of Automotive Studies
Clean Energy Automotive
  Engineering Center
Tongji University
Shanghai, China

Nada Zamel
PhD candidate in Mechanical and
  Mechatronics Engineering
University of Waterloo
Waterloo, Ontario, Canada

Jianlu Zhang
Institute for Fuel Cell Innovation
National Research Council of
  Canada
Vancouver, British Columbia,
  Canada

Jiujun Zhang
Institute for Fuel Cell Innovation
National Research Council
  of Canada
Vancouver, British Columbia,
  Canada

1

PEM Fuel Cell Principles and Introduction to Contamination Issues

Shanna Knights

CONTENTS

1.1 Introduction

Fuel cells have emerged as a vital alternative energy solution to reduce societal dependence on internal combustion engines and lead acid batteries. Fuel cells promise significantly improved energy efficiency with zero or low greenhouse gas emissions, and they are expected to play a key role in the hydrogen economy. While these benefits provide reduced fuel consumption and environmental impacts, fuel cells also offer several other advantages that facilitate their use in a wide range of applications. Advantages over competing technologies include, for example, quiet operation, improved power density compared to batteries, and modularity, i.e., similar efficiencies for small and large units. Commercial use of fuel cells for some applications is a reality today; however, key challenges remain before achievement of wide-spread implementation of fuel cells across a multitude of uses. Some of these challenges will be naturally overcome with increased fuel cell use and production volumes. Government incentives are enabling demonstration projects to reduce the initial market reticence for unproven technology. The operating experience being gained has highlighted opportunities for improved designs to be implemented in successive iterations of fuel cells. As production volumes increase, manufacturing processes arc being improved, and costs are being reduced.

Governments and private industry around the world, including, but not limited to, Japan, the European Union, Canada, and the United States, are supporting fuel cell research and development. Motivation for the support provided by the U.S. Department of Energy (DOE) hydrogen program for research, development, and demonstration of several types of fuel cells, is the desire to transition to hydrogen fuel to meet at least a portion of the U.S. energy needs. The broad range of feedstocks and methods for hydrogen production, such as nonrenewables like natural gas reforming and coal by gasification, and renewables, such as solar, wind, biomass, geothermal energy, and nuclear energy, is attractive and will reduce U.S. dependence on foreign oil requirements and significantly reduce greenhouse gas (GHG) emissions [1]. Similar motivation drives other countries, with additional drivers also coming into play, such as the desire to reduce the need for building new large power plants through the use of fuel cells for distributed power generation or in-home co-generation, and the need for increased reliability of the telecommunications grid, through the use of fuel cells for back-up power.

Fuel cells are currently at the early commercial stage for some applications. There are increasing numbers of units deployed in field trials for an increasing number of applications. The polymer electrolyte membrane (PEM) fuel cell is one of the most common types of fuel cells under development today. (They also are commonly referred to as *proton exchange membrane* fuel cells based on the key characteristic of the solid electrolyte membrane to transfer protons from the anode to the cathode.) With the experience gained through both the field trial demonstrations and through fuel cell research, the key challenges have been identified. The two largest challenges for fuel cells include cost and durability. These two areas are interrelated, and it becomes more difficult to increase or maintain durability while reducing cost of the materials and designs. This book deals with contamination of the fuel cell, the impacts of which contribute to reduced durability. The area of contamination is becoming more important at this early commercialization stage as cheaper fuels, systems, and component materials are required. Introduction of nondesirable species into the fuel cell, or movement of species from one fuel cell component to another, can result in contamination with negative effects. There are various modes for the introduction of the contaminants, as well as several types of effects of the contaminants, which will be discussed in general terms in this chapter, and in more detailed, specific terms in the remainder of this book.

1.1.1 What Is a Fuel Cell?

The most basic description of a fuel cell is that of an electrochemical cell that has reactants supplied from an external source. An electrochemical cell consists of two electrodes, an anode and a cathode, to which reactants (oxidant and reductant, normally referred to as the fuel) are supplied and then react chemically to either produce (as in the case of a fuel cell) or consume (as in the case of water electrolysis) electrical energy. This can be contrasted to a battery, which is also an electrochemical cell, but in which the reactants are contained within the cell. Once the reactants are consumed in the energy-producing reaction, the battery must either be recharged through application of an external electrical energy supply or replaced with a new battery. In contrast, a fuel cell has reactants supplied externally and will operate continuously as long as reactants are available, or the materials used in construction degrade or fail. In this sense, fuel cells can also be considered as an energy-converting device, converting chemical energy into electrical energy. The most common oxidant is the oxygen contained in air, as this is readily available. The most common fuel for PEM fuel cells (PEMFC) is hydrogen, which may be supplied as almost pure hydrogen or as a major component (50 to 70%) of a reformed fuel stream. Alternatively, a hydrogen-containing component may be used, such as methanol in a direct methanol fuel cell.

1.1.2 Benefits of Fuel Cells

1.1.2.1 Energy Efficiency

A key benefit of fuel cells is that of theoretically high-energy efficiency when compared to conventional chemical combustion systems. When used in combined heat and power generation systems, particularly for the high temperature fuel cells, very high efficiencies can be achieved.

The free energy change of a chemical reaction, as designated by Gibbs free energy, ΔG, is the maximum useful work obtainable or the maximum amount of electricity that can be gained from the reaction. The free energy for an electrochemical reaction is given by

$$\Delta G = -nFE \tag{1.1}$$

or for the case when all reactants and products are in their standard states,

$$\Delta G^0 = -nFE^0 \tag{1.2}$$

In terms of the reaction energy, one term, called the reaction enthalpy, ΔH, is used to describe the heat energy delivered by the chemical reaction. To measure the change in the order of the system during the reaction, another term, called the reaction entropy, ΔS, is also defined [2].

The free energy of the reaction can be related to ΔH, ΔS, and the reaction temperature (T) by equation 1.3:

$$\Delta G = \Delta H - T\Delta S \tag{1.3}$$

Equation 1.3 indicates that, for an electrochemical reaction, part of the reaction energy (ΔH) is used to generate electrical energy (ΔG), and the other part is used to produce heat ($T\Delta S$). In a fuel cell system, the most useful energy is electrical energy, while the heat produced is sometimes not desired. Therefore, electrical efficiency (or the reversible or thermodynamic efficiency), η_{rev}, can be defined as the ratio of the maximum electrical energy from the cell reaction to the reaction enthalpy. This represents the theoretical upper limit for fuel cell electrical efficiency.

$$\eta_{rev} = \frac{-\Delta G}{-\Delta H} \tag{1.4}$$

The conventional method of calculating energy efficiency utilizes the enthalpy value, based on the maximum energy you could obtain from a fuel by burning it. The Gibbs energy, at –237.3 kJ/mol, is smaller than the enthalpy value, at 286 kJ/mol, such that a hydrogen fuel cell operating at 25°C has a maximum theoretical efficiency of 83%, assuming the product water is at a liquid state. For comparison, the maximum theoretical efficiency for a combustion engine at 500°C is 58% based on heat rejection at 50°C [3].

Alternatively, the thermal cell voltage is defined as

$$E_H^0 = \frac{-\Delta H}{nF} \tag{1.5}$$

The ideal efficiency for complete conversion of electrochemical energy conversion is

$$\eta_{th} = \frac{E^0}{E_H^0} \tag{1.6}$$

where E^0 is the thermodynamic cell voltage.

The thermal cell voltage at 25°C is 1.48 V for liquid product, again resulting in a maximum theoretical efficiency of 83% for 1.23 V/1.48 V.

The maximum efficiency for a fuel cell will depend on the operating temperature and whether the products are liquid or gaseous. The maximum PEM fuel cell reaction with liquid product has a reversible efficiency of 83%, whereas it rises to 95% with gaseous product.

For power production, of course, the actual cell voltage will be less than E^0, and this value will be used to calculate the load efficiency. For example, to achieve an approximately 50% efficiency requires a cell operating at a voltage of 0.74 V.

The remaining 50% of the energy is produced as heat, and can be used for water or space heating applications, such as in a combined heat and power (CHP) system. This can result in much higher fuel efficiencies compared to the electrical output alone. However, the overall efficiency must also take into account fuel losses and parasitic losses, such as power requirements for fuel and air compressors.

1.1.2.2 Greenhouse Gas Reduction

An important aspect of fuel cell introduction is the reduction in the amount of GHG emissions. Many studies have been completed that show the positive benefit of fuel cells. The amount of greenhouse gas reduction is strongly dependent on the fuel source. The ultimate greenhouse gas reduction will result from the use of renewable energy to produce the hydrogen. In this case, zero greenhouse gases are produced. Even with the use of conventional fuels, the high efficiency of fuel cells can result in a net savings of greenhouse gases.

A study on the GHG emissions reductions predicted for hydrogen fuel cell vehicles in the United States was conducted, with the finding that 50% reduction was achievable for fuel cells when using natural gas feedstock, compared to comparable gasoline vehicles. These reductions are lower for hydrogen delivery by truck, or if produced by electrolysis using grid power [4].

An analysis of two important early markets, forklift propulsion systems and distributed power generation, was conducted at Argonne National Laboratory [5] to compare the potential energy and environmental implications of fuel cell systems to incumbent technologies. The study results indicated lower GHG emissions compared to those associated with the

U.S. grid electricity and alternative distributed combustion technologies. When compared to the ICE (internal combustion engine)-powered forklifts, fuel cell forklifts offer significant advantages in reducing GHG emissions. Battery-powered forklifts that rely on the average U.S. electricity mix produce almost as much GHG emissions as the ICE engines. Savings for both fuel cell- and battery-powered forklifts are realized by the use of renewable energy sources, such as wind power. The use of natural gas converted to hydrogen by steam reforming for fuel cells provides similar GHG levels to battery forklifts based on the California energy mix, which includes a significant portion of nonfossil fuels.

The analysis for distributed power generation indicates that fuel cell technologies will produce lower GHG emissions than those produced by the U.S. grid mix of technologies and all other distributed generation technologies. This reduction in lower pollutant emissions than combustion generators, combined with the opportunity for reduced dependence on petroleum oil, indicates the potential for fuel cells to readily penetrate the distributed electricity market.

The Institute of Energy Research Systems Analysis and Technology Evaluation in Germany conducted a study on the use of fuel cells for CHP residential systems [6]. Only 100 units of primary energy were required for the fuel cell system compared to 153 units of primary energy for conventional separate heat and power generation to produce 30 units of electricity and 60 units of heat. Based on this, and an analysis of possible market penetration levels of the fuel cell CHP units, a significant reduction of GHG from 1 to nearly 4% of the emissions in the residential sector of the Organization for Economic Cooperation and Development (OECD) was predicted. A study conducted on the potential for stationary fuel cell markets in Germany [7] concluded that the main driver for large-scale commercialization in the country is dependent on the introduction of ambitious CO_2 reduction targets. Under this scenario, a large potential for fuel cell commercialization exists. In the medium term, fuel cells can play an important role in efficiently providing heat and electricity from fossil fuels.

In another study, this one on Canadian implementation of a hydrogen economy, with the longer term implementation of renewable hydrogen sources, e.g., wind and solar, the use of fuel cells is an integral part of such a scenario that will result in significant savings of GHG emissions to meet the thermal needs of Canadian households across all provinces, and savings in most Canadian provinces for electrical requirements, such as lighting and operating appliances [8].

1.1.3 Brief History of Fuel Cells

Although fuel cells are considered a modern, leading edge technology, Sir William Robert Grove first demonstrated them in the nineteenth century [9]. Grove built what he called a gas battery in 1839, a device that combined hydrogen and oxygen to produce electricity. He immersed the ends of two platinum electrodes in sulfuric acid and each of the other two ends in separate sealed containers of oxygen and hydrogen, and combined several of these cells in a series circuit to produce electricity. His invention generated significant interest at the time, but this was eclipsed by the soon-to-be-discovered steam engine, and the arrival of the era of cheap fossil fuels. Fuel cell development continued on and off for the next century, with new electrolytes and approaches discovered by a number of different researchers, including precursors to the molten carbonate and solid oxide fuel cell devices of today. During World War II, British engineer Francis Bacon investigated the use of alkali electrolyte, and settled on the use of potassium hydroxide. Although there is no record of fuel cells being used during the war, these advancements led to renewed interest, and in 1959, Bacon developed a 5 kW stationary fuel cell using potassium hydroxide as the electrolyte.

The first fuel cell-powered vehicle in the world was the Allis Chalmers 15 kW fuel cell-powered tractor, built in 1959 by a team led by Harry Ihrig. It was powered by an alkaline fuel cell system, and in October 1959, it successfully ploughed a field of alfalfa in West Allis, Wisconsin. After the demonstration, the Allis Chalmers Manufacturing Company donated the tractor to the Smithsonian.

Fuel cell technology research and development was significantly accelerated by the inclusion of the technology by NASA for the first prolonged manned flight into space. The orbiter needed a source of electricity that could last eight days in space, and both batteries and photovoltaics of the time were too big and heavy. An added advantage was that the astronauts could also consume the fuel cell's water by-product. The problem of cost was not an issue for the space program, and the spacecraft was already carrying liquid hydrogen and oxygen, so the fuel supplies were not a problem. In the 1950s, General Electric Company (GE) introduced the use of a solid ion-exchange membrane as the electrolyte, and developed a method to deposit a Pt (platinum) catalyst onto the membrane. The Gemini missions used fuel cells based on proton exchange membranes (PEM). However, the subsequent Apollo missions and today's Space Shuttle program rely on alkaline fuel cells for electricity and drinking water once in orbit. United Technologies Corporation (UTC) was established in 1958 as a division of Pratt & Whitney, and has been supplying fuel cells to the space program since that time. UTC also first commercialized large-scale (200 kW) stationary power generation systems for use as co-generation power plants in hospitals, universities, and large office buildings.

As a result of the successes in the space program, industry began to recognize in the 1960s the commercial potential of fuel cells for other areas, but encountered technical barriers and high investment costs. Fuel cells were not economically competitive with existing energy technologies. Ballard Power Systems, originally founded to conduct research and development on high-energy lithium batteries, began development work on fuel cells in the 1980s, with initial work funded by the Canadian government. The company decided to focus on PEM fuel cells and fuel cell systems in 1989. In 1993, Ballard succeeded in building a fuel cell-powered bus for demonstration purposes. Ballard engaged in development and demonstration projects in a number of areas, both motive and stationary. In 2000, Ballard opened the first fuel cell volume manufacturing facility. The company's first commercial product, the Nexa™ power module, the world's first volume-produced PEM fuel cell, was launched in 2001.

Fuel cell technology is currently seeing large growth, with a variety of fuel cell products established in niche and luxury market sectors over the past three years. According to *Fuel Cell Today*, commercialization of the fuel cell industry is here, with 2007 being a significant year in terms of increase in manufacturing capability, decreasing costs, and increasing number of original equipment manufacturers (OEMs) involved. Over the three years prior to 2008, industry experienced an annual growth rate of 59% in fuel cell units delivered, with some 12,000 new units shipped during 2007 [10].

1.1.4 Types of Fuel Cells

There are several different types of fuel cells, which are classified primarily by the electrolyte used. This in turn drives requirements around operating temperature range, oxidant and fuel used, types of catalyst, materials requirements, and tolerance to contaminants. Each type of fuel cell may be best suited for certain applications, and will have particular technology challenges to overcome. Table 1.1 from the U.S. DOE H_2 program provides a comparison of the main fuel cell types [11].

1.1.4.1 Polymer Electrolyte Membrane (PEM)

1.1.4.1.1 Hydrogen as Fuel

PEM fuel cells have emerged as the most common type of fuel cell under development today. As stated above, they also are commonly referred to as *proton exchange membrane* fuel cells based on the key characteristic of the solid electrolyte membrane to transfer protons from the anode to the cathode. The solid electrolyte avoids problems caused by liquid electrolytes used in other systems, and the temperature range of <100°C enables rapid start-up under low temperature operation, with operation possible down to subfreezing temperatures. The lower temperature also allows a wider range of materials to be used and enables relatively easy stack design in terms of sealing issues and material selection. This type of fuel cell is the most feasible for use under transportation applications.

The highly acidic membrane necessitates the use of very stable highly reactive catalysts, with platinum (Pt) being the only one in use today sufficiently active to achieve required performances. The fuel used can either be pure hydrogen

TABLE 1.1

Comparison of Fuel Cell Technologies

Fuel Cell Type	Common Electrolyte	Operating Temp.	System Output	Electrical Efficiency	Combined Heat and Power (CHP) Efficiency	Applications	Advantages
Polymer Electrolyte Membrane (PEM)	Solid organic polymer polyper-fluorsulfonic acid	50–100°C 122–212°F	<1k W–100 kW	53–58% (transportation) 25–35% (stationary)	70–90% (low-grade waste heat)	• Backup power • Portable power • Small distributed generation • Transportation • Specialty vehicles	• Solid electrolyte reduces corrosion and electrolyte management problems • Low temperature • Quick startup
Alkaline (AFC)	Aqueous solution of potassium hydroxide soaked in a matrix	90–100°C 194–212°F	10 kW– 100 kW	60%	>80% (low-grade waste heat)	• Military • Space	• Cathode reaction faster in alkaline electrolyte, leads to higher performance • Can use a variety of catalysts
Phosphoric Acid (PAFC)	Liquid phosphoric acid soaked in a matrix	150–200°C 302–392°F	50 kW–1 MW (250 kW module typical)	>40%	>85%	• Distributed generation	• Higher overall efficiency with CHP • Increases tolerance to impurities in hydrogen

	Electrolyte	Operating Temperature	Power	Efficiency		Applications	Advantages
Molten Carbonate (MCFC)	Liquid solutions of lithium, sodium, and/ or potassium carbonates soaked in a matrix	600–700°C 1112–1292°F	<1 kW–1 MW (250 kW module typical)	45–47%	>80%	• Electric utility • Large distributed generation	• High efficiency • Fuel flexibility • Can use a variety of catalysts • Suitable for CHP
Solid Oxide (SOFC)	Yttria stabilized zirconia	600–1000°C 1202–1832°F	<1 kW– 3 MW	35–43%	<90%	• Auxiliary power • Electric utility • Large distributed generation	• High efficiency • Fuel flexibility • Can use a variety of catalysts • Solid electrolyte reduced electrolyte management problems • Suitable for CHP • Hybrid/GT cycle

Source: U.S. Department of Hydrogen Energy Program. 2008. Comparison of fuel cell technologies. http://www1.eere.energy.gov/hydrogenandfuelcells/fuelcells/pdfs/fc_comparison_chart.pdf (accessed May 29, 2009).

or a hydrogen containing stream, typically produced from a reformed fuel, such as natural gas, methanol, or other fuels like kerosene, propane, etc.

1.1.4.1.2 Methanol as Fuel

Direct methanol fuel cells (DMFCs) are a subset of the PEM fuel cell class, operating with very similar materials including the same membrane electrolyte. A DMFC oxidizes the methanol directly in the anode, where it is introduced as a mixture of methanol and water. This eliminates the need for a reforming step and allows the use of the high energy density methanol as fuel, which has an energy density much higher than that of hydrogen.

The reactions are

Anode reaction: $CH_3OH + H_2O \rightarrow 6H^+ + 6e^- + CO_2$ (1.7)

Cathode reaction: $\frac{3}{2}O_2 + 6H^+ + 6e^- \rightarrow 3H_2O$ (1.8)

Overall reaction: $CH_3OH + \frac{3}{2}O_2 \rightarrow 2H_2O + CO_2$ (1.9)

The anode catalyst typically used is PtRu (platinum/ruthenium) in order to reduce the overpotential for methanol oxidation, which is significantly higher than that for hydrogen oxidation, resulting in an overall lower cell voltage and lower efficiency. DMFC applications are typically small electronics and portable power applications, where the advantage of having a liquid fuel that is easily transportable and easy to recharge outweighs the reduced performance. Specific challenges with DMFC include crossover of the methanol to the cathode, resulting in fuel loss and reduced performance, lower performance due to the anode kinetics, and increased degradation rates due to the unstable Ru catalyst on the anode, which can cross over to the cathode resulting in both cathode and anode performance degradation.

1.1.4.2 Alkaline Fuel Cell (AFC)

Francis Bacon first developed the AFC during the 1930s. The AFC offers the advantage of faster cathode reaction rates and, therefore, higher energy efficiency, and the ability to use a wide range of cathode catalysts compared to the acidic electrolyte in PEM fuel cells [12]. The electrochemical reactions in an alkaline fuel cell can be expressed by equation (1.10) through equation (1.12).

Anode reaction: $2H_2 + 4OH^- \rightarrow 4H_2O + 4e^-$ (1.10)

Cathode reaction: $$O_2 + 2H_2O + 4e^- \rightarrow 4OH^-$$ (1.11)

Overall reaction: $$2H_2 + O_2 \rightarrow 2H_2O$$ (1.12)

The liquid electrolyte is typically potassium hydroxide in a mobilized or immobilized aqueous solution (30 to 45 wt%). One of the main challenges for alkaline fuel cells is the potential for poisoning of the electrolyte with any carbon dioxide present in the gas streams, by the conversion of potassium hydroxide (KOH) to potassium carbonate (K_2CO_3). The result of this is blocking of the pores in the cathode with K_2CO_3 and loss of electrolyte ionic conductivity. The advantages of high performance have made the alkaline fuel cell the best option for the space program, where operation on pure oxygen is not an issue. The potential for low cost may make it attractive for other markets, with improvements in electrolyte management and low-cost cell components that have occurred in recent years. The electrolyte system may be static, which can be replaced on a regular basis, or circulating to reduce the poisoning effect. In addition, a scrubber may be incorporated on the inlet air stream to remove CO_2.

1.1.4.3 Phosphoric Acid Fuel Cell

The phosphoric acid fuel cell (PAFC) uses liquid phosphoric acid as an electrolyte contained in a polytetrafluoroethylene (PTFE)-bonded silicon carbide matrix, with porous carbon electrodes containing a platinum catalyst [13]. The phosphoric acid cell typically operates at 150 to 200°C to achieve sufficient electrolyte conductivities. This provides advantages over PEM in terms of tolerance to impurities, which typically desorb very rapidly at these temperatures. The use of the liquid electrolyte contained in a matrix represents challenges with respect to loss of the electrolyte over time. However, this can largely be mitigated with stack design, such that the stacks can operate for 40,000 h without replenishment of the original electrolyte inventory. The increased temperature compared to PEM can also result in higher degradation rates and more difficulty in materials selection, such as, for example, sealing materials. PAFCs are typically used in stationary power generation where the stack operates continuously and where the fuel is a reformed gas stream, with the advantage over PEM of tolerance to increased levels of carbon monoxide. The advantage of the increased temperature also offers higher efficiency in combined heat and power generation.

1.1.4.4 Molten Carbonate Fuel Cell

Molten carbonate fuel cells (MCFCs) are one of the two most common high temperature fuel cells, operating at 600 to 700°C. The key advantage is the ability to reform fuel, such as natural gas *in situ*, without a separate reforming step required, and to use a wide variety of reformed fuels. The issue of carbon monoxide poisoning common to low temperature fuel cells is not an issue at the higher temperatures. An MCFC uses a molten carbon electrolyte containing carbonate ions (CO_3^{2-}), which consist of either of two carbonate salts mixtures—lithium carbonate and potassium carbonate—or lithium carbonate and sodium carbonate [14]. The fuel cell reactions are

Anode reaction: $\qquad CH_3^{2-} + H_2 \rightarrow H_2O + CO_2 + 2e^-$ $\qquad\qquad$ (1.13)

Cathode reaction: $\qquad CO_2 + \frac{1}{2}O_2 + 2e^- \rightarrow CO_3^{2-}$ $\qquad\qquad$ (1.14)

Overall reaction: $\quad H_{2(g)} + \frac{1}{2}O_{2(g)} + CO_{2(cathode)} \rightarrow H_2O_{(g)} + CO_{2(anode)}$ $\qquad$ (1.15)

Because the CO_2 is consumed in the cathode and produced at the anode, it is necessary to transfer the CO_2 produced back to the cathode. This represents an increased system complexity that is not present in other fuel cell types.

Advantages for MCFCs include the ability to use standard materials for construction, such as stainless steel sheets, and the use of nickel-based catalysts on the electrodes. Another advantage common to the high temperature class of fuel cells is the resulting fast electrode reactions, eliminating the need for the use of precious metal catalysts and increasing the fuel cell efficiency. The by-product heat from an MCFC can be used to generate high-pressure steam that can be used in many industrial and commercial applications. The combined high electrical and heat recovery efficiency results in overall high efficiencies for high temperature fuel cells. An MCFC is most suitable for constant power applications due to the relatively long time period required to raise the temperature to the operating condition. However, the higher operating temperature increases material demands in terms of corrosion stability and life of cell components.

1.1.4.5 Solid Oxide Fuel Cell

The solid oxide fuel cell (SOFC) is based on a thin layer of solid ceramic electrolyte of yttria-stabilized zirconia operating at 600 to 1000°C, which transfers the oxygen ion (O^{2-}) from the cathode to the anode. The high temperature is necessary to achieve sufficient ionic conductivity [14]. The electrochemical reaction in an SOFC can be expressed as equation (1.16) to equation (1.18).

Anode reaction: $2H_2 + 2O^{2-} \rightarrow 2H_2O + 4e^-$ (1.16)

Cathode reaction: $O_2 + 4e^- \rightarrow 2O^{2-}$ (1.17)

Overall reaction: $2H_2 + O_2 \rightarrow 2H_2O$ (1.18)

There are two SOFC configurations in use: planar (flat panel) and tubular. The planar generally produces higher performance, but the tubular provides advantages in terms of simpler sealing.

The SOFC has the significant advantages common to the high temperature fuel cells: increased fuel flexibility, including internal fuel reforming in the case of natural gas; the use of a variety of catalysts; tolerance to impurities; and high efficiency for combined heat and power applications. The disadvantages of long start-up times and material selection and design to withstand the high temperatures are also present.

1.2 PEM Fuel Cell Description

1.2.1 Fuel Cell Reactions

The overall reaction for the PEM fuel cell is identical to the combustion reaction of oxygen and hydrogen, but by separating the half reactions, the chemical energy can be converted to electrical energy. The oxidation of hydrogen on the anode produces protons and electrons. The protons are transported through the membrane to the cathode where they react with oxygen and with electrons arriving through the external circuit, to produce water and heat. The reaction is driven by the electrochemical energy stored in the reactants. The electrons traveling through the external circuit produce a current or electrical energy. The electrochemical reactions in a PEM fuel cell can be expressed as equation (1.19) to equation (1.21).

Anode reaction: $2H_2 \rightarrow 4H^+ + 4e^-$ (1.19)

Cathode reaction: $O_2 + 4H^+ + 4e^- \rightarrow 2H_2O$ (1.20)

Overall reaction: $2H_2 + O_2 \rightarrow 2H_2O$ (1.21)

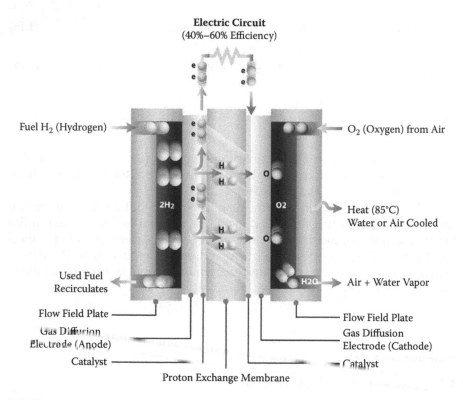

FIGURE 1.1
PEM fuel cell schematic. (From http://www.ballard.com)

1.2.2 Fuel Cell Schematic

The fuel cell is depicted schematically in Figure 1.1.

1.2.3 Fuel Cell Components

1.2.3.1 Proton Exchange Membrane

The heart of the PEM fuel cell is the proton exchange membrane, which transports protons from the anode to the cathode. The membrane also serves to separate the fuel and oxidant gas phases and electronically insulates the cathode from the anode. The most typical membrane is a sulfonated perfluorinated polymer. The Nafion® family of membranes made by DuPont is representative of this class, and is based on a sulfonated tetrafluoroethylene-based fluoropolymer-copolymer, with the chemical structure represented in Figure 1.2. The sulfonic acid (SO_3H) groups on the side chains allow the protons or other cations to "hop" from one acid site to another, in the presence of water. The exact mechanism of the proton movement is an area of significant research. An active area of research is the development of hydrocarbon-based

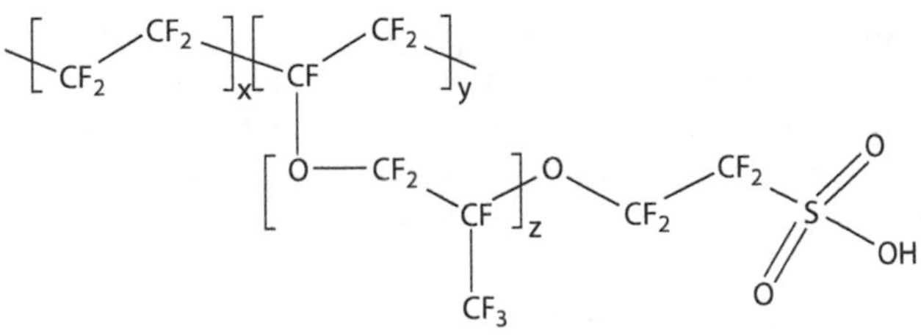

FIGURE 1.2
Structure of Nafion®.

membranes, in order to eliminate or reduce the use of fluorine, which has cost and environmental disadvantages.

Although the fuel cell reactions produce water, it is generally insufficient to maintain the required humidification level essential to maintain the proton conductivity. The passage of the dry reactant gases, as well as the electro-osmotic drag (as the protons move from the anode to the cathode, they "drag" water molecules with them), can result in an under-humidified state. Maintaining fully humidified membranes is most easily accomplished by humidifying the inlet gas streams. Reducing the membrane humidification requirements can reduce system cost, volume, and complexity. Reduced humidification can be tolerated through MEA (membrane electrode assembly) and cell design to keep the membrane hydrated, and some membranes are more tolerant to small reductions in humidification levels. However, the ability to operate with nonhumidified gases remains an immense challenge for fuel cells. In addition to the performance loss due to the reduced conductivity, drying of the membrane can lead to significant membrane damage, as discussed below in section 1.7.2.

1.2.3.2 Catalyst, Catalyst Layer

On either side of the membrane are the anode and cathode catalyst electrode layers. The electrochemically active catalyst sites require the three-phase interface: (1) Pt catalyst surface, electrically connected to the external path to provide electron transport paths; (2) ionomer or electrolyte contact to transport protons; and (3) reactant gas phase access.

The catalysts are required to promote the desired reactions to occur at an appreciable rate. The catalysts used in PEM fuel cells are typically Pt-based due to the high stability and reactivity of Pt. Pt alloys may also be introduced to further increase kinetic activity, improve stability, and improve tolerance to contaminants on the anode for use in reformed fuel. The high cost of Pt necessitates a maximum utilization of the Pt. For this reason, Pt is typically in the form of very small particles of approximately 2 to 8 nm diameters, supported on larger carbon particles. The carbon particles provide a high surface area support structure to enhance the dispersion of the catalyst particles as well as providing an electrical and thermal pathway from the reaction sites toward the external circuit. A similar proton exchange material to that used in the membrane (in this case, generally referred to as ionomer) is mixed into the catalyst layer as small particles to provide proton pathways. The mix of carbon particles and ionomer must provide sufficient porosity in the catalyst layer for reactant gas access to the catalyst reaction sites.

1.2.3.3 Gas Diffusion Layers Including Microporous Layers

On either side of each catalyst layer are the gas diffusion layers (GDLs) or gas diffusion media (GDM), both terms being used in literature. These are typically made of hydrophobic carbon fiber paper or cloth, termed the *substrate*, with a hydrophobic microporous layer (MPL) applied to the catalyst side, made of carbon particles with a hydrophobic binder. The hydrophobicity is typically achieved through application of PTFE to the substrate and by mixing the PTFE with the carbon particles in a carbon ink to be coated on the substrate. The GDLs serve a number of functions:

- Transport of the reactant gases to the catalyst layers, allowing the gases to diffuse under the flowfield landings
- Removal of product water
- Transport of electrons from the catalyst layer to the flowfield landings
- Heat conduction
- Mechanical support of membrane and catalyst layers to span flow-field channels

The combination of the membrane, the electrodes, and the GDLs is called the *membrane electrode assembly* (MEA).

1.2.3.4 Bipolar or Flowfield Plates

The bipolar or flowfield plates are assembled on either side of the GDLs, the entire unit comprising the unit cell of the fuel cell stack. The plates are typically made out of carbon- or graphite-polymer composite materials or metals. The functions of the plates are to transport the incoming gases to the GDL surface, and remove the exhaust gases and product water; transfer electrons and heat; and provide mechanical structure. The flows of reactants through the flowfields and on either side of the MEA are shown schematically in Figure 1.3. The plates must be electrically and thermally conductive, chemically stable in the fuel cell environment, mechanically robust, and free of contamination. The use of metal plates is of particular concern with respect to contaminant introduction into the fuel cell, and considerable effort is being expended in academia and industry to develop metal plate coatings that are stable in the acidic, high-potential fuel cell environment, and do not leach contaminants.

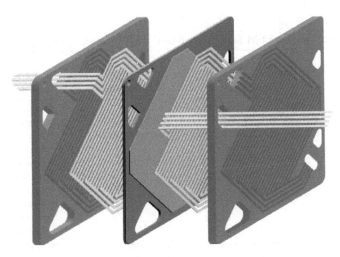

FIGURE 1.3
Schematic of Ballard fuel cell with flowfield plates on either side of membrane electrode assembly.
Flows of reactants through flowfields on bipolar plates shown. (From http://www.ballard.com)

1.2.3.5 Fuel Cell Stack

The unit cells are typically stacked such that the cathode of one cell is electrically connected to the anode of the next. The gases flow through the cells in a parallel fashion, while the electrical connections are in series. In this way, the voltages from each cell add up to create the required stack voltage. The operating voltage range of a single cell is typically 0.6 to 1.0 V, with the total stack voltage the sum of each cell in series, with numbers of cells (depending on the application and cell design) ranging from a few to a few hundred.

1.3 Applications

There are a number of key applications for which fuel cells are being developed as replacements to internal combustion engines on the basis of fuel efficiency and reduced GHG emissions as well as to batteries on the basis of improved storage density, operating characteristics, and fueling/recharging benefits. The automotive application for PEM fuel cells has driven significant development activities, with most of the major car companies having fuel cell activities and demonstration projects [16]. However, there are nearer term markets where PEM fuel cells provide significant advantages, and are likely to provide the early commercialization experience for fuel cells [17]. The near term markets identified for analysis in a study conducted by Battelle under contract to the U.S. DOE are outlined in Table 1.2.

TABLE 1.2

Near Term Markets for PEM Fuel Cells, Based on Information in [17]

Near Term Market Application	Description	Key Industries	Key Characteristics and Requirements
Backup Power	Standby or emergency power to ensure uninterrupted service	Financial services tele-communications	• Reliability and availability • Longer runtimes than batteries • Low operations and maintenance requirements • Reduced emissions as compared to generators
Grid Independent Power	Continuous, stand-alone power to operations that are not connected to the grid	Off-grid sites, chemical plants or other sites with available hydrogen production	• High reliability • Fuel availability • High efficiency • Low maintenance costs • Long lifetime, (>6,000 h/year) • Good load-following characteristics
Specialty Vehicles	Specialized equipment and vehicles, such as forklifts, industrial movers, and motorized scooters	Indoor and outdoor environments Typically used to transport people or goods	• <100 kW • Long runtimes (2,000 to 5,000 h/year) • Low emissions • Easy startup
Portable Power	Continuous power to meet the complete energy needs of small devices	Small electronic products Portable phones, cameras, computers, and security devices	• <1 kW • 1,000 to 5,000 h/year • Energy density, efficiency, and hydrogen storage are critical
Auxiliary Power	Alternate source of power serving specific requirements in portable, mobile onroad transportation, and offroad transportation applications	Trucks, locomotives, airplanes, boats, or military vehicles	• 4 to 30 kW • 20,000 and 35,000 h lifetime

Source: Based on Mahadevan, K. et al. 2007. Identification and characterization of near-term direct hydrogen proton exchange membrane fuel cell markets. Battelle. http://www1. eere.energy.gov/hydrogenandfuelcells/pdfs/pemfc_econ_2006_report_final_0407.pdf (accessed May 29, 2009).

1.4 Fuel Cell Performance

1.4.1 Open Circuit Voltage and the Nernst Equation

The open circuit voltage (OCV) is the voltage provided by a single fuel cell under conditions of no electrical load. Under these open circuit conditions, the true electrochemical equilibrium at slow discharge of current approaching zero represents the thermodynamic cell voltage. For an electrode reaction, the cell voltage value at equilibrium is characterized by the Nernst equation, providing the electrode potential as a function of the bulk concentration of the reactants and products [18].

In general form, the Nernst equation is

$$E = E^0 - \frac{RT}{nF}\ln(Q) \tag{1.22}$$

where
E = equilibrium cell voltage (theoretical OCV value)
E^0 = the equilibrium voltage with all species at their standard states
R = the gas constant (8.3144 J/Kmol)
T = the absolute temperature, in K
n = the stoichiometric coefficient of the electrons in the half-reactions
F = Faraday's constant (96487 C/mol)
Q = the reaction quotient

The reaction quotient, Q, is the ratio (product of all product activities)/ (product of all reactant activities), with each activity raised to the power of that species' stoichiometric coefficient.

For the PEM fuel cell reaction, the Nernst equation takes the form:

$$E = E^0 - \frac{RT}{2F}\ln\left(\frac{1}{P_{H_2}\sqrt{P_{o_2}}}\right) \tag{1.23}$$

where
E^0 = 1.229 V at standard conditions. Product water is assumed to be produced in the liquid form.
P_i = activity of the ith species, expressed as partial pressure, in atmospheres, with the standard state being P = 1 atm.

In practice, the measured OCV is always lower than the theoretical OCV. The major contributions to this difference include mixed cathode potential due to PtO_x formation on the catalyst surface, and fuel crossover from the anode to the cathode, reducing surface concentration of O_2. Other factors also can contribute to this difference including leakage currents moving the open circuit voltage away from true open circuit or no load conditions, and the presence of other species, including the fuel cell materials as well as contaminants.

1.4.2 Overpotential

The actual cell voltage achieved under a current load will be less than the theoretical equilibrium voltage due to voltage losses in the cell or "polarization" of the electrodes, where the potentials at each electrode approach each other or exhibit "overpotential," thereby resulting in a reduced cell voltage. Contributions to the polarization include activation losses and concentration or mass transport losses. The additional cell loss includes ohmic loss due to current flowing through the cell and between the electrodes.

The cell voltage can be described as

V = Open circuit voltage – cathode activation losses – anode activation
    losses – ohmic losses – (concentration and mass transport losses)   (1.24)

The cell voltage as a function of current density is referred to as the *polarization curve*. This is illustrated in Figure 1.4, with most of the possible contributing losses schematically shown.

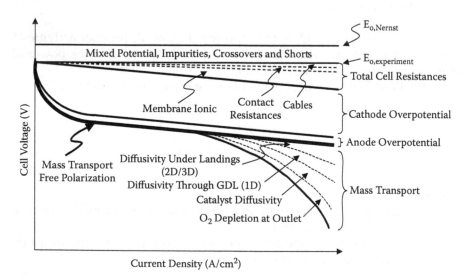

FIGURE 1.4
Illustrative fuel cell polarization with losses noted. (From Ye, S. 2007. Electrocatalysts for PEM fuel cells: Challenges and opportunities. *14th National Meeting of Chinese Society of Electrochemistry,* Yangzhou, China.)

1.4.3 Activation Polarization

Activation polarization is electrode polarization caused by sluggish electrode kinetics. It is associated with the activation energy barrier that must be overcome by the reactants. It is dependent on electrode characteristics and dominates the total overpotential at low currents.

The cathode kinetics are considerably slower than the anode kinetics, resulting in a large cathode overpotential and a relatively small anode overpotential. While the following equations apply equally to both electrodes, the anode activations losses are generally ignored and, in practice, these are usually applied only to the cathode. (See section 1.4.6 for further discussion on anode activation losses.)

There are several excellent texts that provide the development of electrochemical thermodynamic equations governing fuel cell behavior [18,20]. Some of the basic equations will be described here. Please refer to the noted references for a more in-depth understanding.

1.4.4 Exchange Current Density, i_o, and Rate Constant, k^0

The exchange current, i_o, represents the balanced faradaic activity occurring at a net current of zero. The exchange current is equal in magnitude to either the cathodic or anodic current at equilibrium of an electrochemical half-cell reaction. The exchange current is proportional to k^0

$$i_o = nFAk^0C_o^{(1-\alpha)}C_R^{\alpha} \tag{1.25}$$

Where k^0 refers to the standard rate constant, and alpha (α) is the electron transfer coefficient. The standard rate constant is the value of the rate constants for the forward and backward reactions, at equilibrium, where they are equal. The deviation of these rate constants at potentials differing from equilibrium is an exponential function of potential difference from equilibrium and the transfer coefficient,

$$k_f = k^0 e^{-\alpha nf(E-E^o)} \tag{1.26}$$

$$k_b = k^0 e^{(1-\alpha)nf(E-E^o)} \tag{1.27}$$

The electron transfer coefficient can be thought of as a symmetry factor between the forward and backward reactions. In many systems, alpha can be approximated by ½. The exchange current is often normalized to the exchange current density, $j_0 = i_0/A$.

1.4.5 Tafel Equation

The behavior of electrode systems, in absence of reactant transport limitations, typically displays an exponential relationship between the current and the overpotential, η. This is represented by the Tafel equation:

$$i = a'e^{\eta/b}$$

(1.28)

Or, in the more typical form,

$$\eta = a + b\log i$$

(1.29)

Where b is referred to as the Tafel slope and can be determined from the plot of overpotential as a function of log(i). Using the exchange current density, the Tafel equation for the cathodic reaction in the fuel cell is provided by

$$\eta = \frac{RT}{\alpha nF}\ln i_o - \frac{RT}{\alpha nF}\ln i$$

(1.30)

1.4.6 Butler–Volmer Equation

For an electrochemical half-cell reaction, the overall relationship describing the complete current-potential characteristic is given by the Butler–Volmer equation:

$$i_o = nFAk^0C_o(0,t)e^{-\alpha nf(E-E^o)} - C_R(0,t)e^{(1-\alpha)nf(E-E^o)}$$

(1.31)

Or, in its simpler form:

$$i = i_o[e^{-\alpha nf\eta} - e^{(1-\alpha)nf\eta}]$$

(1.32)

It is important to understand that the forward and backward reactions can be driven to large values, even with a small k^0, at significant deviations from E^0. This is important for the cathode reaction in the fuel cell, which has sluggish kinetics that must be overcome by large overpotentials on the cathode.

1.4.7 Charge Transfer Resistance, R_{ct}

For very small overpotentials, the exponential function can be represented by a linear equation, and the ratio of overpotential to current is referred to as the *charge transfer resistance*, R_{ct}.

$$R_{ct} = \frac{RT}{nFi_o}$$

(1.33)

The charge transfer resistance is a convenient indicator of kinetic facility, which approaches zero for highly kinetically active systems.

1.4.8 Concentration Polarization or Mass Transport Losses

As reactants are consumed at the catalyst surfaces, new reactants will diffuse in and a concentration gradient will quickly be established between the bulk fluid and the reaction sites. The result is a reduction in reactant concentration at the catalyst surface. In addition, the reactant concentration in the flowfields decreases along the length of the flowfield as the reactants are consumed and the product water vapor is produced. The equilibrium voltage as a function of gas composition is provided by the Nernst equation (equation (1.23)) above. While these phenomena occur at both electrodes, in reality, since the hydrogen diffusion rate is sufficiently fast and pure hydrogen is often used on the anode, the hydrogen concentration polarization is usually neglected. The change in cell voltage due to the decreased oxygen concentration at the cathode catalyst surface is the concentration polarization. This is also often interchangeably referred to as mass transport losses, referring to the significant contribution that the slow oxygen mass transport plays in this voltage loss. Concentration polarization becomes increasingly significant as the load increases.

A review of the typical shape of the polarization curve indicates a region where mass transport losses begin to dominate as the oxygen concentration is reduced at the electrode surface at the higher current density (see Figure 1.4).

The oxygen transport is based on several mechanisms. The gaseous reactants enter the fuel cell typically through the gas channels, and then diffuse through the GDL and into the catalyst layer. The gases diffuse into the catalyst layer either to the catalyst surface where they react or are dissolved into the ionomer in the catalyst layer. The gas phase mass transport in the GDL is due to diffusion governed by Fick's law, in which the rate of diffusion is a function of the concentration gradient, the thickness of the GDL, and the diffusion coefficient for the species [21].

As the pore size decreases, the Knudsen diffusion model may become important. This model describes diffusion when the gas molecules collide more often with the pore walls than with each other, and it is these molecule–pore wall interactions that start to dominate the rate of diffusion. Knudsen diffusion should be considered when pore diameters are typically less than 100 nm under atmospheric conditions, and is important in the catalyst layer and possibly in the GDL [21].

1.4.9 Ohmic Losses

The ohmic polarization refers to voltage losses due to the movement of charge through the cell. Ohmic polarization obeys Ohm's law: $\eta_{ohm} = IR$, and thus is linear with respect to increasing current density.

The cell resistance is a summation of the electronic resistance through each of the fuel cell components, the interfacial resistance between each component, and the ionic (or protonic) resistance of the proton transport through the ionomer networks in the catalyst layers and through the membrane or other electrolytes.

1.4.10 Anode Polarization

The processes occurring at the anode are the same as those described for the cathode. However, the hydrogen oxidation reaction is extremely facile and occurs with a much lower overpotential than that of the cathode oxygen reduction reaction. Gasteiger et al. [22] provide an analysis of the anode overpotential and the effect of anode Pt loadings. Due to the extremely fast rate of the reaction, the kinetic parameters are difficult to measure accurately. Based on fuel cell testing of the effect of anode catalyst loading, an extremely high reaction rate is likely, and overpotential would be as low as ~1 to 10 mV at loadings of 0.05 to 0.4 mg Pt/cm². For this reason, the anode overpotential is often neglected in the analysis of fuel cell performance.

However, the situation can change significantly in the presence of contaminant species. For example, the presence of small amounts of carbon monoxide in the fuel stream can cover ~90% of the anode catalyst surface [23], and result in significant anode overpotentials and decreased cell performance [24].

1.5 Fuel Cell Diagnostics

A suite of many different diagnostic approaches have been developed to study fuel cells for performance, durability, and component functionality [25]. These include electrochemical, electrical, chemical, and mechanical tests, both on an operating fuel cell and in *ex situ* measurements. Some of the most commonly used *in situ* tests conducted on operating fuel cells are described below in this section, along with reference to applicability to contamination testing.

1.5.1 Polarization Curve

The most commonly used performance diagnostic is the polarization curve. This consists of recording the cell voltage response to various current or current density levels. As the current is increased from zero at open circuit voltage state, the electrodes begin to polarize, resulting in a lower cell voltage. This is an essential operating characteristic for application of fuel cells, as it dictates the power that is available. The shape of the curve provides information as to the magnitude of the polarization losses, categorized as kinetic, ohmic, and mass transport, which was discussed in section 1.4. The U.S. Fuel Cell Council has developed a standard polarization procedure [26]. In studying contamination, polarizations are typically run prior to introduction of the contamination, and may be run after exposure to the contaminant for a specified length of time to understand performance impacts.

1.5.2 Steady State Testing

Likely the second most used diagnostic is steady state testing, and it is used to study durability of the fuel cell. This is the simplest test and can be accomplished quite easily. The fuel cell is typically set at a constant load or current or, in some cases, at a constant voltage, and allowed to operate for a specified period of time or until a failure criteria is reached, while the change in voltage or current is monitored. Diagnostics, such as polarization curves or some of the other tests described in the following sections, may be intermittently conducted to characterize rate and type of degradation. This is a very common test in contamination testing; the cell or stack is typically run for a short length of time to establish baseline performance, then the contaminant is introduced into one of the inlet reactant streams and the performance degradation is monitored. The contaminant may be stopped and the test continued to determine if there is spontaneous recovery of any performance losses or other diagnostics may be conducted, e.g., to detect the presence of adsorbed species on the catalyst.

1.5.3 Duty Cycle Testing

The simulation of the fuel cell duty cycle may be quite complex, depending on the application to be simulated. A study conducted by Sandia National Laboratories (Albuquerque, New Mexico) provides an example of the efforts involved in characterizing the duty cycle for a robotic vehicle through the use of a data acquisition system [27]. Although duty cycle testing is a valuable tool used by fuel cell developers and is essential for validation of the technology, it is a technique rarely run by most fuel cell researchers. This technique requires automated test stations and a good knowledge of the intended duty cycle. Although the effect of contaminants on the performance over the duty cycle is the ultimate understanding goal, the high cost and complexity of these tests significantly limits their use.

1.5.4 Oxygen Testing

In addition to the standard polarization curve conducted on the intended oxidant, typically air, oxygen polarization curves are also commonly conducted. The oxygen polarization curve provides a measure of the activation and ohmic losses, with minimal contributions from mass transport or concentration losses, as the diffusion rate of the oxygen is much higher than in the binary gas diffusion with air. This provides a means to separate the kinetic, ohmic, and mass transport losses.

1.5.5 Voltage Decay

The voltage decay diagnostic technique is based on analysis of the transient voltage response after a step change in load from zero to a constant current. A typical procedure involves running the cell at the conditions of interest to stabilize, and then the cell is set to open circuit. The current is stepped from zero to the load of interest, and the transient voltage response is measured. The initial response is based on the fact that the reactants are at a uniform concentration as there is no current being generated. The voltage just after the load is applied (V_{0+}) arises before a diffusion gradient can be established in the electrode, while the reactant concentration at the catalyst layer is the same as the bulk gas concentration. Also, with a uniform distribution of temperature and reactant concentration, V_{0+} should occur with a uniform current distribution. Repeating this test and stepping to different load points can obtain a mass transport free polarization curve. As the reactants begin to be consumed, a diffusion gradient is established, and the subsequent response provides information about mass transport.

1.5.6 AC Impedance

Alternating current (AC) impedance testing differentiates itself from the majority of diagnostic techniques by applying a very small perturbation to the system in the form of an alternating signal. The small signal results in reduced errors introduced by the measurement technique [18]. The theory uses an equivalent electronic circuit consisting of resistors, capacitors, and inductors to represent the impedance to electron flow in an electrochemical cell, such as the slow electrode kinetics and diffusion resistance [28,29]. Relatively simple equivalent circuit models are frequently good approximations of real systems, and the data can be fitted to yield reasonably accurate results. By introducing a perturbation in the direct current (DC) delivered by the fuel cell and measuring both the perturbed current and the voltage response, an AC impedance can be calculated. Electrochemical impedance spectroscopy (EIS) involves the measurement of the response at several frequencies.

Cleghorn et al. [30] provide an example of the use of AC impedance for the assessment of degradation over a lifetime test. In this case, by measuring the AC impedance spectra at different current densities for new and lifetime-tested cells, they found that the resistance to transport, or diffusion processes, has significantly increased between the new cell and the life-tested cell. By applying the technique under different humidification levels, they were further able to ascertain a connection between the presence of liquid water and the increased resistance to diffusion. They also studied the changes in ionic conductivity of the electrodes.

1.5.7 Cyclic Voltammetry

Cyclic voltammetry is commonly used to study fuel cell electrodes and hydrogen crossover. In this technique, a linear sweep potential is applied to one electrode, while the other is held constant. The potential is cycled in a triangular wave pattern, while the current produced is monitored. The shape and magnitude of the current response provides useful quantitative and qualitative information regarding the amount of catalyst that is electrochemically active, the double layer capacitance, hydrogen crossover, and the presence of oxide layers and contaminants. Wu et al. provide a description of this technique with example voltammograms [29].

1.5.8 Single Cell versus Stack Testing

The majority of research tests are conducted on single cells, while fuel cell developers tend to use a mixture of single cells and stack testing. Each type of testing has advantages, as discussed by Schneider et al. [31] Single cells are more cost effective and provide greater control of operating conditions, and, therefore, are more useful for fundamental studies. The stack tests, on the other hand, more closely simulate operation under realistic conditions including the range of conditions experienced by the cells in the stack. The main differences for operation in a stack are due to uneven thermal, electrical, and flow distribution effects between the cells. In addition, the heat loss from a single cell is quite large, and it becomes difficult to simulate the same thermal conditions as a cell would see in a stack. This can cause some difference for contamination studies, as temperature effects are very important for adsorption and desorption phenomena.

1.5.9 Accelerated Durability Testing

Due to the long lifetimes required for many fuel cell applications, it is not practical to rely solely on testing under normal time conditions. Applications, such as stationary power generation at 40,000 h and bus applications at 20,000 h, would take years to test and require considerable quantities of fuel. The development and design cycle requires more rapid turnaround in a make–break–fix cycle involving determination of failure modes, followed by technology development, and, finally, implementation of mitigating materials and designs. Therefore, as fuel cells approach sufficient maturity to last the required lifetimes, accelerated testing techniques are being developed by many fuel cell developers and material suppliers, and are an important component of fuel cell testing strategy.

Accelerated testing techniques may be used for rapid screening of different materials and designs for down-selection of improved durability, or they may be used to predict actual lifetimes. Wu et al. [29] and Zhang et al. [32] present overviews of general accelerated stress test (AST) methods used in fuel cell lifetime testing. Accelerated testing techniques can take a number of forms. The most common is identification of a key stressor that contributes significantly to one or more failure modes. An AST is normally designed to increase this stressor in a defined and quantitative manner. For example, accelerated voltage cycling tests are developed to simulate voltage cycling expected under actual duty cycle operation. These may be accelerated in time, while maintaining similar voltages expected under operation, or they may additionally be accelerated in magnitude in order to further reduce the test time. Additional stressors may be applied, such as increased temperature and changes in humidification levels, to further accelerate the failure mechanism. An important consideration is that the failure mechanism occurring in the accelerated test procedure must closely mimic that observed in duty cycle operation. Diagnostic performance testing and failure analysis techniques are employed to compare the characteristics of the failures between real and accelerated conditions.

1.6 Fuel Cell Operating Conditions

The operating conditions for a fuel cell are highly dependent on the application and have a significant impact on the performance and durability. They will also have a significant impact on contamination effects. The conditions used in fuel cells today are typically those required by the fuel cells, but system developers desire different conditions to enhance system simplicity and reduce costs. The key drivers today are toward higher temperature and lower humidity conditions.

1.6.1 Temperature

The temperature of operation has one of the strongest influences on durability and contamination effects. It can also have a strong influence on performance when coupled with humidification effects. Typical fuel cell operation is in the range of 60 to 80°C, but the overall drive in the fuel cell industry is to run at higher temperatures in order to reduce system cooling requirements and catalyst contamination issues.

The relationship between temperature and performance is complex. An increased temperature drives increased kinetic activity for the oxygen reduction reaction (ORR); however, the reversible cell voltage decreases with increasing temperature. The Tafel slope increases in the low current density region with little effect seen in the high current density region. The exchange current density increases at higher temperatures [23]. Increased temperatures can also result in a loss of water in the membrane and in the catalyst layer ionomer, resulting in decreased ionomer conductivity and decreased performance. The fuel cell may need to operate over a wide temperature range, from subzero to close to or over 100°C. The low temperatures also impact performance due to increased ohmic losses caused by reduced proton conductivity in the membrane as the temperature decreases.

Increased temperature generally results in reduced adsorption of contaminants and, therefore, increased tolerance to the presence of contaminants. Zhang et al. [23] summarize this for one of the most common fuel contaminants, carbon monoxide, where fractional coverage of CO on platinum catalyst is shown to decrease from close to 1 at 50°C to close to 0 at 170°C, for 1 ppm CO. (The temperature effect is also discussed in chapter 2.) However, increased temperatures can significantly reduce durability. Many of the components degrade according to mechanisms exhibiting Arrhenius behavior, in particular, the membrane and gasketing materials, resulting in an approximate doubling in rate for every 10°C rise in temperature.

1.6.2 Pressure

The operating pressure for the fuel cell influences performance, parasitic system energy, humidification requirements, stack and system design, and seal and membrane durability considerations. The performance of the cell is improved at higher pressures according to the Nernst relationship (see section 1.4.1). On the other hand, this improved performance will be offset by increased parasitic system energy to compress the inlet gases to the required pressures. The optimization of these competing effects will depend on the application. Additional trade-offs also exist. Increased pressures result in improved ability to remove water droplets from the flowfields and improve uniform gas distribution throughout the cells and stack; however, higher operating pressures will increase the seal and stack design and material requirements to ensure safe, durable, leak-free operation.

An added consideration for contamination issues is the use of a filter in the air or fuel circuits. This will result in an increased pressure drop, impacting system design and parasitic load, and, for this reason, typically is not desired. Nevertheless, with the current state of the technology, filters may be used when the advantages of stable, noncontaminated performance outweigh the drawbacks.

1.6.3 Humidity

Water management is arguably the most critical component of fuel cell operation. The currently available membranes and catalyst layer ionomers require a high water content to maintain proton conductivity. However, the presence of liquid water is detrimental to performance, as it tends to impede the oxygen transport to the catalyst layer. In addition, water droplets in the flow fields can result in uneven gas flows resulting in flow sharing effects and, in some cases, low cell voltages due to oxygen starvation in those cells.

Water is also a critical transport mechanism for contaminants, many of which require a water matrix to solubilize. The effect of water on gaseous contaminants is fairly complex, and further understanding is required to control the water to avoid erroneous data and conclusions when conducting contamination studies [33]. Once a cell is contaminated, however, operating with high water content can aid the removal of the contaminants through a flushing behavior.

The membrane is typically negatively impacted by low humidity conditions, with the acceleration of mechanical failure observed. On the other hand, the catalyst layer durability is generally higher under drier conditions. The dominant degradation mechanisms of Pt dissolution and carbon corrosion are both accelerated under wetter conditions.

1.7 Fuel Cell Degradation Modes

The durability of fuel cell systems remains a key challenge for fuel cell commercialization. Stationary applications require the longest lifetimes in terms of operational hours due to the high daily usage rate. The U.S. DOE has published the minimum commercialization target of 40,000 h with <10% rated power degradation over lifetime, to be achieved by 2011. PAFCs have demonstrated this lifetime, whereas PEM fuel cells have demonstrated ca. 30,000 h in the laboratory. Automotive applications, on the other hand, require approximately 5,000 operational hours, but the duty cycle is much more demanding, and, therefore, is arguably equal to or even more challenging than the requirements of stationary power generation. The DOE published target of 5,000 h has been met in duty cycle lab demonstrations [34], but actual usage in the field is still to be determined.

Schmittinger and Vahidi [35] provide a list of degradation rates observed in laboratory tests and for various achieved lifetimes based on a review of literature, with a range of 600 to 26,300 h for the lifetimes and 0.5 to 120 µV/h voltage degradation rates. This large range of durability represents significant variation in test mode, operating conditions, materials, and designs. It should be noted that not all tests listed were operated to failure. Wu et al. [29] provide a similar list, with the additional classification of steady state lifetime tests, showing degradation rates of 1 to 20 µV/h and accelerated durability tests with degradation rates of up to 800 µV/h.

Significant research into failure modes and advances in the understanding of the mechanisms as well as mitigation has been achieved in recent years as complied in several review documents [25,35–40]. The degradation modes described in the following section are some of the most critical challenges to long-life operation. Contamination is considered an important durability issue on its own, but is discussed here only as it relates to the other degradation modes. Contamination degradation is covered more fully in section 1.8 and in the remainder of this book.

1.7.1 Cathode Degradation

Degradation of the cathode catalyst is the main contributor to performance degradation in the fuel cell in the absence of contamination effects. This will be discussed in terms of degradation of the catalyst metal itself as well the corrosion of the carbon catalyst support.

1.7.1.1 Catalyst Degradation (Pt Dissolution, Agglomeration, Ostwald Ripening)

The catalyst in the PEM fuel cell consists of small Pt or Pt-alloy particles of 2 to 8 nm, deposited on larger carbon particles. The main degradation modes include Pt catalyst particle growth and Pt migration. Both of these degradation mechanisms are believed to be closely associated with the dissolution of Pt. A Pt oxidation/dissolution process occurs due to voltage cycling and high potentials on the cathode that results in movement or migration of the Pt in a dissolved state. This process was modeled by Darling and Meyers [41], who found that the Pt surface oxidation under air/air potentials is a significant contributor to the Pt stability, acting to protect it from dissolution at higher potentials. However, at potentials between the H_2/air potential and the air/air potential, the Pt is quite soluble. Pt loss becomes accelerated when cycling between normal operating voltage and air/air potentials, such as would occur during start-up and shutdown. At least a portion of this dissolved Pt often ends up as a band of Pt in the membrane, parallel to the membrane surface. It has been found that the location of this Pt band corresponds to the location in the membrane where the mixed potential, calculated based on the gas partial pressures, changes dramatically. The dissolved Pt ions migrating through the membrane precipitate when they first encounter a sufficiently reducing atmosphere caused by hydrogen flux toward the cathode [42–44].

The remaining Pt particles tend to grow in size, resulting in loss of active catalyst surface area. The Pt particles' surface area loss can be attributed to a combination of hypothesized mechanisms, which can be summarized as: (1) Ostwald ripening occurs when small Pt particles dissolve and then redeposit on the surface of larger particles, leading to particle growth; (2) the dissolved Pt may redeposit elsewhere, forming new particles; or (3) the small particles may move across the carbon surface and coalesce to form larger particles [37].

In addition to these chemical mechanisms, the Pt particles can also be lost from contact with the carbon support or ionomer through processing of the catalyst ink during manufacturing, or through structural changes to the catalyst layer during operation caused by degradation of the ionomer and carbon supports.

1.7.1.2 Cathode Corrosion

Carbon is used extensively in fuel cells due to the favorable properties of high electrical and thermal conductivity, the relatively low cost, the absence of damaging degradation products when compared to metals, and the ability to form various structures from high surface area particles for the carbon support and in the microporous layer, to carbon or graphitic fibers for the gas diffusion substrate, to graphite or carbon-polymer composite flow field plates. Dicks provides an overview of the role of carbon in fuel cells [45].

Although carbon is thermodynamically unstable in the fuel cell environment, the slow kinetics of the corrosion reaction (equation (1.34)) allow quite stable operation under controlled conditions, such as steady state operation.

$$C + 2H_2O \rightarrow CO_2 + 4H^+ + 4e^- \quad E = 0.207V_{RHE} \tag{1.34}$$

Combined with the presence of the Pt catalyst, which accelerates this reaction, measurable CO_2 production occurs above 1.0 V. While this indicates that slow corrosion will occur at open circuit potential, the rate is still relatively low, but is enough to contribute to degradation over longer lifetimes. However, the key driver for corrosion of the carbon support is the increased potentials that typically occur during shutdown or startup, when the cathode potential can go up to 1.5 V or higher [46].

These high cathode potentials occur due to the absence of hydrogen in a localized area of the anode. Without the strong influence of hydrogen to maintain the normal anode potential, the local potential will be influenced by the other species present, which is typically oxygen, resulting in an increased anode potential. The oxygen present on the anode may either cross over from the cathode or may occur during shutdown due to an influx of air from the exhaust manifold. Because the fuel cell plates are highly conductive, they will act to maintain a fairly constant cell voltage over the entire active area of the cell. The hydrogen-starved part of the cell is driven to maintain the same voltage as the normal part of the cell, thus driving the cathode potential to that of the anode potential plus the cell voltage, resulting in the very high potentials specified above (see Figure 1.5).

This mechanism only occurs when hydrogen is present in part of the anode to maintain the high cell voltage in the "normal" region, to act as a driving force for the high potential conditions in the starved region of the cell. The mechanism also will occur whether the cell is under load or not, and, in fact, is the most damaging when there is no external current flowing.

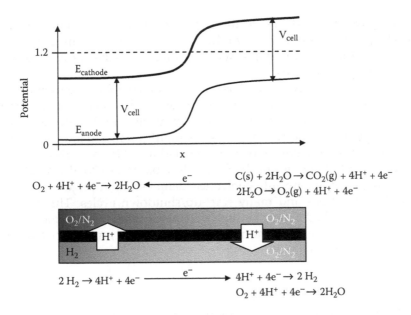

$$O_2 + 4H^+ + 4e^- \rightarrow 2H_2O \xleftarrow{\quad e^- \quad}$$

$$C(s) + 2H_2O \rightarrow CO_2(g) + 4H^+ + 4e^-$$
$$2H_2O \rightarrow O_2(g) + 4H^+ + 4e^-$$

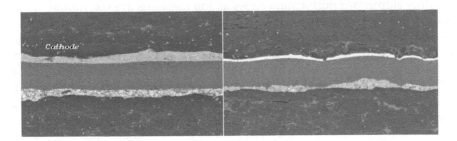

$$2H_2 \rightarrow 4H^+ + 4e^- \xrightarrow{\quad e^- \quad} 4H^+ + 4e^- \rightarrow 2H_2$$
$$O_2 + 4H^+ + 4e^- \rightarrow 2H_2O$$

FIGURE 1.5
Schematic of the fuel cell environment and electrochemical potentials during a partial fuel starvation. (From Lauritzen, M.V., et al. 2007. *J. New Mat. Electrochem. Sys* 10:143–145. With permission.

FIGURE 1.6
MEA cross section picture comparison of MEAs at 0 and 30 h of corrosion cycling. (Left): 0 h, prior to corrosion occurring; (right): 30 h, after corrosion occurred; cathode catalyst layer thinned by 65%. (From Young, A.P., et al. 2009. *J. Electrochem. Soc.* 156 (8):B913–B922. With permission.)

Under current load, the cell voltage, and thus the driving force for damage, is reduced.

As the carbon support corrodes, the connection to the Pt catalyst particles is severed, resulting in loss of effective Pt. The structure of the cathode catalyst layer collapses, as evidenced by thinning of the layer in scanning electron microscope (SEM) cross sections (Figure 1.6). The carbon becomes much more hydrophilic due to the presence of oxygen species on the carbon surface, and the collapsed structure reduces oxygen diffusion to the remaining catalyst particles, resulting in significant performance degradation. The performance loss due to the presence of any contaminant species adsorbed on the cathode catalyst will be increased as the surface area of the cathode catalyst decreases.

The issue of carbon corrosion has received considerable attention in recent years. There are several drivers for this: (1) the cost drivers for commercialization require the use of high performance catalysts with less durable carbon catalyst supports, (2) the need for system simplification and low cost prevents additional control systems to be implemented to avoid the carbon corrosion conditions, and (3) the use of the fuel cells subjected to "real world" conditions as opposed to carefully controlled demonstration projects, with very dynamic duty cycles and many start-up/shutdown cycles. This increased attention has resulted in new or improved measurement techniques and several studies and reviews on the high cathode potential and associated carbon corrosion mechanism [39,40,48–51].

1.7.2 Membrane Degradation

The thin polymer electrolyte membrane at the heart of the fuel cell is prone to degradation, and failure of this membrane can result in the end of the useful fuel cell's lifetime. Either extreme thinning or hole formation can result in excessive gas transfer between electrodes, signaling the end-of-life due to either significant performance loss as the reactants are consumed or displaced, or safety concerns due to mixing of combustible gases. It is widely understood that the failure of the membrane is due to a combination of chemical and mechanical stresses, with several steps involved. Prevention of any of the contributing mechanisms can significantly improve lifetime. Collier et al. [52] provide a review of the membrane degradation literature and classify the failures to a combination of electrochemical/chemical, mechanical, and thermal stressors.

1.7.2.1 Chemical Degradation

The first step in the chemical degradation mechanism is the production of hydrogen peroxide, which may be produced as a by-product of the oxygen reduction reaction on the cathode, or may be produced chemically by crossover of either hydrogen or oxygen to the opposite electrode. The hydrogen peroxide reacts with metal ion contaminants (M^{n+}) acting as Fenton's catalysts to produce very reactive hydroperoxy and peroxy radicals, as described by equation (1.35) and equation (1.36).

$$H_2O_2 + M^{2+} \rightarrow M^{3+} + \cdot OH + OH^-$$ (1.35)

$$\cdot OH + H_2O_2 \rightarrow H_2O + HOO\cdot$$ (1.36)

The most common, highly reactive Fenton's catalyst in the fuel cell is typically iron, the presence of only small amounts of which is required to initiate this reaction. Iron is typically present as a contaminant in some of the component materials and also may be introduced from the system piping or components. Pozio et al. [53] presented an example of the impact of iron contamination due to the use of stainless steel endplates, which resulted in significantly higher membrane degradation than for the same design using aluminum endplates.

The hydroperoxy radical is highly reactive and attacks the membrane polymer at susceptible spots in the polymer chain. The perfluorinated backbone is chemically resistant; however, nonfluorinated end groups [54], as well as the ether linkage on the side chain [55], represent attack points in the polymer (see section 1.2.3.1 for chemical structure of a typical membrane). As the polymer is degraded, fluoride is released from the polymer structure, and can be measured in the exhaust water as a useful indicator of the amount of degradation occurring. A reduction in the nonfluorinated end groups, as shown in Figure 1.7, results in significantly improved membrane durability as demonstrated by results from DuPont for experimental membrane candidates, where the measured fluoride emission rates have been significantly reduced [54], but does not completely eliminate this degradation mode. This chemical attack results in loss of ionomer material and weakens the mechanical structure, thereby making the membranes more prone to mechanical degradation and failure.

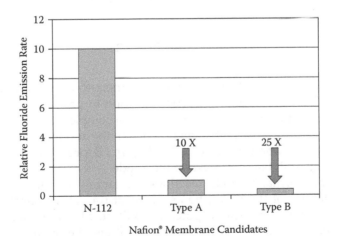

Nafion® Membrane Candidates

FIGURE 1.7
Reduction in fluoride emissions for developmental Nafion® membranes made using DuPont's proprietary protection strategies. (From Curtin, D.E, R.D. Lousenberg, T.J. Henry, et al. 2004. Advanced materials for improved PEMFC performance and life. *J. Power Sources* 131:41–48. With permission.)

1.7.2.2 Mechanical and Thermal Degradation

The membrane may fail due to thinning, rips, tears, or holes. Uneven compression or a rough electrode or GDL surface structure can result in mechanical stress on the membrane and eventually failure. Reduced humidification has been observed to result in significantly reduced membrane lifetime [56], and low membrane water content can make the membrane brittle and fragile [57]. However, the most dominant mechanical failure is believed to be due to the in-plane tensile and compressive stresses imposed by the membrane swelling and shrinking cycle resulting from wet and dry operating regimes, which can occur over the operational duty-cycle [58–60].

Thermal stresses can further accelerate both the chemical and mechanical degradation mechanisms.

1.7.3 Anode Degradation

The most significant anode degradation mechanism, aside from contamination, is that initiated by fuel starvation. In the absence of sufficient fuel on the anode, in a cell (or cells) within a stack, to supply the protons required to maintain the current, the anode potential will rise until protons can be produced from another source. This rise in anode potential, above that of the cathode potential, is manifested as a negative cell voltage, also referred to as *cell voltage reversal*. The secondary source of protons is typically water, which will undergo electrolysis (oxidation) above 1.2 V on the anode, according to reaction (equation (1.37)), resulting in the evolution of oxygen and protons on the anode.

$$H_2O \rightarrow O_2 + 2H^+ + 2e^- \tag{1.37}$$

The protons will pass through the membrane and combine with oxygen at the cathode in the normal reduction reaction to produce water (reverse of reaction; equation (1.37)). A polarization of a complete fuel starvation (no hydrogen) with humidified nitrogen flowing on the anode, with the cell connected to a power supply, is presented in Figure 1.8 for an anode with a 4 mg Pt/cm^2 loading of Pt black catalyst on the anode.

Because the normal cathode reactions are largely unaffected, a typical cathode potential of 0.7 V can be assumed at 0.4 A/cm^2 current density. Therefore, the resulting anode potential will be >1.5 V. If the catalyst is supported on carbon, the carbon support will also be oxidized, in a reaction similar to that described for the cathode carbon corrosion (see section 1.7.1.2), and significant degradation will occur. This scenario only occurs when there are enough cells in the stack that are operating normally to produce the power required to drive the abnormally operating cells to the high

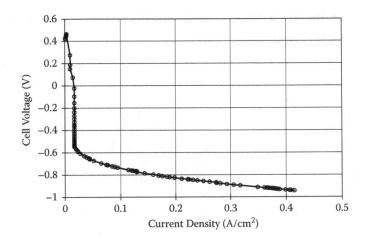

FIGURE 1.8
Fuel starvation polarization. Humidified anode/cathode feed streams: nitrogen/air. 3 bara, 75°C, 4 mg/cm² Pt on each of cathode and anode. (From Knights, S.D., et al. 2004. *J. Power Sources* 127, 127–134. With permission.)

anode potential. The electrolysis reaction can only be sustained for a limited period of time, until either the water supply is depleted or the degradation occurring significantly impacts the anode's ability to function. At this point the cell voltage will go further negative and the anode carbon corrosion is accelerated. Significant degradation can occur, depending on the amount of time spent in fuel starvation.

This failure mechanism can have significant impact on the ability of the anode to tolerate adsorbed contaminants. Similar to the impact of carbon corrosion on the cathode, the reduced electrochemically active catalyst surface area becomes very sensitive to the presence of contaminants. This is very important, for example, for operation on reformate where even small amounts of carbon monoxide can result in significant performance loss.

1.8 Introduction to Contamination Issues in Fuel Cells

Contamination of PEM fuel cells is an important durability and operational issue. Introduction of contaminants in the fuel or oxidant streams, or through fuel cell component materials composition, can result in negative performance effects and increase degradation modes. This book will cover the issues and understanding associated with contamination of PEM fuel cells. A brief overview of contamination issues, along with reference to the material in this book, is included in this section.

1.8.1 Contamination Sources and Chemical/Electrochemical Reactions

1.8.1.1 Cathode Contaminants

The cathode contaminants' sources and general mechanisms are described in chapter 2, and a more detailed discussion on mechanisms, experimental results, and mitigation is provided in chapter 3.

The major air contaminants are carbon oxides (CO and CO_2), sulfur oxides (SO_2 and SO_3), nitrogen oxides (NO_2 and NO), volatile organic carbons (VOCs), and ammonia (NH_3). In the absence of a filtration system, these contaminants can impact the fuel cell performance.

The CO_x in the air is mainly in the form of CO_2, which comes from both natural and manmade sources. This has no direct poisoning effect on the PEM fuel cell, and is in low enough quantities that the dilution effect on the oxygen is negligible.

Sulfur oxides are the most common contaminant of concern for fuel cell operation. The source of SO_2 in the air is almost entirely from human sources, including the main sources of electricity generation, fossil fuel combustion, and industrial processes. SO_2 can strongly adsorb on the Pt surface, blocking active sites for the ORR and leading to performance degradation at concentrations as low as 1 ppm. High potentials are required to remove the adsorbed sulfur species, and recovery can be slow or incomplete. Hydrogen sulfide can also poison the cathode when present in the air stream with significant performance loss observed in the ppm range due to strong H_2S adsorption on the Pt surface. Performance recovery by operation on neat air is slow and incomplete, whereas voltage cycling results in substantial performance recovery.

Nitrogen oxides are formed during combustion processes, industrial processes, fertilizer use, electricity generation, and from other sources. NO_x in the air will result in performance degradation to the fuel cell in the ppb to ppm range. However, the adsorption on the catalyst sites is quite weak and the performance is typically easily recovered with cessation or reduction in the NO_x concentrations. However, very long exposures to NO_x may result in permanent performance loss.

The class of chemicals designated as VOCs includes a wide range of carbon-based molecules of sufficient vapor pressure to be present in the air, such as aldehydes and ketones. The most common VOC is methane, the primary component of natural gas. There are various sources, both natural and human, of VOCs. The response of the fuel cell to VOCs will vary significantly, depending on the molecules in question, but can be significant. For example, benzene and toluene at the ppm level have both been found to significantly affect performance, with the dominant effect believed to be due to adsorption on the catalyst surface resulting in kinetic losses. A semiempirical model for toluene contamination based on kinetic losses is described in chapter 3, section 3.8.

Ammonia (NH_3) comes from several sources, including waste and fertilizers, natural decay processes, combustion, sewage treatment plants, industrial processes, and vehicle exhaust. The main degradation mechanism is the exchange of the ammonium ions for the H^+ ions in the membrane necessary to carry the proton charge. The resulting loss in membrane conductivity will decrease the fuel cell performance. The NH_3 contamination also can impact the catalyst layer conductivity and the cathode and anode kinetics. Ammonia will negatively affect the cell performance at concentrations as low as 1 ppm. The performance loss can spontaneously recover when the contaminant is no longer in the feed stream, but only for low degradation levels. Prolonged exposure to high concentrations will not fully recover in a practical time frame.

1.8.1.2 Anode Contaminants

Hydrogen is presently primarily produced from the reforming of fossil fuels, with approximately half from the reforming of natural gas. Other key sources include water electrolysis and hydrogen by-product from industrial sources. A general introduction to anode contaminants sources is provided in chapter 2. A more detailed discussion of the various hydrogen production processes and the expected fuel compositions, as well as anode contaminant mechanisms, experimental results, and mitigation strategies, are given in chapter 4.

The major hydrogen fuel contamination sources include carbon monoxide and carbon dioxide (CO_x), hydrogen sulfide (H_2S), and ammonia (NH_3).

Carbon monoxide results in significant performance loss, even at low concentrations (low ppm level), due to adsorption of CO on either bare Pt sites or Pt-H sites. The performance degradation is increased at lower temperatures. Carbon dioxide also will result in a poisoning effect, primarily due to the production of CO from the CO_2 on the Pt surface due to the well-known reverse water–gas shift (RWGS) reaction or from electroreduction. However, much higher concentrations of CO_2 (% level) than that of CO can be tolerated. Although there is a strong performance impact in the presence of CO_x, these species are relatively easy to clean off the catalyst surface through chemical and electrochemical oxidation reactions and the performance generally recovers fairly rapidly if the CO_x concentration is reduced or eliminated. The performance impact of CO can also be reduced through the use of Pt alloy catalysts on the anode, the most typical one being PtRu. The use of an air or oxygen bleed, in which a small amount of air or O_2 is fed into the fuel stream, is a highly effective means to counteract much of the performance loss due to the CO and allows continual operation with ppm levels of CO present in the fuel.

Hydrogen sulfide has the largest impact of the most common fuel cell contaminants and will negatively affect fuel cell performance in the ppb levels. H_2S is present in natural gas, a common feed stream for hydrogen production, is a by-product of many industrial processes, and is produced by living organisms. H_2S adsorbs dissociatively on the Pt surface and blocks the active sites for H_2 oxidation. Again, the adsorption process is strongly affected by temperature with lower temperatures resulting in increased coverage on the Pt surface. The formation of Pt-S on the catalyst surface is very stable and difficult to remove, although some recovery is possible when the anode is exposed to increased potentials.

Ammonia in the hydrogen fuel originates from the hydrogen production process. The impact on fuel cell performance is as described for the cathode contamination, and is similar whether it is introduced in the cathode or anode.

1.8.1.3 Anionic Contaminants

Halide ions are the most important anion contaminants and may be introduced from the environment or through component materials, such as chlorine used to synthesize catalysts or the fluorine degradation product from the membrane. The presence of halide ions can increase Pt dissolution rates, thereby resulting in reduced lifetimes.

1.8.1.4 Membrane Contamination

The membrane contaminants sources and general mechanisms are described in chapter 2. Chapter 5 provides a discussion on membrane properties and degradation mechanisms, and a more detailed discussion on membrane contamination sources, mechanisms, and mitigation.

The main contaminants for the membrane are cationic species, such as metal ions, which may come from contaminated air and fuel streams when moisture is present, metal fuel cell components, balance-of-plant components, or nonmetal contaminated component materials. Other organic and inorganic materials can also contaminate the membrane, but the effects of these are less well documented. Component materials supplying contaminants may include the platinum catalyst or alloying metals, such as ruthenium or cobalt, which may leach out into the membrane; the raw material source for the carbon materials (in the catalyst support, microporous layer, gas diffusion layer, or plate materials) may also have inherent metal or other chemical impurities; and seal and gasketing materials, such as silicone, can decompose and contaminate the membrane. All of the membrane contaminants can also impact the ionomer materials present in the catalyst layers.

The sulfonic acid groups in the proton exchange membrane have a high affinity for many cationic species. Replacement of the protons by the contaminant cations will result in reduced conductivity of the membrane and performance loss. In addition, the membrane morphology and structure can be affected. An additional degradation issue associated with contaminant metal ions is the occurrence of Fenton's reactions to produce peroxy and hydroperoxy radicals, which in turn will chemically attack the membrane polymer structure. This mechanism is described in section 1.7.2.1 of this chapter and in chapter 5. The most common Fenton's metal of concern commonly found in the MEA is iron.

Of the most common air and fuel gaseous contaminants, ammonia is the most significant membrane contaminant, as the cationic species NH^{4+} will interfere with the proton conduction mechanism and result in decreased membrane conductivity. Other gaseous contaminants may also affect the membrane and ionomer functionality through pH effects and decomposition products; these are discussed in chapter 5.

1.8.2 Fuel Cell Contamination Modeling

Contamination modeling is an important aspect of fuel cell development. It is required to interpolate and extrapolate experimental results to expected conditions in real-world operation, as it is impractical to test all combinations of reactant concentrations and fuel cell operating conditions. Modeling also assists in the development and validation of hypothesized contamination mechanisms. Model development for the anode is more extensive than that for the cathode contamination. The majority of the modeling deals with the kinetic effects associated with adsorption of contaminant species on the cathode and anode catalysts.

1.8.2.1 Cathode Contamination Modeling

Chapter 6 provides an overview of cathode contamination modeling in the literature and presents a model by the authors of that chapter. The oxygen reduction mechanism is described in some detail to provide a basis for the modeling approach. Several kinetic models reported in literature employ either an associative or dissociative mechanism. Based on literature evidence, the authors determined that the associative mechanism provided better simulation agreement with experimental and theoretical calculations. A cathode contamination model is presented, which assumes the first electron transfer is the rate-determining step, for a general contaminant, P. The model is validated against experimental results for toluene contamination of the cathode, in the range of 1 to 10 ppm toluene. Performance loss can be predicted as a function of toluene concentration and current density at the ppb range, which is representative of expected indoor and outdoor conditions. From the predicted effects, allowable toluene levels can be set. For example, to achieve a performance loss of <10 mV at 1 A/cm², the allowable toluene concentration should be set to less than 100 ppb. The model also provides an analysis of expected Pt surface coverage by adsorbed toluene and oxygen coverage as a function of overpotential or current density and toluene concentration. At low overpotential, the toluene surface coverage is quite low, regardless of toluene concentration. However, as the overpotential is increased, the toluene coverage increases significantly. For example, at 1 A/cm² and 750 ppb toluene, the surface coverage is estimated to be 73%.

Contamination models in the literature tend to be either empirical models or competitive adsorption models. Empirical models for air side contaminants include SO_2, NO_2 (kinetic effect), and NH_3 (ohmic effect). There is a nonempirical model for NO_2 reported in literature based on competitive adsorption kinetics of O_2 and NO_2 on Pt catalysts.

1.8.2.2 Anode Contamination Modeling

Chapter 7 provides a comprehensive literature review and discussion on fuel cell anode contamination modeling. An introduction to fuel contamination is provided, including a hydrogen fuel quality specification for transportation hydrogen.

Carbon monoxide is the most understood poisoning phenomena due to extensive studies. The goal of this chapter (7) is to provide a full understanding of the CO poisoning phenomenon, sufficient to enable the reader to numerically simulate CO poisoning as well as its mitigation methods. Empirical models are usually developed and an example of this is described. Mathematical models, which include the fundamental physical and chemical properties of the system under study, are more useful for parametric studies, such as, for example, the effects of the catalyst layer structure. Several approaches to the mathematical models are briefly described.

The adsorption, desorption, and electro-oxidation reactions of hydrogen and carbon monoxide are discussed. The hydrogen oxidation reaction requires the dissociation of the hydrogen molecule onto bare platinum sites. The CO molecules will also adsorb on the Pt sites, and require oxidation at higher electrode potentials in the range of 0.6 to 0.9 V for removal. Since this potential does not readily occur on the anode, the hydrogen oxidation occurs on a reduced number of Pt sites, resulting in increased anode overpotential.

Several mitigation methods for carbon monoxide poisoning are described. The introduction of oxygen into the fuel stream relies on the heterogeneous oxidation of carbon monoxide by oxygen on the catalyst surface. This can be a very effective method of mitigating the effects of CO poisoning. The use of CO tolerant binary catalysts is presented, along with an extensive list of references for the most common catalyst types. The reaction kinetics and associated model for a PtRu/C are also provided. Additional mitigating methods include advanced reformer design to reduce the CO concentration entering the fuel cell; the use of high temperature membranes, such as phosphoric acid-doped polybenzimidazole (PBI); and using a two-layer structure that relies on the different diffusion rates of H_2 and CO.

A complete model of CO poisoning is presented, including the catalyst layer as well as transport in the GDL and flow channel. An example of validation of the models based on the polarization curve is presented.

The comprehensive overview of the modeling approaches includes a discussion of the model approaches used to investigate the actual kinetics including the use of either Langmuir or Tempkin kinetics. The CO poisoning effect also depends on the concentration of H_2, with a greater effect for more dilute H_2 streams. The modeling results for several effects on CO poisoning are presented, including the effects of temperature on both steady state and transient CO poisoning, pressure, transient CO exposure, oxygen bleeding, CO tolerant catalysts, and CO poisoning on the liquid water distribution in the cell.

The poisoning effect of CO_2 is mainly due to the formation of CO by the RWGS reaction and the electro-reduction of CO_2 by Pt hydrides at low potentials. Review of the CO_2 reaction kinetics and performance effects are discussed.

The presence of H_2S causes significant performance loss. A comprehensive overview of approaches to study and model the poisoning effects of H_2S is presented. H_2S poisoning occurs kinetically similar to CO poisoning, and modeling approaches are also similar. Mitigation methods discussed for H_2S include:

- Use of H_2S-tolerant catalysts, which have not been established to be beneficial with the exception of increased Pt loading, which will have cost implications.
- Use of transient concentrations of pure hydrogen, which results in partial recovery
- Use of transient voltage pulses, which provides significant to almost full recovery when high voltages are applied.

A brief summary of ammonia poisoning results in the literature is provided. However, little modeling work has been completed and no models are described. While there is consensus that there is an effect of ammonia on proton conductivity, other effects on the anode and crossover to the cathode are not uniformly supported by the existing literature.

The effects of mixtures of contaminants have not been extensively studied and little experimental data is available. There is a need for experimental studies and modeling in this area. Chapter 7 closes with a discussion on critical needs in fuel cell contamination modeling.

1.8.2.3 Membrane Contamination Modeling

Chapter 8 provides a comprehensive review of membrane cationic contamination with macroscopic physics-based mathematical models. The introduction of cationic species into the electrolyte membrane can displace the proton associated with a sulfonate group. In general, the presence of cationic contaminants will not significantly affect performance for many hours of operation. However, cation contaminants tend to be stable in the membrane and build up over time.

Three main effects of cationic contamination include a loss in fuel cell performance; an increase in high frequency cell resistance, which does not wholly account for the performance losses; and a decrease in limiting current. In addition to the effect on the membrane conductivity, the presence of contaminant cation species changes the membrane properties, including membrane water content and electro-osmotic drag. The most studied species is ammonia.

The membrane structure assumptions for the basis of membrane models include various combinations of homogeneous and porous structures for vapor and liquid phases. The authors in chapter 8 present a membrane model based on a homogeneous vapor phase of the membrane as being sufficient in predicting performance of cation contamination. The model includes relationships describing water content, water diffusion flux, conductivity, and water electro-osmotic drag. The effects of multiple cations are included, with ammonium ion contamination as an example. The effect of ionic potential across the membrane is added, resulting in migration of charged ions. Ion diffusion, combined with a gradient in water electrochemical potential, results in combined ion and water fluxes. A discussion on concentrated solution theory, which allows the treatment of all species interacting with each other, is included. The transport equations are developed with the continuity equation, with no sources or sinks inside the membrane, only at the boundaries.

The general behavior of the membrane with cation contamination, with and without the water effects, is presented and discussed based on predictions from the model. The model is extended to include the potential from the metal on each side into and through the ionomer in the electrodes. Because the work is focused on the membrane, this aspect is restricted to only include a hydrogen pump cell with hydrogen electrodes on both the anode and cathode sides.

The model results are compared to experimental data based on cesium contamination of a fuel cell operated in hydrogen pump mode.

1.8.3 Impurity Mitigation Strategies

Chapter 9 provides an overview of mitigation strategies that are divided into three sections: (1) mitigation of hydrogen impurities in hydrogen production sources, (2) mitigation of impurities in the fuel cell system, and (3) reducing the impact of impurities on fuel cells.

A summary table of hydrogen production processes is presented in the first section on mitigation of impurities in hydrogen production sources. A detailed discussion of the various processes, including a general description of each process, thermodynamic and kinetic considerations, and advantages and disadvantages, is followed by an extensive overview of hydrogen purification technologies for removal of hydrogen sulfide, carbon monoxide, and carbon dioxide.

Three technologies are described for hydrogen sulfide removal. Chemical absorption with metal oxide or metal hydroxide, such as the reaction of H_2S with CaO to produce CaS and water, is an effective method. Partial oxidation, such as the Claus reaction, converts hydrogen sulfide through reaction with oxygen into elemental sulfur and water, and is typically used in fuel processing industries. An Fe^{3+}/Fe^{2+} redox system also can be used for partial oxidation. Decomposition of H_2S recovers both H_2 and S, but requires thermal or electrical energy to carry out the reaction, which occurs at temperatures in excess of $1500°C$, and must be followed by a rapid cooling to prevent the reverse reaction. Metal- or metal oxide-based cycles also can be used to accomplish decomposition in a thermochemical cycle.

Carbon monoxide and carbon dioxide separation from hydrogen is most commonly accomplished via four categories of removal technologies. Adsorption, including temperature swing adsorption and pressure swing adsorption, can be used to remove CO and CO_2. Absorption (physical or chemical) of CO_2 can produce pure CO_2. Membrane separation, e.g., palladium (Pd)-based membrane, is very effective for purification of H_2 down to low levels of CO, such as ~10 ppm. Cryogenic separation based on partial condensation will separate H_2 from impurities with higher boiling points.

In the second section on mitigation of impurities at the fuel cell system level, a general description of a fuel cell system is provided, along with a list of possible contamination sources in the fuel cell stack and system. The strategies for reducing or eliminating the contamination sources fall into four categories. The proper selection and screening of materials is important to avoid trace contaminants, such as anion (F^-, Cl^-) and cation impurities, metallic corrosion products, and organic polymer decomposition products. Prevention of contaminants from air or fuel supplies entering the system can be accomplished through the use of filtration systems and through hydrogen purification systems. Continuous removal of contaminants from recirculating fluids, such as the use of ion exchange filters in water loops, is important to prevent contaminant buildup. Periodic decontamination operations can be used to flush contaminants out of the system that bypass the other mitigation methods. For example, CO poisoning can be recovered via periodic air bleedings; startup/shutdown processes can purge contaminants; periodic potential cycling or oxygen introduction can oxidize and clean adsorbed contaminants; deionized water or acidic, basic, or chelating solutions can be washed through the cell depending on the nature of the contaminant; and cyclic oxidant starvation can reduce species and allow them to be removed.

The third section of chapter 9 includes strategies for mitigation of impurity impacts. Discussion on fuel side contamination mitigation includes strategies for carbon monoxide mitigation, including oxidant bleeding, the use of CO tolerant catalysts, and optimizing operating conditions, such as temperature, current cycling, and fuel starvation. Hydrogen sulfide mitigation strategies have not been well developed as compared to CO strategies, and many of the same ones have been attempted with much less success. Reintroducing neat hydrogen only results in partial recovery; increased temperature was not sufficient to prevent H_2S adsorption on the catalyst; and the use of CO tolerant catalysts did not provide improved tolerance to H_2S. Current cycling, however, was found to provide some measure of performance recovery from H_2S contamination.

Mitigation of air side contamination is focused on chloride ion, sulfur containing compounds, and nitrogen oxides. Attempted mitigation of the presence of Cl^-, introduced in the catalyst synthesis procedure or through the humidifier, including flushing with hot water, was unsuccessful. Sulfur compounds of concern include SO_2, H_2S, and COS. Sulfur adsorbed onto the catalyst surface can be removed through voltage cycling, flushing with water, and exposure to pure air. NO_2 contamination in the fuel cell can be mitigated through introduction of clean air or by voltage cycling.

Ammonia contamination can affect both the air and fuel sides, regardless of from which side the contamination enters. Introduction of acid into the fuel cell to displace the ammonium ion from the proton exchange sites was unsuccessful in promoting performance recovery. Introduction of an air bleed was also of no impact. The cell will slowly recover performance after the contamination introduction is stopped, but the extent of this recovery will depend on the amount of contamination present and the cell might not fully recover even after several days.

1.9 Summary

Contamination is a significant durability challenge affecting the operation of fuel cells. The main contaminants of concern and their sources have been identified for the anode, cathode, and membrane. The contamination mechanisms for the most common contaminants are thoroughly discussed in this book, providing a detailed overview of the current understanding and gaps in knowledge. Modeling of fuel contaminants is more extensive than that of cathode contaminants. The mechanisms of the oxygen reduction reaction, the hydrogen oxidation reaction, and the membrane transport processes are discussed to provide a base set of model relationships. To these are added the contamination reactions to provide predictive modeling capability of the effects of contaminants on performance and behavior of the fuel cell. A significant gap in contamination modeling is that of the combined effects of multiple species. A number of contamination mitigation strategies have been investigated, with varying degrees of success. Although some contaminants, such as carbon monoxide, can be dealt with fairly readily in the fuel cell at low concentrations, others, such as sulfur compounds, are more effectively removed prior to entry. Significant progress has been made in fuel cell contamination characterization and understanding, but much work remains to be completed in order to develop a comprehensive strategy for fuel cell operation under commercial applications.

References

1. Marcinkoski, J., J.P. Kopasz, and T.G. Benjamin. 2008. Progress in the US DOE fuel cell subprogram efforts in polymer electrolyte fuel cells. *Int. J. Hydrogen Energy* 33:3894–3902.
2. Vielstich, W. 2003. Ideal and effective efficiencies of cell reactions and comparison to carnot cycles. In *Handbook of fuel cells volume 1*, ed. H.A. Gasteiger, A. Lamm, and W. Vielstich, 27, Chichester, England: John Wiley & Sons Ltd.
3. World Energy Council. 2009. Fuel Cells. http://www.worldenergy.org/focus/fuel_cells/377.asp (accessed May 29, 2009)
4. Unnasch, S. 2006. Greenhouse gas analysis for hydrogen fuel cell vehicles. *Fuel Cell Seminar 2006*. Honolulu, Hawaii.
5. Elgowainy, A., L. Gaines, and M. Wang. 2009. Fuel-cycle analysis of early market applications of fuel cells: Forklift propulsion systems and distributed power generation. *Int. J. Hydrogen Energy*, 34:3557–3570.
6. Birnbaum, U., M. Haines, J.F. Hake, and J. Linssen. 2008. Reduction of greenhouse gas emissions through fuel cell combined heat and power applications. *17th World Hydrogen Energy Conference*, Brisbane, Australia.
7. Krewitt, W., J. Nitsch, M. Fischedick, et al. 2006. Market perspectives of stationary fuel cells in a sustainable energy supply system—long-term scenarios for Germany. *Energy Policy* 34:793–803.
8. Lubis, L.I., I. Dincer, G.F. Naterer, and M.A. Rosen. 2009. Utilizing hydrogen energy to reduce greenhouse gas emissions in Canada's residential sector. *Int. J. Hydrogen Energy* 34:1631–1637.
9. Smithsonian Natural History Museum. 2008. Collecting the history of fuel cells. http://americanhistory.si.edu/fuelcells/ (accessed May 29, 2009)
10. Adamson, K.A., J. Butler, M. Hugh, and P. Doran. 2008. Fuel cells: Commercialization. *Fuel Cell Today*, Industry Review 2008:2–15.
11. U.S. Department of Hydrogen Energy Program. 2008. Comparison of fuel cell technologies. http://www1.eere.energy.gov/hydrogenandfuelcells/fuelcells/pdfs/fc_comparison_chart.pdf (accessed May 29, 2009)
12. Cifrain, M., and K. Kordesch. 2003. Hydrogen/oxygen (air) fuel cells with alkaline electrolytes. In *Handbook of fuel cells*, vol. 1, ed. H.A. Gasteiger, A. Lamm, and W. Vielstich, 267–279, Chichester, England: John Wiley & Sons Ltd.
13. King, J.M., and H.R. Kunz. 2003. Phosphoric acid electrolyte fuel cells. In *Handbook of fuel cells*, ed. H.A. Gasteiger, A. Lamm, and W. Vielstich, 287–300, West Sussex: John Wiley & Sons Ltd.
14. Yokokawa, H., and N. Sakai. 2003. History of high temperature fuel cell development. In *Handbook of fuel cells*, vol. 1, ed. H.A. Gasteiger, A. Lamm, and W. Vielstich, 219–266, Chichester, England: John Wiley & Sons Ltd.
15. http://www.ballard.com/ (accessed May 29, 2009)
16. Wee, J-H. 2007. Applications of proton exchange membrane fuel cell systems. *Renewable Sustainable Energy Rev.* 11:1720–1738.
17. Mahadevan, K., K. Judd, H. Stone, et al. 2007. Identification and characterization of near-term direct hydrogen proton exchange membrane fuel cell markets. Battelle. http://www1.eere.energy.gov/hydrogenandfuelcells/pdfs/pemfc_econ_2006_report_final_0407.pdf (accessed May 29, 2009)

18. Bard, A.J., and L.R. Faulkner. 1980. *Electrochemical methods.* Toronto: John Wiley & Sons.
19. Ye, S. 2007. Electrocatalysts for PEM fuel cells: Challenges and opportunities. *14th National Meeting of Chinese Society of Electrochemistry,* Yangzhou, China.
20. Bagotsky, V.S, E.M. Belgsur, E.J. Cairns, et al. 2003. *Handbook of fuel cells,* vol. 1. ed. Gasteiger, A. Lamm, and W. Vielstich. Chichester, England: John Wiley & Sons Ltd.
21. Weber, A., R. Darling, J. Meyers, and J. Newman. 2003. Mass transfer at two-phase and three-phase interfaces. In *Handbook of fuel cells,* vol. 1, ed. H.A. Gasteiger, A. Lamm, and W. Vielstich, 47–69, Chichester, England: John Wiley & Sons Ltd.
22. Gasteiger, H.A., J.E. Panels, and S.G. Yan. 2004. Dependence of PEM fuel cell performance on catalyst loading. *J. Power Sources,* 127:162–171.
23. Zhang, J., Z. Xie, J. Zhang, et al. 2006. Review, high temperature PEM fuel cells. *J. Power Sources,* 160:872–891.
24. Cheng, X., Z. Shi, N. Glass, et al. 2007. A review of PEM hydrogen fuel cell contamination: Impacts, mechanisms, and mitigation. *J. Power Sources* 165: 739–756.
25. Wu, J., X.Z. Yuan, J.J. Martin, et al. 2008. A review of PEM fuel cell durability: Degradation mechanisms and mitigation strategies. *J. Power Sources* 184:104–119.
26. McNeil, D.A. 2005. Establishing a standardized single cell testing procedure through industry participation, consensus and experimentation. *Electrochemical Society ECS 207th Meeting,* Quebec City, Canada.
27. Maish, A., E. Nilan, and P.M. Baca. 2002. *Characterization of fuel cell duty cycle elements.* Albuquerque, NM: Sandia National Laboratories.
28. Yuan, X., H. Wang, J.C. Sun, and J. Zhang. 2009. AC impedance technique in PEM fuel cell diagnosis—A review. *Int. J. Hydrogen Energy* 32: 4365–4380.
29. Wu, J., X.Z. Yuan, H. Wang, et al. 2008. Diagnostic tools in PEM fuel cell research: Part I electrochemical techniques. *Int. J. Hydrogen Energy.* 33:1735–1746.
30. Cleghorn, S.J.C., D.K. Mayfield, D.A. Moore, et al. 2006. A polymer electrolyte fuel cell life test: 3 years of continuous operation. *J. Power Sources* 158:446–454.
31. Schneider, J., M. Rigolo, S. Knights, et al. 2007. An automotive perspective on durability protocol challenges from single cells to fuel cell vehicle systems. *Fuel Cells Durability & Performance 2007 – Real World Solutions to the Most Significant Challenges Facing Fuel Cells Commercialization.* Miami, FL.
32. Zhang, S., X. Yuan, H. Wang, et al. 2009. A review of accelerated stress tests of MEA durability in PEM fuel cells. *Int. J. Hydrogen Energy* 34:388–404.
33. Van Zee, J.W., J. St. Pierre, K. Punyawudho, and M. Jung. 2008. PEMFC performance studies: The effects of interactions between trace contaminants and water. *Fuel Cells Durability & Performance 2008.* Las Vegas, NV.
34. Chiem, B., P. Beattie, and K. Colbow. 2007. The development and demonstration of technology on the path to commercially viable PEM fuel cell stacks. http://www.electrochem.org/meetings/scheduler/abstracts/214/1090.pdf (accessed May 30, 2009)
35. Schmittinger, W., and A. Vahidi. 2008. A review of the main parameters influencing long-term performance and durability of PEM fuel cells. *J. Power Sources* 180:1–14.

36. Wilkinson, D.P., and J. St-Pierre. 2003. *Handbook of fuel cells: Fundamentals, technology and applications*, vol. 3. ed. W. Vielstich, H. Gasteiger, and A. Lamm. 611–626. Chichester, England: John Wiley & Sons Ltd.

37. Borup, R., et al. 2007. Scientific aspects of polymer electrolyte fuel cell durability and degradation. *Chem. Rev.* 107: 3904–51.

38. Yousfi-Steiner, N., et al. 2008. A review on PEM voltage degradation associated with water management: Impacts, influent factors and characterization. *J. Power Sources* 183:260–274.

39. Yu, X., and S. Ye. 2007. Recent advances in activity and durability enhancement of Pt/C catalytic cathode in PEMFC: Part II: Degradation mechanism and durability enhancement of carbon supported platinum catalyst. *J. Power Sources* 172:145–154.

40. Yu, X., and S. Ye. 2007. Recent advances in activity and durability enhancement of Pt/C catalytic cathode in PEMFC: Part I. Physico-chemical and electronic interaction between Pt and carbon support, and activity enhancement of Pt/C catalyst. *J. Power Sources* 172:133–144.

41. Darling, R.M., and J.P. Meyers. 2005. Mathematical model of platinum movement in PEM fuel cells. *J. Electrochem. Soc.* 152:A242–A247.

42. Ohma, A., S. Suga, S. Yamamoto, and K. Shinohara. 2007. Membrane degradation behavior during open-circuit voltage hold test. *J. Electrochem. Soc.* 154(8):B757–B760.

43. Bi, W., G.E. Gray, and T.F. Fuller. 2007. PEM fuel cell Pt/C dissolution and deposition in Nafion® electrolyte. *Electrochem. Solid-State Lett.* 10(5):B101–B104.

44. Peron, J., Y. Nedellec, D.J. Jones, and J. Roziere. 2008. The effect of dissolution, migration and precipitation of platinum in Nafion®-based membrane electrode assemblies during fuel cell operation at high potential. *J. Power Sources* 185:1209–1217.

45. Dicks, A.L. 2006. The role of carbon in fuel cells. *J. Power Sources* 156:128–141.

46. Lauritzen, M.V., P. He, A. Young, et al. 2007. Study of fuel cell corrosion processes using dynamic hydrogen reference electrodes. *J. New Mat. Electrochem. Sys.* 10:143–145.

47. Young, A.P., J. Stumper, and E. Gyenge. 2009. Characterizing the structural degradation in a PEMFC cathode catalyst layer: carbon corrosion. *J. Electrochem. Soc.* 156 (8):B913–B922.

48. Baumgartner, W.R., P. Parz, S.D. Fraser, et al. 2008. Polarization study of a PEMFC with four reference electrodes at hydrogen starvation conditions. *J. Power Sources* 182:413–421.

49. Maass, S., F. Finsterwalder, G. Frank, et al. 2008. Carbon support oxidation in PEM fuel cell cathodes. *J. Power Sources* 176:444–451.

50. Shao, Y., G. Yin, and Y. Gao. 2007. Understanding and approaches for the durability issues of Pt-based catalysts for PEM fuel cell. *J. Power Sources* 171:558–566.

51. Wang, J., G. Yin, Y. Shao, et al. 2007. Effect of carbon black support corrosion on the durability of Pt/C catalyst. *J. Power Sources* 171:331–339.

52. Collier, A., H. Wang, X.Z.Yuan, et al. 2006. Degradation of polymer electrolyte membranes. *Int. J. Hydrogen Energy* 31:1838–1854.

53. Pozio, A., R.F. Silva, M. De Francesco, and L. Giorgi. 2003. Nafion degradation in PEFCs from end plate iron contamination. *Electrochim. Acta* 48:1543–1549.

54. Curtin, D.E, R.D. Lousenberg, T.J. Henry, et al. 2004. Advanced materials for improved PEMFC performance and life. *J. Power Sources* 131:41–48.

55. Marsil, K., M.K. Kadirov, A. Bosnjakovic, and S. Schlick. 2005. Membrane-derived fluorinated radicals detected by electron spin resonance in UV-irradiated Nafion and Dow ionomers: Effect of counterions and H_2O_2. *J. Phys. Chem.* B 109:7664–7670.

56. Knights, S.D., K.M. Colbow, J. St-Pierre, and D.P. Wilkinson. 2004. Aging mechanisms and lifetime of PEFC and DMFC. *J. Power Sources* 127, 127–134.

57. LaConti, A.B., M. Hamdan, and R.C. McDonald. 2003. Mechanisms of membrane degradation. In *Handbook of fuel cells*, vol. 3, ed. H.A. Gasteiger, A. Lamm, and W. Vielstich. Chichester, England: John Wiley & Sons Ltd.

58. Mathias, M.F., R. Makharia, and H.A. Gasteiger. 2005. Two fuel cell cars in every garage. *Electrochem. Soc. Interface* (Fall) 2005:24–35.

59. Tang, H., S. Peikang, S.P Jiang, et al. 2007. A degradation study of Nafion proton exchange membrane of PEM fuel cells. *J. Power Sources*, 170:85–92.

60. Budinski, M.K., C.S. Gittleman, C.L. Lewis, et al. 2006. Method of operating a fuel cell stack. Int. Patent Application W0/2007/053560 A2.

2

Fuel Cell Contaminants: Sources and
Chemical/Electrochemical Reactions

Jianlu Zhang, Zheng Shi, Hui Li, and Jiujun Zhang

CONTENTS

2.1 Introduction

Proton exchange membrane (PEM) fuel cells are prospective alternative power sources for mobile and stationary applications due to their high-energy conversion efficiency and environmental friendliness [1,2], and have been intensively developed worldwide in recent years. However, this technology faces many challenges, including high cost, low performance, and low durability/reliability. Indeed, durability is one of the most important issues and can be adversely affected by impurities in the fuel cell system. Fuel and air feed streams are major sources of impurities, and fuel cell components or accessories can introduce others impurities, which cause performance degradation and sometimes permanent damage to the membrane electrode assembly (MEA).

The effect of impurities on fuel cells, often referred to as *fuel cell contamination*, has been identified as one of the most important issues in fuel cell operation and applications. Studies have shown that the component most affected by contamination is the MEA [3]. Three major effects of contamination on the MEA have been identified [3,4]: (1) the kinetic effect, which involves poisoning of the catalysts or a decrease in catalytic activity; (2) the conductivity effect, reflected in an increase in the solid electrolyte resistance; and (3) the mass transfer effect, caused by changes in catalyst layer structure, interface properties, and hydrophobicity, hindering the mass transfer of hydrogen and/or oxygen.

In order to establish a fundamental understanding of contamination mechanisms, develop effective contamination control strategies, and then improve the reliability/durability of PEM fuel cells, it is important to recognize the sources and their possible effects on fuel cell operation. In this chapter, we review the origins of contaminants and their chemistries, with some description of their effects on PEM fuel cell operation.

2.2 Anode Contaminants/Impurities and Their Basic Chemistry

Contaminants/impurities at the anode are mainly brought in by the fuel feed stream. Impurities in the hydrogen fuel, such as CO, H_2S, NH_3, organic sulfur–carbon, as well as carbon–hydrogen compounds, are primarily from the manufacturing process, in which natural gas or other organic fuels are reformed to produce hydrogen. In this section, several major anode contaminants, such as CO_x, H_2S, and NH_3, will be discussed.

2.2.1 CO_x and Their Basic Chemistry

2.2.1.1 Sources of CO and CO_2

2.2.1.1.1 H_2 Production Processes

CO_x typically indicates carbon monoxide (CO) and carbon dioxide (CO_2). CO_x, especially CO, is one of the key contaminants in PEM fuel cells [5–9]. CO_x in the fuel stream derives mainly from H_2 production processes. Currently, the hydrogen used for fuel cell operation is usually produced by one of the following processes: reforming of natural gas or gasoline [10,11], steam reforming of naphtha [12], oxidation and steam reforming of methanol and propane [13], partial oxidation of gasoline [11], or thermal decomposition of hydrocarbons [11]. Hydrogen produced by these processes inevitably contains carbon oxides and sulfur compounds. For example, in methanol oxidation and steam reforming processes, CO and/or CO_2 can be formed by the following reactions [13]:

$$CH_3OH + \frac{3}{2}O_2 = CO_2 + 2H_2O \qquad (2.1)$$

$$CH_3OH + H_2O = CO_2 + 3H_2 \qquad (2.2)$$

$$CH_3OH = CO + 2H_2 \qquad (2.3)$$

or by direct methanol partial oxidation [14]:

$$CH_3OH + \frac{1}{2}O_2 = CO + H_2O + H_2 \qquad (2.4)$$

2.2.1.1.2 Carbon Corrosion

Another source of carbon oxide impurities in fuel cells is carbon corrosion. Carbon, which is employed as an electrode support in both anode and cathode, can undergo oxidation reactions and produce CO_x [15–20]:

$$C + 2H_2O \rightarrow CO_2 + 4H^+ + 4e^- \, (0.207\,V \text{ vs. NHE}) \qquad (2.5)$$

$$C + H_2O \rightarrow CO + 2H^+ + 2e^- \qquad (2.6)$$

$$CO + H_2O \rightarrow CO_2 + 2H^+ + 2e^- \qquad (2.7)$$

Studies [16–18] have shown that these carbon corrosion processes can be accelerated in PEM fuel cells by harsh operating conditions, such as high voltage, high temperature, low humidity, and start/stop cycles.

Carbon monoxide also can be produced inside a fuel cell on the platinum surface by the reduction of CO_2 [21]. This reaction requires the presence of hydrogen atoms adsorbed on the platinum (Pt) catalyst surface [7,22,23], and can be expressed as

$$CO_2 + 2Pt-H \Rightarrow Pt-CO + H_2O + Pt \tag{2.8}$$

2.2.1.1.3 Other Sources of CO

CO can also arise from CO_2 conversion, the major routes of this being through the reverse water gas shift reaction (RWGSR, equation (2.9)) and the electrochemical equivalent of the RWGSR (equation (2.11)) [24]. From thermodynamics, it has been calculated that at 80°C, for a reformate containing 25% CO_2 in hydrogen, 100 to 200 ppm of CO can be generated by the RWGSR [23,24].

$$CO_2 + H_2 \rightarrow H_2O + CO \tag{2.9}$$

$$2Pt + 2H^+ + 2e^- \rightarrow 2(Pt-H_{ads}) \tag{2.10}$$

$$2(Pt-H_{ads}) + CO_2 \rightarrow Pt-CO + H_2O + Pt \tag{2.11}$$

2.2.1.2 Chemistry of CO

2.2.1.2.1 CO Adsorption

The effect of CO_x contamination on fuel cell performance is shown in Figure 2.1. Evidently, even a small amount of CO_x (and only a trace amount of CO) in the fuel cell feed stream causes a dramatic decrease in performance [9,25], and a higher CO concentration can cause an even larger drop. In order to mitigate the effect of CO_x contamination on fuel cell performance, it is important to understand the contamination mechanism.

CO is known to poison H_2 electro-oxidation on a Pt surface, through adsorption of CO on either bare Pt sites or Pt-H sites [23]. CO adsorbs on Pt surfaces through two modes, linear and bridge [26], as illustrated in Figure 2.2 [27].

CO adsorption on Pt can be described by the following reactions:

$$CO + Pt \rightarrow Pt-CO \tag{2.12}$$

$$2CO + 2(Pt-H) \rightarrow Pt-CO + H_2 \tag{2.13}$$

As shown in equation (2.12), CO adsorption on a Pt surface is associative, while H_2 adsorption, as shown in equation (2.14), is dissociative:

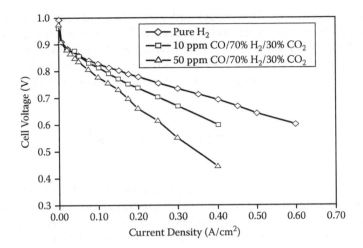

FIGURE 2.1
Cell performance of an MEA when the anode fuel was H_2, 10 ppm CO/70% H_2/30% CO_2, and 50 ppm CO/70% H_2/30% CO_2, respectively. Anode, Pt-Ru = 0.60 mg/cm²; cathode, Pt = 1.7 mg/cm²; Nafion® 1135 membrane. (From Qi, Z. et al. 2002. *J. Power Sources* 111:239–247. With permission.)

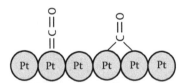

FIGURE 2.2
CO adsorption on Pt surface. (From Zhang, J. et al. 2006. *J. Power Sources* 160:872–891. With permission.)

$$2Pt + H_2 \rightarrow 2(Pt-H) \qquad (2.14)$$

Because the bond between Pt and CO is much stronger than that between Pt and H, CO can block the active Pt sites for H adsorption and also competitively displace the adsorbed H from Pt. The preferential adsorption of CO molecules at the active sites on a platinum surface also lowers the reactivity of the remaining uncovered sites through dipole interactions and electron capture. Therefore, any factors that can promote CO coverage will also promote CO poisoning of the Pt surface [6,28]. CO coverage depends on the catalyst surface state (surface roughness) and the atmosphere of the electrode/electrolyte interface.

Adsorption of CO on Pt is associated with a high negative entropy, indicating that adsorption is strongly favored at low temperatures, and disfavored at high temperatures [26,29–31]. Dhar, et al. [30] found that the CO adsorption isotherm on a Pt surface during H_2 electro-oxidation was a standard Temkin isotherm at 298 K, and the CO coverage (θ_{CO}) could be expressed as

$$\theta_{CO} = 19.9\exp(-7.69 \times 10^{-3}\,T) + 0.085\ln\left(\frac{[CO]}{[H_2]}\right) \qquad (2.15)$$

Equation (2.15) shows that CO coverage increases with CO concentration in the feed stream, and decreases with temperature and H_2 concentration. A thermodynamic analysis of the temperature requirements for CO tolerance was conducted by Yang et al.[32] who reported that the increased tolerance to CO was related to the thermodynamics of adsorption of CO and H_2 on Pt. CO adsorbs associatively on Pt below 500 K, whereas H_2 adsorbs dissociatively.

The fractional coverages (θ) of CO and H_2 on the surface of the catalyst are given by equation (2.16) and equation (2.17):

$$\theta_{CO} = \frac{K_{CO}P_{CO}}{1 + K_{CO}P_{CO} + K_H^{1/2}P_{H_2}^{1/2}} \qquad (2.16)$$

$$\theta_H = \frac{K_H^{1/2}P_{H_2}^{1/2}}{1 + K_{CO}P_{CO} + K_H^{1/2}P_{H_2}^{1/2}} \qquad (2.17)$$

where K_{CO} and K_H are the equilibrium constants for adsorption, and P_{CO} and P_{H_2} are the partial pressures of CO and H_2 in the gas phase, respectively. Because hydrogen adsorption is less exothermic than CO, and H_2 adsorption requires two adsorption sites, increasing the temperature causes a beneficial shift toward higher H_2 coverage, at the expense of CO coverage.

Figure 2.3 shows plots of equilibrium coverage of CO on Pt at CO concentrations ranging from 1 to 100 ppm, with an H_2 pressure of 0.5 bar. It is clear that CO coverage on a Pt surface increases with greater CO concentration, and decreases dramatically with increasing temperature, indicating a higher CO tolerance at higher operating temperatures. Adjemian et al. [33] and Malhorta et al. [34] reported similar curves for CO adsorption. They showed that hydrogen coverage increased from 0.02 monolayers at 350 K (77°C), to 0.39 monolayers at 400 K (127°C).

CO poisoning is strongly related to how long the anode electrode is exposed to CO. As shown in Figure 2.4, the cell voltage decrease becomes more significant with prolonged CO exposure.

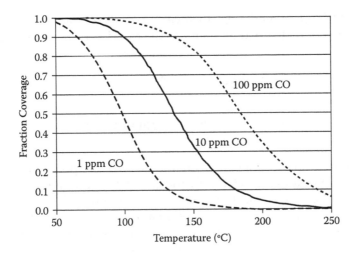

FIGURE 2.3

CO coverage on platinum as a function of temperature and CO concentration. The partial pressure of H_2 is 0.5 bar. (From Yang, C. et al. 2001. *J. Power Sources* 103:1–9. With permission.)

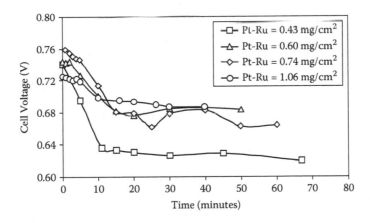

FIGURE 2.4

Cell voltage change with time after 10 ppm CO/70% H_2/30% CO_2 was switched to 50 ppm CO/70% H_2/30% CO_2 for MEAs with different anode Pt-Ru loadings. Current density = 200 mA/cm²; Nafion® 1135 membrane. (From Qi, Z. et al. 2002. *J. Power Sources* 111:239–247. With permission.)

2.2.1.2.2 CO Electro-Oxidation

Oxidizing adsorbed CO on a Pt surface can mitigate the CO poisoning effect. In the presence of water, adsorbed CO can be eliminated through the reactions in equation (2.18) and equation (2.19). The water is oxidized on the free Pt sites, forming adsorbed OH (equation (2.18)), and then the adsorbed CO is eliminated (electrochemically oxidized) through reacting with the adsorbed OH (equation (2.19)). However, when Pt sites are poisoned, water cannot be readily adsorbed on them, so the reaction is ineffective for eliminating CO.

$$Pt + H_2O \rightarrow Pt - OH_{ads} + H^+ + e^-$$
(2.18)

$$Pt - CO_{ads} + Pt - OH_{ads} \rightarrow 2Pt + CO_2 + H^+ + e^-$$
(2.19)

CO electro-oxidation depends on the adsorption potential [23,35–37], operating temperature [23,26,28,31], and catalyst type at the anode. Bellows et al. [23] studied the adsorption potential dependence of CO electro-oxidation. As shown in Figure 2.5, adsorption at the higher potential of 450 mV shows a pronounced single oxidation peak at 900 mV without measurable oxidative current below 800 mV, while adsorption at the lower potentials of 100 mV shows a small oxidation peak beginning at about 400 mV and centered near 500 mV, with a second and larger oxidation peak centered near 700 mV. These findings suggest that CO adsorbs onto bare Pt at 450 mV and adsorbs onto a surface covered with Pt-H at 100 mV. However, it is not clear why CO adsorbed onto Pt-H is more electroactive than CO adsorbed onto bare Pt.

The effect of temperature on CO oxidation is illustrated in Figure 2.6. It can be seen that increasing temperature shifts the electro-oxidation features

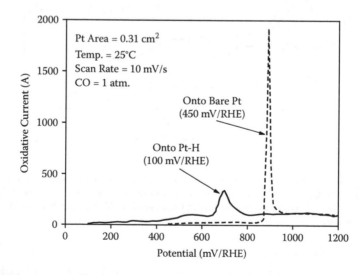

FIGURE 2.5
Linear scanning voltammetry for CO electro-oxidation at 10 mV/s on polycrystalline Pt at 25°C in 1 N H_2SO_4. (From Bellows, R.J. et al. 1996. *Ind. Eng. Chem. Res.* 35:1235–1242. With permission.)

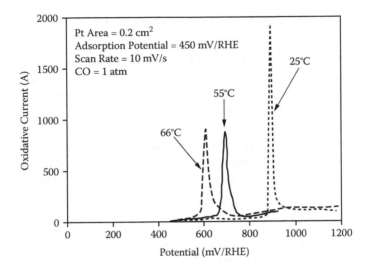

FIGURE 2.6
Linear scanning voltammetry for CO electro-oxidation at 10 mV/s on polycrystalline Pt in 1 N H_2SO_4. CO adsorbed at 450 mV. (From Bellows, R.J. et al. 1996. *Ind. Eng. Chem. Res.* 35:1235–1242. With permission.)

to progressively lower potentials [23]. For example, at 66°C, CO adsorbs at 450 mV and displays an oxidation peak at about 620 mV, without significant oxidation before 500 mV. Bellows et al. [23] concluded that at higher temperatures, CO adsorbed at potentials in the Pt-H region was more easily oxidized than that adsorbed in the double-layer region. The combination of low adsorption potentials and higher temperatures, thus, is more effective for oxidizing CO.

The type of anode catalyst has a strong effect on the severity of CO poisoning, since the catalyst affects the kinetics of CO adsorption (equation (2.12) and equation (2.13)) and CO oxidation (equation (2.18) and equation (2.19)). Based on these mechanisms, many CO-tolerant electrocatalysts have been developed by Pt alloying, such as PtRu (platinum/ruthenium) [24,38], PtSn (platinum/tin) [39–41], and PtMo (platinum/molybdenum) [42–44]. Generally, alloying Pt with a second element can enhance the catalytic activity of the Pt through one or more of the following effects:

Bifunctional effect [45–49]: Water activation is first initiated by the second component (M) to form M-OH, which then reacts with an adjacent CO adsorbed on the Pt atom to complete CO oxidation and clear the Pt surface for hydrogen oxidation. This mechanism can be expressed by equation (2.20) and equation (2.21):

$$M + H_2O \rightarrow M-OH + H^+ + e^- \tag{2.20}$$

$$Pt-CO+M-OH \rightarrow Pt+M+CO_2+H^++e^- \qquad (2.21)$$

Electronic effect [50–53]: The second component alters the electronic properties of the catalytically active Pt surface, reducing the stability of the Pt-CO bonding, so the adsorption of CO on Pt is weakened and the adsorption of hydrogen is promoted.

Ensemble effect: The dilution of the active Pt component with a catalytically inert metal changes the distribution of the active Pt sites, thereby opening different reaction pathways for hydrogen adsorption and oxidation. One example of such an inert metal is Pd-Ag alloy.

2.2.1.3 Chemistry of CO_2

Although the acidic membrane electrolyte in PEMFCs can tolerate the presence of CO_2, experimental results [5,22,24,54,55] suggest that CO_2 nonetheless has a significant poisoning effect on cell performance, especially at higher current densities and low temperatures. The poisoning is mainly due to the *in situ* production of CO from CO_2 on the platinum surface, and the subsequent adsorption of CO on the active platinum sites [7,22,24,26,55]. The *in situ* production of CO occurs mainly through either the reverse water gas shift reaction or the electrochemical reduction of CO_2, as indicated in equation (2.9) and equation (2.11), respectively.

In addition, the presence of both CO and CO_2 has a synergetic poisoning effect on cell performance, which is evident from the aggravated poisoning effect when CO_2 is added to a PEMFC via a CO-containing fuel stream [5]. The presence of CO_2 can also decrease H_2 partial pressure, thus reducing fuel cell performance [5].

Adsorbed CO_2 can be reduced via equation (2.11). At a potential below 0.4 V versus RHE, CO_2 reduction is favored because there is sufficient $Pt-OH_{ads}$ promoting CO_2 electroreduction. Beden et al. [56] investigated the nature of reduced CO_2 by infrared (IR) spectroscopy and found that the reduction of CO_2 involved reaction with adsorbed hydrogen. They also found that reduced CO_2 yielded a significant amount of highly coordinated adsorbed CO species rather than the linearly bonded CO species formed by direct CO adsorption. However, the exact structure and adsorption form of these surface species have not yet been clarified; they could be linear, bridge, or triple-bonded CO [56–61]. Studies have also found CO/COOH radicals [21], $COOH_{ads}$, as well as COH_{ads} [58,62–64] during CO_2 reduction.

2.2.2 H_2S and Its Basic Chemistry

Like CO, H_2S is a major contaminant in the fuel streams of PEM fuel cells. Modeling and experimental studies [65–74] have indicated that trace amounts of H_2S can cause dramatic degradation in fuel cell performance. For example,

5 ppm H_2S in H_2 can result in a 96% loss in fuel cell performance at 50°C within 12 hours [70]. Even with as little as 10 ppb H_2S in H_2, 42.3% of a Pt surface would be covered by H_2S in 1,000 hours [74].

H_2S occurs naturally in the environment (e.g., in marshes, swamps, sulfur springs, decaying organic matter, and volcanic gases). It is also produced by living organisms, human beings included, through the digestion and metabolism of sulfur-containing materials. In addition, H_2S is a by-product of many industrial processes, including paper manufacturing, sewage treatment, landfills, and concentrated animal feed operations [75,76]. H_2S is found in petroleum and natural gas as well, the latter containing up to 28% hydrogen sulfide. Indeed, anthropogenic hydrogen sulfide arises principally as a by-product in the purification of natural gas and the refinement of crude oil [76]. H_2S in the PEMFC fuel stream mainly comes from reformate gas or the reforming of natural gas [10,73].

H_2S in the fuel stream adsorbs dissociatively on the Pt surface and blocks the active sites for H_2 oxidation. As a result, it retards the reaction kinetics and degrades PEM fuel cell performance. Early research into H_2S poisoning on Pt was conducted in aqueous solutions, and two forms of chemisorbed sulfur were identified [77,78], which are strongly and weakly bound to Pt, respectively. In aqueous solutions, the adsorptions of H_2S, HS^-, and H^+ occur through the following reactions [79,80]:

$$Pt + H_2S \rightarrow Pt - H_2S_{ads} \tag{2.22}$$

$$Pt - HS^-_{ads} + H^+ \rightarrow Pt + H_2S_{ads} \tag{2.23}$$

$$Pt + HS^- \rightarrow Pt - HS^-_{ads} \tag{2.24}$$

$$Pt + H^+ \rightarrow Pt^+ - H_{ads} \tag{2.25}$$

where Pt^+ represents an equivalent positive charge on the Pt surface. Equation (2.24) and equation (2.25) can result in an electrochemical potential, which may cause the following dissociative adsorption [80]:

$$2Pt + H_2S \rightarrow Pt - SH_{ads} + Pt - H_{ads} \tag{2.26}$$

According to a potential step study, the dissociative potentials of H_2S on Pt were about 0.4 V at 90°C, 0.5 V at 60°C, and 0.6 V at 30°C [68]. Following dissociative adsorptions, oxidation of the adsorbed SH and H_2S occurred, forming platinum sulfide [80]:

$$Pt - SH_n \Leftrightarrow Pt - S + nH^+ + ne^- (n = 1 or 2) \tag{2.27}$$

The adsorption rate, coverage, and stability of the H_2S species were found to be strongly affected by temperature [68,70]. Mohtadi et al. [70] studied the rate of Pt-S formation at different temperatures and found that this rate was 69% lower at 50°C than that at 90°C. It also has been shown that adsorbed H_2S and SH are highly unstable on Pt, whereas adsorbed S and H are the most stable intermediates on Pt [81]. The strong bond between sulfide and platinum makes it impossible for the fuel cell to recover [68,71] by flowing pure hydrogen alone. In order to recover the fuel cell performance, a potential higher than 0.9 V versus NHE is required to oxidize the adsorbed sulfur through the following reactions [71,78]:

$$Pt-S+3H_2O \rightarrow SO_3+6H^++6e^-+Pt \tag{2.28}$$

$$Pt-S+4H_2O \rightarrow SO_4^{2-}+8H^++6e^-+Pt \tag{2.29}$$

Thus, similar to CO, H_2S coverage on a Pt surface is affected by concentration, exposure time, fuel cell current, and temperature. Normally, H_2S coverage rises with increasing concentration, exposure time, and fuel cell current density, along with decreasing operating temperature [65,68,70].

2.2.3 NH$_3$ and Its Basic Chemistry

NH_3 is another typical contaminant in PEM fuel cell anodes. Trace amounts of NH_3 in a fuel cell system will cause significant fuel cell degradation [82–86]. NH_3 in the fuel stream originates mainly from hydrogen production and storage, due to several causes:

1. NH_3 can be formed in the fuel-reforming process by the reaction of N_2 in air, if the reformer is operated with air (autothermal reforming and partial oxidation).
2. NH_3 may also be present as a residual component if it is used as a feedstock for hydrogen production by thermal splitting [87].
3. NH_3 also can be formed catalytically from traces of N_2 in H_2 when AB_5 metal hydride alloys are used for hydrogen storage [88,89].
4. NH_3 can be introduced into the fuel system when it is used as a tracer gas in natural gas distribution.

As ammonia is alkaline, it is absorbed by a PEM fuel cell's acidic membrane (e.g., Nafion® membrane), where it forms ammonium. Partial or full exchange of H^+ by NH_4^+ or other cations in the membrane will reduce membrane conductivity and thereby decrease fuel cell performance. As shown in Table 2.1 [84], the conductivity of a fully exchanged NH_4^+-form Nafion 105 membrane is only about one-quarter that of the protonic-form membrane.

TABLE 2.1

Conductivity of Nafion® 105 Membrane in Various Cationic Forms

Cation	Conductivity (S cm⁻¹)	Conductivity (S cm⁻¹)
NH_4^+	0.106	0.032
NH_4^+	0.108	0.033
K^+	0.113	0.021
H^+	0.133	0.133

Source: Uribe, F.A. et al. 2002. *J. Electrochem. Soc.* 149:A293–A296. With permission.

Conductivities from resistance measurements in deionized water at 10 kHz and 25°C:

- Membrane pretreated by immersion in 0.1 M solution for 1 h
- Membrane pretreated by immersion in 1.0 M solution for 66 h
- Immersion in $(NH_4)_2SO_4$ solution
- Immersion in NH_4OH solution
- Immersion in KOH solution

Membrane conductivity in PvEM fuel cell systems decreases with increasing NH_3 concentration, as reported by Uribe et al. [84,86]. The membrane contamination process can be described by the following reactions [85]:

$$NH_3(gas) \rightarrow NH_3(membrane) \qquad (2.30)$$

$$NH_3(membrane) + H^+ \rightarrow NH_4^+ \qquad (2.31)$$

Similarly, the Nafion ionomer in a catalyst layer can also be neutralized by NH_3, leading to a NH_4^+-form Nafion ionomer [84,86]. This will further decrease fuel cell performance.

Studies have shown that the decrease in membrane conductivity can explain only part of the total loss in fuel cell performance, typically 5 to 15%. Uribe et al. [84] and Soto et al. [86] attributed some performance loss to decreased ionomer conductivity in the catalyst layer. However, other studies [85,89] indicated that NH_3 could also affect the kinetics of the hydrogen oxidation reaction (HOR) and the oxygen reduction reaction (ORR). Halseid et al. [85] found that the HOR in a symmetric H_2/H_2 cell was affected by NH_3 and they suggested that the adsorbed species could partially block the anode catalyst surface. They also found that the presence of NH_3 could shift the potential of the hydrogen adsorption process on Pt in aqueous solutions [85]. This catalyst poisoning effect was not observed by Uribe et al [82,83].

Similar to H_2 crossover in PEM fuel cells, NH_3 can also cross from the anode (or cathode) side over to the cathode (or anode) side. Diffusion of NH_3 in the membrane is rapid [85,89]. For instance, for a typical membrane thickness (l) of 10 to 100 μm and a diffusivity of 10^{-10} m^2/s, the estimated characteristic time constant for NH_3 diffusion is of the order l^2/D_{NH_3}, or 1 to 100 s [85,89]. NH_3 that has diffused from the anode can decrease fuel cell performance by affecting the ORR at the cathode. Szymanski et al. [83] studied the effect of NH_3 on phosphoric acid fuel cell performance and found that the performance loss occurred primarily at the cathode, indicating that NH_3 did affect the ORR. However, the mechanism is not clear. Halseid et al. [85] proposed the following NH_3 contamination mechanisms: (1) electrochemical oxidation of NH_3, leading to the formation of adsorbed species that block active sites on the cathode, (2) formation of mixed potential on the cathode due to oxidation of NH_3 occurring simultaneously with the ORR, and (3) reduced proton activity in the cathode catalyst layer due to NH_3, which therefore affects the ORR. The Tafel slope of the ORR, particularly at low current densities, can also be affected by NH_3 adsorption, including intermediate species as well as N-O species formed at high potentials [90].

Adsorbed NH_3 (NH_4^+) can be oxidized electrochemically at the cathode according to the following reaction [83]:

$$NH_4^+ \rightarrow \frac{1}{2}N_2 + 4H^+ + 3e^- \qquad (2.32)$$

Its reaction rate is strongly potential dependent, and can be increased by a factor of 60 with a potential increase from 0.65 to 0.85 V [83]. Higher temperature operation may also increase the rate of NH_3 oxidation, as suggested by Halseid et al. [85].

2.3 Cathodic Contaminants/Impurities and Their Basic Chemistry

In practical applications, a PEM fuel cell cathode will be exposed to air, and all of the impurities in that air may enter the fuel cell system if it has no effective filtration. The major air contaminants are CO_x (CO_2, CO), SO_x (SO_2, SO_3), NO_x (NO_2, NO), and NH_3. In this section, we will discuss each of these cathodic contaminants in detail.

2.3.1 CO_x and Their Basic Chemistry

2.3.1.1 Sources of CO_x

Carbon dioxide is the main CO_x in a fuel cell's cathodic feed stream. The CO_2 in the atmosphere comes primarily from the following sources [91–94]:

1. *Natural sources.* The Earth's natural volcanic activity can generate sources of carbon dioxide, such as geothermal vents and natural wells. Depending on the location, carbon dioxide concentrations of over 90% can be found in underground wells [93].

2. *Fermentation.* The conversion of sugars, starches, and other carbohydrates into alcohols or acids by fermentation usually produces gaseous carbon dioxides as by-products. The purity of carbon dioxide produced in this way can be very high.

3. *Respiration.* Carbon dioxide is produced in the bodies of humans and other animals by the combining of oxygen and sugars, and then emitted through breathing. Respiration also occurs in plants as part of their energy generation. It is well known that photosynthesizing plants consume CO_2, but they also produce it through respiration.

4. *Decay.* The decaying of plants and of animal bodies through bacterial and other decomposition processes produces CO_2.

5. *Industrial processes.* Industrial processes are some of the key sources of carbon dioxide. CO_2 is usually produced as a by-product of various nonenergy-related industrial activities. Four main types of industrial processes generate CO_2:

   a. Production and consumption of mineral products, such as cement, lime, and soda ash

   b. Production of metals, such as iron, steel, aluminum, zinc, and lead

   c. Production of chemicals (e.g., ammonia, petrochemicals, and titanium dioxide)

   d. Consumption of oil products in feedstocks and other end-uses

6. *Fossil fuel combustion.* Fossil fuel combustion is another major source of CO_2. Carbon is one of the main elements in fossil fuels, and is converted to CO_2 when fossil fuels combust. Petroleum (oil), natural gas, and coal are the main fossil fuels consumed by humans. Electricity generation, industrial production, and the use of internal combustion engines are the primary activities that burn fossil fuels and thereby generate CO_2.

7. *Deforestation.* With the development of agriculture and industry, humans have burned vast amounts of forest land. Burning forests not only produces CO_2, but also reduces the amount of CO_2 being consumed by photosynthesis.

2.3.1.2 Chemistry of CO_x

CO_2 in the cathodic feed stream has no direct poisoning effect on PEM fuel cells. Instead, it causes fuel cell performance degradation by diluting the oxygen. However, due to the low concentration of CO_2 in air, its dilution effect is much smaller than that of nitrogen in the air stream.

2.3.2 SO_x and Their Basic Chemistry

Sulfur oxides, mainly SO_2, are common contaminants in PEM fuel cell air streams. About 99% of the sulfur dioxide in the air comes from human sources [95,96], the main one being industrial activities that process sulfur-containing materials, e.g., electricity generation from coal, oil, or gas that contains sulfur. Some mineral ores also contain sulfur, and sulfur dioxide is released when they are processed. Sulfur dioxide is also present in motor vehicle emissions as a result of fuel combustion, and, thus, occurs in high concentrations in urban areas with heavy traffic.

Figure 2.7 shows the 2002 national summary of sulfur dioxide emissions by source sectors prepared by the U.S. Environmental Protection Agency (EPA), with input from numerous state and local air agencies, from Native Americans, and from industry [95]. As the figure indicates, SO_2 comes primarily from electricity generation, fossil fuel combustion, and industrial processes.

When the air stream of the fuel cell contains acidic impurities, such as SO_x, the pH inside the MEA decreases, resulting in free acids within it and causing potential shifts.

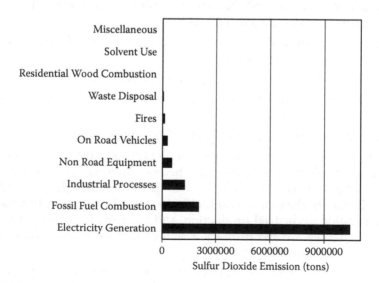

FIGURE 2.7

National sulfur dioxide emissions by source sector in 2002. (From http://www.epa.gov/air/emissions/so2.htm)

Like H_2S, SO_2 can strongly adsorb on the Pt surface, competing with oxygen adsorption and leading to performance degradation [3,97–100]. The SO_2 adsorbed on Pt can be reduced electrochemically to sulfur with the formation of an SO intermediate, as shown in following reactions [98,101]:

$$Pt + SO_2 + 2H^+ + 2e^- \rightarrow Pt-SO + H_2O \tag{2.33}$$

$$Pt-SO + 2H^+ + 2e^- \rightarrow Pt-S + H_2O \tag{2.34}$$

Studies [77,78,101] have shown that two forms of sulfur adsorb on the Pt surface, one strongly and one weakly bound to Pt through the bridge mode and the linear mode, respectively.

The adsorbed sulfur on the Pt catalyst surface can be removed by oxidation at higher electrode potentials, as suggested by Loucka [78], according to reaction (2.28) and reaction (2.29). Mangun et al. [102] studied the adsorption of SO_2 on activated carbon and suggested that, due to the presence of delocalized electrons at the edge sites of carbons where oxygen could chemisorb, the adsorbed SO_2 could react with this oxygen to form SO_3 that then reacted with water to form H_2SO_4. These processes are described by the following reactions [97,102]:

$$2(Pt-SO_2) + O_2 \rightarrow 2(Pt-SO_3) \tag{2.35}$$

$$Pt-SO_3 + H_2O \rightarrow H_2SO_4 + Pt \tag{2.36}$$

Thus, fuel cell performance could be recovered by removing sulfur from the Pt catalyst surface. However, studies [97,103] showed that the fuel cell performance could not be recovered completely. The reason could be as suggested by Garsany et al. [99,100], who studied the impact of SO_2 on the ORR on Pt/C and Pt_3Co/C electrocatalysts, and found that with SO_2, the ORR pathway could be changed from 4-electron to 2-electron. When the sulfur coverage was larger than 0.14, the hydrogen peroxide formation was enhanced, which might have caused irreversible damage to the MEA [99].

2.3.3 NO_x and Their Basic Chemistry

Similar to SO_x, NO_x are also important fuel cell contaminants. NO_x derive from several sources [104–106]. The primary sources are motor vehicles, electrical utilities, and other industrial, commercial, and residential sources that burn fuels. Other NO_x arise from human activities that include biomass burning, waste disposal, solvent use, and fertilizer use. NO_x are formed when fuel is burned at high temperatures, as in a combustion process. Fossil

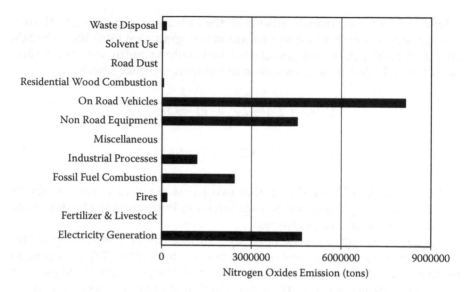

FIGURE 2.8
National nitrogen oxides emissions by source sector in 2002. (From http://www.epa.gov/air/emissions/nox.htm)

fuel combustion is responsible for over half the total global emissions of nitrogen oxides, and up to two-thirds of the emissions from human activities. Although the contribution from coal combustion alone is not known, it is evident that oil combustion in the transport sector commonly contributes over 50% of emissions from human activities in individual countries. During coal combustion, nitrogen present in both the coal and the combustion air is converted to NO_x.

Figure 2.8 shows the 2002 national summary of nitrogen oxides emissions by source sector, prepared by the U.S. EPA with input from numerous state and local air agencies, from Native Americans, and from industry [104]. The figure clearly shows that NO_x come mainly from on-road vehicle emissions, nonroad equipment, fossil fuel combustion, electricity generation, and industrial processes.

When NO_x are present in the air and/or in fuel cell cathode, NO is easily oxidized to NO_2. Mohtadi et al. [98] performed a cyclic voltammetry (CV) study on a fuel cell after it was exposed to NO_2 and found no oxidation peaks corresponding to adsorbed NO_2 species on the Pt surface. This indicated that the poisoning effects of NO_2 do not appear to be a catalyst poisoning issues; instead, they suggested [98] that the ionomer and/or the catalyst–ionomer interface could be affected by NO_2 exposure through the following reaction:

$$NO_2(g) + 8H^+ + 7e^- \rightarrow NH_4^+ + 2H_2O \qquad (2.37)$$

Thus, NO_2 would compete with O_2 for electrons on the cathode and the NH_4^+ thus formed would further contaminate the fuel cell, causing performance degradation by poisoning the Nafion membrane and ionomer in the catalyst layer, as discussed in section 2.2.3.

Jing et al. [97] and Rodes et al. [107] observed an NO_2 oxidation peak on a CV scanning curve. This indicated that NO_x could adsorb on the Pt surface and affect the ORR. The observation of increased kinetic resistance in the ORR after exposure to NO_x [108] provided further support for this mechanism.

However, NO_x adsorption is very weak and NO_x can be oxidized electrochemically, therefore, fuel cell performance loss due to NO_2 contamination can be almost fully recovered [97,98,108]. The mechanism of performance recovery might be the oxidation of NH_4^+ according to reaction (2.32), or the oxidation of NO_2 expressed in reaction (2.38) [97] or in reaction (2.39) and reaction (2.40) [108]:

$$NO_2 + H_2O \rightarrow NO_3^- + 2H^+ + e^- \qquad (2.38)$$

$$2NO_2 + H_2O \rightarrow NO_3^- + NO_2^- + 2H^+ \qquad (2.39)$$

$$2NO_2^- + O_2 \rightarrow 2NO_3^- \qquad (2.40)$$

2.3.4 VOCs and Their Basic Chemistry

Volatile organic compounds (VOCs) are common in the atmosphere, as VOCs have high enough vapor pressures under normal conditions to significantly vaporize and enter the air. A wide range of carbon-based molecules, such as aldehydes, ketones, and other light hydrocarbons, are referred to as VOCs [109].

VOCs can be found both indoors and outdoors. The most common is methane, the primary component of natural gas. Major worldwide sources of atmospheric methane include wetlands, ruminants (such as cows), rice agriculture, landfills, and the burning of biomass (such as wood) [109].

Untreated exhaust emissions from vehicles that run on compressed natural gas have the highest VOC emissions, in addition to other chemical emissions [110]. Formaldehyde, toluene, and limonene are a few of the most common VOCs emitted in indoor environments by consumer products, such as cleaning solvents, paints, and printers [111]. As an important biological source, trees also can emit large amounts of VOCs, especially isoprene and terpenes. Another significant source of VOC emissions is crude oil tanking, during both offloading and loading of tanks. Office buildings also contain many sources of VOCs, including new furnishings, wall coverings, and office equipment, such as photocopy machines that can off-gas

VOC particles into the air [111]. In battlefields, chemical warfare can release significant amounts of VOCs, such as sarin, hydrogen cyanide (HCN), and so forth [112].

These VOCs can cause disastrous degradation of PEMFC performance. The effect of benzene contamination on cell performance is dependent on the operating current density. In testing, performance drops were not fully recovered after the benzene was shut off and the fuel cell had been purged with pure air for 10 minutes. However, the performance was recoverable by operating at open circuit potential [36]. Toluene contamination in the air can affect fuel cell performance as well, mainly resulting in performance degradation in the kinetics of the fuel cell reactions [4,113,114]; this form of contamination also results in easy flooding at the cathode side, possibly by changing the hydrophilicity of the cathode catalyst layer [4,113,114]. Fuel cell cathode contamination caused by chemical warfare gases is disastrous, and the degraded performance is only partly recoverable. Contamination with sarin and sulfur mustard is even more disastrous than with HCN. So far, the contamination mechanisms of chemical warfare agents are not clear.

2.3.5 NH$_3$ and Its Basic Chemistry

NH$_3$, which exists commonly in air, has several sources [115–117]:

1. The most common sources are waste and fertilizers. Large amounts of ammonia can be released into the air near farms and industries. Farms have high levels of ammonia due to animal waste storage and the use of ammonia-containing fertilizers.

2. Decaying plants or animals, coal or wood fires, as well as marshes, all release small amounts of ammonia into the air.

3. Sewage treatment plants may release large amounts of ammonia.

4. High levels of ammonia can also be released from industrial sites that store ammonia or use it as a refrigerant, if the chemical leaks or is spilled.

5. Ammonia can be emitted from on-road vehicles. In some countries, three-way catalysts are installed in automobiles to control nitrogen oxide pollutants. These catalysts operate by constantly oscillating the air-to-fuel ratio in the engine between oxidating and reducing conditions, to simultaneously control hydrocarbon, carbon monoxide (CO), and nitrogen oxides (NO$_x$) emissions. Under reducing conditions, the catalyst may also produce ammonia.

As discussed in section 2.2.3, NH$_3$, as a contaminant in a fuel cell system, can cause significant fuel cell performance degradation. NH$_3$ may be introduced into the fuel cell system via either the fuel stream or the air stream. Due to the fast diffusion rate of NH_3 in the membrane [85,89], it is probably not important whether it enters the fuel cell system from the anode or cathode. Therefore, the effect of NH_3 contamination from either side is the same.

2.4 Membrane Contaminants/Impurities and Their Basic Chemistry

The foreign cationic ions, such as alkali metals, alkaline earth metals, transition metals, and rare earth metals, can cause membrane contamination and then fuel cell performance degradation. The sources of these cationic ions are as follows:

1. Metal ions can come from fuel cell components. For example, the ions of Fe, Cr, Ni, and Mo can come from fuel cell end plates or stainless steel bipolar plates [118,119].
2. Metal ions can come from fuel cell accessories, such as coolants, humidifiers, deionized water, reactant pipelines, etc.
3. Metals ions also can come from electrocatalysts in fuel cells. For example, some unalloyed metals can be released from multicomponent catalysts, and some metal ions can also be released from Pt-based alloy catalysts [120–122].

Due to the high affinity of many cationic species to the sulfonic acid groups in Nafion [123], contaminant cations will decrease the ionic conductivity of a membrane by replacing the membrane's H^+ [124–127], thereby decreasing fuel cell performance. The effect is more pronounced for cations with higher valence than those with lower valence because Nafion displays greater affinity for many cations, and particularly strong bonding with di- and trivalent cations. In general, the affinity of sulfonic acid PEMs for cations increases as the charge and size of the cation increases [126,128].

Trace metals accelerate peroxide degradation by catalyzing the radical formation reaction, as described by reaction (2.41):

$$H_2O_2 + M^{2+} \rightarrow M^{3+} + {}^\bullet OH + OH^- \tag{2.41}$$

LaConti et al. [129] proposed a decomposition mechanism of hydrogen peroxide in the presence of impurity cations, such as Fe^{2+}:

$$H_2O_2 + Fe^{2+} \rightarrow HO^\bullet + OH^- + Fe^{3+} \tag{2.42}$$

$$Fe^{2+} + HO^{\bullet} \rightarrow Fe^{3+} + OH^{-} \tag{2.43}$$

$$H_2O_2 + HO^{\bullet} \rightarrow HO_2^{\bullet} + H_2O \tag{2.44}$$

$$Fe^{2+} + HO_2^{\bullet} \rightarrow Fe^{3+} + HO_2^{-} \tag{2.45}$$

$$Fe^{3+} + HO_2^{\bullet} \rightarrow Fe^{2+} + H^{+} + O_2 \tag{2.46}$$

The hydroxyl radical, formed via this process, is considered one of the most reactive chemical species known, and it could degrade Nafion polymer chains and release fluoride.

Impurity cations can also affect the oxygen reduction kinetics at platinum electrodes covered with perfluorinated ionomer [130]. The reduced kinetics for the ORR could be ascribed to a change in the electric double layer at the platinum–ionomer interface. Study has also shown that the diffusion coefficients of both oxygen and hydrogen in the membrane were decreased by the presence of impurity cations, an effect that could be attributed to changes in the polymer flexibility [130].

Modeling and experimental studies [131–133] have also indicated that when other cations (except Li^+) replaced protons in the polymer, less water was coordinated and the amount of water in the cell was affected. For example, the exchange of Ca^{2+} for H^+ can cause a 19% reduction in membrane water content. Less water may result in a greater concentration of H_2O_2 and, therefore, a greater concentration of destructive hydroxyl radicals [131,134,135]. Moreover, low water content might accelerate mechanical degradation due to membrane dehydration.

2.5 Anions and Their Chemistry

Anions, such as halide ions (F^-, Cl^-), SO_4^{2-}, and NO_3^-, also can be introduced into a fuel cell system during operation. Halide ions are the most important contaminants among all anions, and can enter fuel cells in different ways.

In practical operation of PEM fuel cells, especially in coastal or marine areas, airborne salts can be introduced into the cathode during air intake. Depending on weather conditions, the natural NaCl in the air in coastal areas can be as high as ca. 800 $nmol/m^3$ [136,137]. This is an important issue for PEM fuel cell operation.

Another source of halogen contamination is halide-containing educts (e.g., Cl^-) used to synthesize high surface-area fuel cell catalysts [138–141], which are not always completely removed after synthesis. Chloride is also likely to be present in the water contained in PEM fuel cell feed streams [142].

The proton exchange membrane can be a source of fluoride ions as well [143]. Hydroxyl radicals, formed via crossover gases or reactions of hydrogen peroxide with Fenton-active contaminants (e.g., Fe^{2+}), could attack the backbone of Nafion, causing the release of fluoride anions; these anions in turn promote corrosion of the fuel cell plates and catalyst, and release transition metals into the fuel cell [143]. Transition metal ions, such as Fe^{2+}, then catalyze the formation of radicals within the Nafion membrane, resulting in a further release of fluoride anions. On the other hand, transition metal ions also can cause decreased membrane and ionomer conductivity in catalyst layers, as discussed in section 2.4 of this chapter.

The chloride ion is considered one of the most severe anion contaminants in fuel cell systems [142–145]. Yadav et al. [146] studied the effect of halogen ions on Pt dissolution under potential cycling in a 0.5 M H_2SO_4 solution, and found that F^- ions did not have any significant influence on Pt dissolution, while the effect of Cl^- was strongly dependent on its concentration and the morphology of the Pt samples. In the test, Pt dissolution was induced with a chloride ion concentration of as low as 10 ppm. Regardless of the thickness and particle size of the Pt sample, Pt dissolution was more enhanced at and above 100 ppm Cl^-. Matsuoka et al. [145] investigated the effects of four anions (Cl^-, F^-, SO_4^{2-}, and NO_3^-) on PEM fuel cell operation. F^-, SO_4^{2-}, and NO_3^- did not show significant effects on cell performance, while Cl^- caused visible performance degradation as well as a 30% loss in the electrochemical surface area of Pt. Their study also showed that Cl^- caused the formation of a Pt band inside the membrane by promoting Pt dissolution. In addition, another study has shown that Cl^- can affect the ORR on a Pt/carbon fuel cell catalyst by adsorbing on the Pt surface. Even trace amounts of chloride (i.e., 10^{-4} M Cl^- or ca. 1 ppm) could drastically decrease the catalyst activity for the ORR and enhance the formation of H_2O_2 as well, by changing the ORR pathway [142].

The adsorption of chloride ion on a Pt electrode depends on the ion concentration, Pt morphology, temperature, and electrode potential. The coverage of Cl^- on Pt decreases when both temperature [147,148] and concentration [147] increase, and when electrode potential decreases [147,148]. At 20°C, the coverage of Cl^- on polycrystalline Pt was found to be ca. 0.5 monolayers at 0.7 V and ca. 0.45 monolayers at 0.6 V, when Cl^- concentration in the electrolyte was 10^{-3} M [147]. Stamenkovic et al. [149] found that Pt(111) is more active than Pt(100) for both the ORR and the HOR when Cl^- is present in the electrolyte. They attributed the differences in catalytic activities to a much stronger interaction of Cl_{ad} with the (100) sites than with the (111) sites, and they proposed that Cl_{ad} strongly adsorbed onto Pt(100) can simultaneously suppress both the adsorption of O_2 and H_2 molecules and the formation of pairs of Pt sites needed to break the O-O and H-H bonds.

2.6 Other Contamination Sources

Other than the contaminant sources mentioned above, some contaminants can also come from coolants and deionized water (e.g., Si, Al, S, K, Fe, Cu, Cl, V, Cr), compressors (e.g., oils), and sealing gaskets (e.g., Si), as summarized by Cheng et al. [150]. In recent studies [151–153], silicon was detected in a fuel cell catalyst layer after long-term operation. Silicon released due to gasket failure will get into the fuel cell catalyst layer, adsorb on the Pt catalyst surface or the interface of the Pt and ionomer, and thereby cause degradation in fuel cell performance.

2.7 Summary

Major PEM fuel cell contaminants, their sources, and their chemical/electro-chemical reactions were reviewed in this chapter. The key contaminants and their principal sources are summarized in Table 2.2 [150].

At the fuel cell anode side, the main contaminants are CO, H_2S, and NH_3. These can originate from the reformate gas and hydrogen production processes. Both CO and H_2S reduce fuel cell performance by strong adsorption on the Pt catalyst surface, thus poisoning the catalyst and retarding H_2 oxidation. NH_3 causes a decrease in membrane conductivity by forming NH_4^+ and then replacing H^+ in the Nafion membrane and ionomer in both anode and cathode catalyst layers.

At the fuel cell cathode side, the prominent contaminants are NO_x (NO and NO_2) and SO_x (SO_2 and SO_3). These mainly arise from electricity generation and fossil fuel combustions. NO_x adsorb weakly on the Pt catalyst surface, and the contamination mechanism is mainly via the formation of NH_4^+, which

TABLE 2.2

Major Contaminants Identified in the Operation of PEM Fuel Cells

Impurity Source	Typical Contaminant
Air	N_2, NO_x (NO, NO_2), SO_x (SO_2, SO_3), NH_3, O_3
Reformate hydrogen	CO, CO_2, H_2S, NH_3, CH_4
Bipolar metal plates (end plates)	Fe^{3+}, Ni^{2+}, Cu^{2+}, Cr^{3+}
Membrane (Nafion®)	Na^+, Ca^{2+}
Sealing gaskets	Si
Coolants, DI water	Si, Al, S, K, Fe, Cu, Cl, V, Cr
Battlefield pollutants	SO_2, NO_2, CO, propane, benzene
Compressors	Oils

Source: Cheng, X. et al. 2007. *J. Power Sources* 165:739–756. With permission.

poisons the Nafion membrane and ionomers in the catalyst layers. SO_x cause fuel cell performance degradation mainly by blocking Pt active sites for the ORR, and they also can change the ORR reaction pathways from 4-electron to 2-electron.

Metal ions mainly come from fuel cell components and system accessories, and cause a decrease in membrane conductivity by replacing H^+. Metal ions also cause membrane decay by catalyzing the formation of hydroxyl radicals, which are the most active chemicals involved in chemical degradation of membranes.

References

1. Steele, B.C.H. and A. Heinzel. 2001. Materials for fuel-cell technologies. *Nature* 414:345–352.
2. Haile, S.M., et al. 2001. Solid acids as fuel cell electrolytes. *Nature* 410:910–913.
3. Knights, S., et al. 2005. Fuel cell progress, challenges and markets. Paper presented at the *Fuel Cell Seminar 2005,* Palm Springs, CA.
4. Li, H., et al. 2008. Polymer electrolyte membrane fuel cell contamination: Testing and diagnosis of toluene induced cathode degradation. *J. Power Sources* 185:272–279.
5. Smolinka, T., et al. 2005. CO_2 reduction on Pt electrocatalysts and its impact on H_2 oxidation in CO_2 containing fuel cell feed gas—A combined *in situ* infrared spectroscopy, mass spectrometry and fuel cell performance study. *Electrochim. Acta* 50:5189–5199.
6. Davies, J.C., et al. 2004. CO desorption rate dependence on CO partial pressure over platinum fuel cell catalysts. *Fuel Cells* 4:309–319.
7. Janssen, G.J.M. 2004. Modelling study of CO_2 poisoning on PEMFC anodes. *J. Power Sources* 136:45–54.
8. Jiménez, S., et al. 2005. Assessment of the performance of a PEMFC in the presence of CO. *J. Power Sources* 151: 69–73.
9. Qi, Z., C. He, and A. Kaufman. 2002. Effect of CO in the anode fuel on the performance of PEM fuel cell cathode. *J. Power Sources* 111:239–247.
10. Dicks, A.L. 1996. Hydrogen generation from natural gas for the fuel cell systems of tomorrow. *J. Power Sources* 61:113–124.
11. Jamal, Y. and M.L. Wyszynski. 1994. On-board generation of hydrogen-rich gaseous fuels—a review. *Int. J. Hydrogen Energy* 19:557–572.
12. Darwish, N.A., et al. 2004. Feasibility of the direct generation of hydrogen for fuel-cell-powered vehicles by on-board steam reforming of naphtha. *Fuel* 83:409–417.
13. Avci, A.K., Z.İ. Önsan, and D.L. Trimm. 2003. On-board hydrogen generation for fuel cell-powered vehicles: the use of methanol and propane. *Top Catal.* 22:359–367.
14. Trimm, D.L. and Z.I. Onsan. 2001. On-board fuel conversion for hydrogen-fuel-cell-driven vehicles. *Catal. Rev.* 43:31–84.

15. Meyers, J.P. and R.M. Darling. 2006. Model of carbon corrosion in PEM fuel cells. *J. Electrochem. Soc.* 153:A1432–A1442.
16. Atanassova, P., et al. 2007. Carbon corrosion effects in fuel cells. Paper presented at the *Gordon Research Conference on Fuel Cells.* 2007. Bryant University, Smithfield, RI.
17. Baumgartner, W.R. 2006. Carbon corrosion in PEMFC electrodes under real operating conditions. http://www.eva.ac.at/publ/pdf/afc_baumgartner.pdf
18. Fullera, T.F. and G. Gray. Carbon corrosion induced by partial hydrogen coverage. http://www.fcbt.gatech.edu/Publications/Carbon_Corrosion_Induced_by_Partial_Hydrogen_Coverage.pdf
19. Tang, H., et al. 2006. PEM fuel cell cathode carbon corrosion due to the formation of air/fuel boundary at the anode. *J. Power Sources* 158:1306–1312.
20. Venkatasubramanian, V., et al. 2008. Tungsten carbide as possible support for Pt in electrochemical Reactions. *Bull. Catal. Soc. Ind.* 7:146–152.
21. Giner, J. 1963. Electrochemical reduction of CO_2 on platinum electrodes in acid solutions. *Electrochim. Acta* 8:857–865.
22. de Bruijn, F.A., et al. 2002. The influence of carbon dioxide on PEM fuel cell anodes. *J. Power Sources* 110:117–124.
23. Bellows, R.J., E.P. Marucchi-Soos, and D.T. Buckley. 1996. Analysis of reaction kinetics for carbon monoxide and carbon dioxide on polycrystalline platinum relative to fuel cell operation. *Ind. Eng. Chem. Res.* 35:1235–1242.
24. Ralph, T.R. and M.P. Hogarth. 2002. Catalysis for low temperature fuel cells. *Platinum Met. Rev.* 46:117–135.
25. Qi, Z., C. He, and A. Kaufman. 2001. Poisoning of proton exchange membrane fuel cell cathode by CO in the anode fuel. *Electrochem. Solid-State Lett.* 4:A204–A205.
26. Li, Q., et al. 2003. The CO poisoning effect in PEMFCs operational at temperatures up to 200°C. *J. Electrochem. Soc.* 150:A1599–A1605.
27. Zhang, J., et al. 2006. High temperature PEM fuel cells. *J. Power Sources* 160:872–891.
28. Kim, J.D., et al. 2001. Characterization of CO tolerance of PEMFC by AC impedance spectroscopy. *Solid State Ionics* 140:313–325.
29. Vogel, W., et al. 1975. Reaction pathways and poisons—II: The rate controlling step for electrochemical oxidation of hydrogen on Pt in acid and poisoning of the reaction by CO. *Electrochim. Acta* 20:79–93.
30. Dhar, H.P., L.G. Christner, and A.K. Kush. 1987. Nature of CO adsorption during H_2 oxidation in relation to modeling for CO poisoning of a fuel cell anode. *J. Electrochem. Soc.* 134:3021–3026.
31. Dhar, H.P., et al. 1986. Performance study of a fuel cell Pt-on-C anode in presence of CO and CO_2, and calculation of adsorption parameters for CO poisoning. *J. Electrochem. Soc.* 133:1574–1582.
32. Yang, C., et al. 2001. Approaches and technical challenges to high temperature operation of proton exchange membrane fuel cells. *J. Power Sources* 103:1–9.
33. Adjemian, K.T., et al. 2002. Silicon oxide Nafion® composite membranes for proton-exchange membrane fuel cell operation at 80–140°C. *J. Electrochem. Soc.* 149:A256–A261.
34. Malhotra, S. and R. Datta. 1997. Membrane-supported nonvolatile acidic electrolytes allow higher temperature operation of proton-exchange membrane fuel cells. *J. Electrochem. Soc.* 144:L23–L26.

35. Grambow, L. and S. Bruckenstein. 1997. Mass spectrometric investigation of the electrochemical behavior of adsorbed carbon monoxide at platinum in 0.2 M sulphuric acid. *Electrochim. Acta* 22:377–383.

36. Kita, H., K. Shimazu, and K. Kunimatsu. 1998. Electrochemical oxidation of CO on Pt in acidic and alkaline solutions: Part I. Voltammetric study on the adsorbed species and effects of aging and Sn(IV) pretreatment. *J. Electroanal. Chem.* 241:163–179.

37. Gutiérrez, C. and J.A. Caram. 1991. Electro-oxidation of dissolved CO on a platinum electrode covered with a monolayer of the chemisorbed CO formerly considered to be a poison. *J. Electroanal. Chem.* 308:321–325.

38. Oetjen, H.F., et al. 1996. Performance data of a proton exchange membrane fuel cell using H_2/CO as fuel gas. *J. Electrochem. Soc.* 143:3838–3842.

39. Lee, D., S. Hwang, and I. Lee. 2005. A study on composite PtRu(1:1)-PtSn(3:1) anode catalyst for PEMFC. *J. Power Sources* 145:147–153.

40. Lim, D.-H., et al. 2009. A new synthesis of a highly dispersed and CO tolerant PtSn/C electrocatalyst for low-temperature fuel cell; its electrocatalytic activity and long-term durability. *Appl. Catal. B: Environ.* 89:484–493.

41. Dupont, C., Y. Jugnet, and D. Loffreda. 2006. Theoretical evidence of PtSn alloy efficiency for CO oxidation. *J. Am. Chem. Soc.* 128:9129–9136.

42. Mukerjee, S., et al. 1999. Investigation of enhanced CO tolerance in proton exchange membrane fuel cells by carbon supported PtMo alloy catalyst. *Electrochem. Solid-State Lett.* 2:12–15.

43. Grgur, B.N., N.M. Markovic, and P.N. Ross. 1999. The electro-oxidation of H_2 and H_2/CO mixtures on carbon-supported Pt_xMo_y alloy catalysts. *J. Electrochem. Soc.* 146:1613–1619.

44. Mukerjee, S., et al. 2004. Electrocatalysis of CO tolerance by carbon-supported PtMo electrocatalysts in PEMFCs. *J. Electrochem. Soc.* 151:A1094–A1103.

45. Watanabe, M. and S. Motoo. 1975. Electrocatalysis by ad-atoms: Part III. Enhancement of the oxidation of carbon monoxide on platinum by ruthenium ad-atoms. *J. Electroanal. Chem.* 60:275–283.

46. Gasteiger, H.A., et al. 1994. Carbon monoxide electro-oxidation on well-characterized platinum-ruthenium alloys. *J. Phys. Chem.* 98:617–625.

47. Friedrich, K.A., et al. 1996. CO adsorption and oxidation on a Pt(111) electrode modified by ruthenium deposition: an IR spectroscopic study. *J. Electroanal. Chem.* 402:123–128.

48. Gasteiger, H.A., N.M. Markovic, and P.N. Ross. 1995a. H_2 and CO electro-oxidation on well-characterized Pt, Ru, and Pt-Ru. 1. Rotating disk electrode studies of the pure gases including temperature effects. *J. Phys. Chem.* 99:8290–8301.

49. Gasteiger, H.A., N.M. Markovic, and P.N. Ross. 1995b. H_2 and CO electro-oxidation on well-characterized Pt, Ru, and Pt-Ru. 2. Rotating disk electrode studies of CO/H_2 mixtures at 62 °C. *J. Phys. Chem.* 99:16757–16767.

50. Giorgi, L., et al. 2001. H_2 and H_2/CO oxidation mechanism on Pt/C, Ru/C and Pt–Ru/C electrocatalysts. *J. Appl. Electrochem.* 31:325–334.

51. Christoffersen, E., et al. 2001. Anode materials for low-temperature fuel cells: A density functional theory study. *J. Catal.* 199:123–131.

52. Buatier de Mongeot, F., et al. 1998. CO adsorption and oxidation on bimetallic Pt/Ru(0001) surfaces—A combined STM and TPD/TPR study. *Surf. Sci.* 411:249–262.

53. Lin, S.D., et al. 1999. Morphology of carbon supported Pt-Ru electrocatalyst and the CO tolerance of anodes for PEM fuel cells. *J. Phys. Chem. B* 103:97–103.

54. Urian, R.C., A.F. Gullá, and S. Mukerjee. 2003. Electrocatalysis of reformate tolerance in proton exchange membranes fuel cells: Part I. *J. Electroanal. Chem.* 554–555: 307–324.

55. Ball, S., et al. 2002. The proton exchange membrane fuel cell performance of a carbon supported PtMo catalyst operating on reformate. *Electrochem. Solid-State Lett.* 5:A31–A34.

56. Beden, B., et al. 1982. On the nature of reduced CO_2: An IR spectroscopic investigation. *J. Electroanal. Chem.* 139:203–206.

57. Arévalo, M.C., et al. 1994. A contribution to the mechanism of "reduced" CO_2 adsorbates electro-oxidation from combined spectroelectrochemical and voltammetric data. *Electrochim. Acta* 39:793–799.

58. Iwasita, T., et al. 1992. On the study of adsorbed species at platinum from methanol, formic acid and reduced carbon dioxide via *in situ* FT-IR spectroscopy. *Electrochim. Acta* 37:2361–2367.

59. Taguchi, S., et al. 1994. Adsorption of CO and CO_2 on polycrystalline Pt at a potential in the hydrogen adsorption region in phosphate buffer solution: An *in situ* Fourier transform IR study. *J. Electroanal. Chem.* 369:199–205.

60. Nikolic, B.Z., et al. 1990. Electroreduction of carbon dioxide on platinum single crystal electrodes: electrochemical and *in situ* FT-IR studies. *J. Electroanal. Chem.* 295:415–423.

61. Huang, H., et al. 1991. *In situ* Fourier transform infrared spectroscopic study of carbon dioxide reduction on polycrystalline platinum in acid solutions. *Langmuir* 7:1154–1157.

62. Sobkowski, J. and A. Czerwinski. 1974. Kinetics of carbon dioxide adsorption on a platinum electrode. *J. Electroanal. Chem.* 55:391–397.

63. Sobkowski, J. and A. Czerwinski. 1975. The comparative study of CO_2+H_{ads} reaction on platinum electrode in H_2O and D_2O. *J. Electroanal. Chem.* 65:327–333.

64. Sobkowski, J. and A. Czerwinski. 1985. Voltammetric study of carbon monoxide and carbon dioxide adsorption on smooth and platinized platinum electrodes. *J. Phys. Chem.* 89:365–369.

65. Shi, Z., et al. 2007. Transient analysis of hydrogen sulfide contamination on the performance of a PEM fuel cell. *J. Electrochem. Soc.* 154:B609–B615.

66. Shah, A.A. and F.C. Walsh. 2008. A model for hydrogen sulfide poisoning in proton exchange membrane fuel cells. *J. Power Sources* 185:287–301.

67. Shi, W., et al. 2007. The effect of H_2S and CO mixtures on PEMFC performance. *Int. J. Energy Res.* 32:4412–4417.

68. Shi, W., et al. 2007. The influence of hydrogen sulfide on proton exchange membrane fuel cell anodes. *J. Power Sources* 164:272–277.

69. Shi, W., et al. 2007. Hydrogen sulfide poisoning and recovery of PEMFC Pt-anodes. *J. Power Sources* 165:814–818.

70. Mohtadi, R., W.K. Lee, and J.W. Van Zee. 2005. The effect of temperature on the adsorption rate of hydrogen sulfide on Pt anodes in a PEMFC. *Appl. Catal. B: Environ.* 56:37–42.

71. Mohtadi, R., et al. 2003. Effects of hydrogen sulfide on the performance of a PEMFC. *Electrochem. Solid-State Lett.* 6:A272–A274.

72. Urdampilleta, I., et al. 2007. PEMFC poisoning with H_2S: Dependence on operating conditions. *ECS Trans.* 11:831–842.

73. Rockward, T., et al. 2007. The effects of multiple contaminants on polymer electrolyte fuel cells. *ECS Trans.* 11:821–829.
74. Garzon, F., et al. 2006. The impact of hydrogen fuel contaminates on long-term PMFC performance. *ECS Trans.* 3:695–703.
75. http://en.wikipedia.org/wiki/Hydrogen_sulfide
76. http://www.earthworksaction.org/hydrogensulfide.cfm#SOURCES
77. Contractor, A.Q. and H. Lal. 1979. Two forms of chemisorbed sulfur on platinum and related studies. *J. Electroanal. Chem.* 96:175–181.
78. Loucka, T. 1971. Adsorption and oxidation of sulphur and of sulphur dioxide at the platinum electrode. *J. Electroanal. Chem.* 31:319–332.
79. Mathieu, M.-V. and M. Primet. 1984. Sulfurization and regeneration of platinum. *Appl. Catal.* 9:361–370.
80. Ramasubramanian, N. 1975. Anodic behavior of platinum electrodes in sulfide solutions and the formation of platinum sulfide. *J. Electroanal. Chem.* 64:21–37.
81. Michaelides, A. and P. Hu. 2001. Hydrogenation of S to H_2S on Pt(111): A first-principles study. *J. Chem. Phys.* 115:8574.
82. Rajalakshmi, N., T.T. Jayanth, and K.S. Dhathathreyan. 2003. Effect of carbon dioxide and ammonia on polymer electrolyte membrane fuel cell stack performance. *Fuel Cells* 3:177–180.
83. Szymanski, S.T., et al. 1980. The effect of ammonia on hydrogen-air phosphoric acid fuel cell performance. *J. Electrochem. Soc.* 127:1440–1444.
84. Uribe, F.A., S. Gottesfeld, and J.T.A. Zawodzinski. 2002. Effect of ammonia as potential fuel impurity on proton exchange membrane fuel cell performance. *J. Electrochem. Soc.* 149:A293–A296.
85. Halseid, R., P.J.S. Vie, and R. Tunold. 2006. Effect of ammonia on the performance of polymer electrolyte membrane fuel cells. *J. Power Sources* 154:343–350.
86. Soto, H.J., et al. 2003. Effect of transient ammonia concentrations on PEMFC performance. *Electrochem. Solid-State Lett.* 6: A133–A135.
87. Chellappa, A.S., C.M. Fischer, and W.J. Thomson. 2002. Ammonia decomposition kinetics over $Ni-Pt/Al_2O_3$ for PEM fuel cell applications. *Appl. Catal. A: Gen.* 227:231–240.
88. Zhu, H. Y. 1996. Room temperature catalytic ammonia synthesis over an AB_5-type intermetallic hydride. *J. Alloys Compd.* 240:L1–L3.
89. Halseid, R., et al. 2008. The effect of ammonium ions on oxygen reduction and hydrogen peroxide formation on polycrystalline Pt electrodes. *J. Power Sources* 176:435–443.
90. Halseid, R., T. Bystron, and R. Tunold. 2006. Oxygen reduction on platinum in aqueous sulphuric acid in the presence of ammonium. *Electrochim. Acta* 51: 2737–2742.
91. http://www.epa.gov/climatechange/emissions/co2_human.html
92. http://www.ghgonline.org/co2bioburn.htm
93. http://www.jmcatalysts.com/ptd/site.asp?siteid=671&pageid=672
94. http://www.seed.slb.com/subcontent.aspx?id=3988
95. http://www.epa.gov/air/emissions/so2.htm, 2002
96. http://www.environment.gov.au/atmosphere/airquality/publications/sulfurdioxide.html
97. Jing, F., et al. 2007. The effect of ambient contamination on PEMFC performance. *J. Power Sources* 166:172–176.

98. Mohtadi, R., W. Lee, and J.W. Van Zee. 2004. Assessing durability of cathodes exposed to common air impurities. *J. Power Sources* 138:216–225.
99. Garsany, Y., O.A. Baturina, and K.E. Swider-Lyons. 2007. Impact of sulfur dioxide on the oxygen reduction reaction at Pt/vulcan carbon electrocatalysts. *J. Electrochem. Soc.* 154:B670–B675.
100. Garsany, Y., O. Baturina, and K. Swider-Lyons. 2007. Impact of SO_2 on the kinetics of Pt_3Co/vulcan carbon electrocatalysts for oxygen reduction. *ECS Trans* 11:863–875.
101. Contractor, A.Q. and H. Lal. 1978. The nature of species adsorbed on platinum from SO_2 solutions. *J. Electroanal. Chem.* 93:99–107.
102. Mangun, C.L., J.A. DeBarr, and J. Economy. 2001. Adsorption of sulfur dioxide on ammonia-treated activated carbon fibers. *Carbon* 39:1689–1696.
103. Uribe, F., et al. 2005. Effect of fuel and air impurities on PEM fuel cell performance, in U.S. *DOE Hydrogen Program: FY 2005 Progress Report*, Washington, D.C. 1046–1051.
104. http://www.epa.gov/air/emissions/nox.htm
105. http://www.raypak.com/lownoxtech.htm
106. http://www.coalonline.org/site/coalonline/content/browser/81423/NOx-emissions-and-control
107. Rodes, A., et al. 1998. Nitric oxide adsorption at Pt(100) electrode surfaces. *Electrochim. Acta* 44:1077–1090.
108. Yang, D., et al. 2006. The effect of nitrogen oxides in air on the performance of proton exchange membrane fuel cell. *Electrochim. Acta* 51:4039–4044.
109. http://en.wikipedia.org/wiki/Volatile_organic_compound
110. Hesterberg, T.W., C.A. Lapin, and W.B. Bunn. 2008. A comparison of emissions from vehicles fueled with diesel or compressed natural gas. *Environ. Sci. Tech.* 42:6437–6445.
111. Bernstein, J.A., et al. 2008. The health effects of nonindustrial indoor air pollution. *J. Allergy Clin. Immun.* 121:585–591.
112. Moore, J.M., et al. 2000. The effects of battlefield contaminants on PEMFC performance. *J. Power Sources* 85:254–260.
113. Li, H., et al. 2008. Durability of PEMFC cathodes exposed to toluene-contaminated air. *ECS Trans.* 16:1059–1067.
114. Li, H., et al. 2009. PEM fuel cell contamination: Effects of operating conditions on toluene-induced cathode degradation. *J. Electrochem. Soc.* 156:B252–B257.
115. Air quality research subcommittee. 2000. Atmospheric ammonia: Sources and fate. In *Air Quality Research Subcommittee Meeting Report*. ed. Committee on the environment and natural resources.
116. http://www.engineeringtoolbox.com/ammonia-health-symptoms-d_901.html
117. http://dhs.wisconsin.gov/eh/air/fs/Ammonia.htm
118. Pozio, A., et al. 2003. Nafion® degradation in PEFCs from end plate iron contamination. *Electrochim. Acta* 48:1543–1549.
119. Agneaux, A., et al. 2006. Corrosion behaviour of stainless steel plates in PEMFC working conditions. *Fuel Cells* 6:47–53.
120. Colón-Mercado, H.R. and B.N. Popov. 2006. Stability of platinum based alloy cathode catalysts in PEM fuel cells. *J. Power Sources* 155:253–263.
121. Sulek, M.S., S.A. Mueller, and C.H. Paik. 2008. Impact of Pt and Pt–alloy catalysts on membrane life in PEMFCs. *Electrochem. Solid-State Lett.* 11: B79–B82.

122. Seo, A., et al. 2006. Performance and stability of Pt-based ternary alloy catalysts for PEMFC. *Electrochim. Acta* 52:1603–1611.
123. Okada, T. 2003. Effect of ionic contaminants. In *Handbook of Fuel Cells: Fundamentals, Technology and Applications.* Vol. 3., 627. ed. W. Vielstich, H.A. Gasteiger, and A. Lamm. New York: John Wiley & Sons, Ltd.
124. Wang, H. and J.A. Turner. 2008. The influence of metal ions on the conductivity of Nafion 112 in polymer electrolyte membrane fuel cell. *J. Power Sources* 183:576–580.
125. Pupkevich, V., V. Glibin, and D. Karamanev1. 2007. The effect of ferric ions on the conductivity of various types of polymer cation exchange membranes. *J. Solid State Electrochem.* 11:1429–1434.
126. Okada, T., et al. 1999. The effect of impurity cations on the transport characteristics of perfluorosulfonated ionomer membranes. *J. Phys. Chem. B* 103:3315–3322.
127. Kelly, M.J., et al. 2005. Contaminant absorption and conductivity in polymer electrolyte membranes. *J. Power Sources* 145:249–252.
128. Kelly, M.J., et al. 2005. Conductivity of polymer electrolyte membranes by impedance spectroscopy with microelectrodes. *Solid State Ionics* 176:2111–2114.
129. Laconti, A.B., M. Mamdan, and R.C. McDonald. 2003. In *Handbook of Fuel Cells,* Vol. 3, 647. ed. W. Vielstich, A. Lamm, and H.A. Gasteiger. New York: John Wiley & Sons.
130. Okada, T., et al. 2001. The effect of impurity cations on the oxygen reduction kinetics at platinum electrodes covered with perfluorinated ionomer. *J. Phys. Chem. B* 105:6980–6986.
131. Okada, T., et al. 1997. Ion and water transport characteristics in membranes for polymer electrolyte fuel cells containing H^+ and Ca^{2+} cations. *J. Electrochem. Soc.* 144:2744–2750.
132. Okada, T. 1999. Theory for water management in membranes for polymer electrolyte fuel cells: Part 1. The effect of impurity ions at the anode side on the membrane performances. *J. Electroanal. Chem.* 465:1–17.
133. Okada, T. 1999. Theory for water management in membranes for polymer electrolyte fuel cells: Part 2. The effect of impurity ions at the cathode side on the membrane performances. *J. Electroanal. Chem.* 465:18–29.
134. Shi, M. and F.C. Anson. 1997. Dehydration of protonated Nafion® coatings induced by cation exchange and monitored by quartz crystal microgravimetry. *J. Electroanal. Chem.* 425:117–123.
135. Collier, A., et al. 2006. Degradation of polymer electrolyte membranes. *Int. J. Hydrog. Ener.* 31:1838–1854.
136. Vogt, R., P.J. Crutzen, and R. Sander. 1996. A mechanism for halogen release from sea-salt aerosol in the remote marine boundary layer. *Nature* 383:327–330.
137. Mikkola, M.S., et al. 2007. The effect of NaCl in the cathode air stream on PEMFC performance. *Fuel Cells* 7:153–158.
138. Kinoshita, K., K. Routsis, and J.A.S. Bett. 1974. The thermal decomposition of platinum(II) and (IV) complexes. *Thermochim. Acta* 10:109–117.
139. Watanabe, M., M. Uchida, and S. Motoo. 1987. Preparation of highly dispersed Pt + Ru alloy clusters and the activity for the electro-oxidation of methanol. *J. Electroanal. Chem.* 229:395–406.
140. Luczak, F.J. and D.A. Landsman. 1987. Ordered ternary fuel cell catalysts containing platinum and cobalt and method for making the catalysts. Patent: US4677092.

141. Stonehart, P. 1997. Platinum alloy catalyst. Patent: US5593934.
142. Schmidt, T.J., et al. 2001. The oxygen reduction reaction on a Pt/carbon fuel cell catalyst in the presence of chloride anions. *J. Electroanal. Chem.* 508:41–47.
143. Halalay, I.C., B. Merzougui, and A. Mance. 2008. Three mechanisms for protecting the fuel cell membrane, plates and catalysts. *ECS Trans.* 16:969–981.
144. Zhao, X., et al. 2005. Effects of chloride anion as a potential fuel impurity on DMFC performance. *Electrochem. Solid-State Lett.* 8:A149–A151.
145. Matsuoka, K., et al. 2008. Degradation of polymer electrolyte fuel cells under the existence of anion species. *J. Power Sources* 179:560–565.
146. Yadav, A.P., A. Nishikata, and T. Tsuru. 2007. Effect of halogen ions on platinum dissolution under potential cycling in 0.5 M H_2SO_4 solution. *Electrochim. Acta* 52:7444–7452.
147. Molina, F.V. and D. Posadas. 1988. Temperature dependence of the adsorption of chloride ion on platinum electrodes. *Electrochim. Acta* 33:661–665.
148. Li, N. and J. Lipkowski. 2000. Chronocoulometric studies of chloride adsorption at the Pt(111) electrode surface. *J. Electroanal. Chem.* 491:95–102.
149. Stamenkovic, V., N.M. Markovic, and P.N. Ross. 2001. Structure-relationships in electrocatalysis: Oxygen reduction and hydrogen oxidation reactions on Pt(111) and Pt(100) in solutions containing chloride ions. *J. Electroanal. Chem.* 500:44–51.
150. Cheng, X., et al. 2007. A review of PEM hydrogen fuel cell contamination: Impacts, mechanisms, and mitigation. *J. Power Sources* 165:739–756.
151. Ahn, S.-Y., et al. 2002. Performance and lifetime analysis of the kW-class PEMFC stack. *J. Power Sources* 106:295–303.
152. Husar, A., M. Serra, and C. Kunusch. 2007. Description of gasket failure in a 7 cell PEMFC stack. *J. Power Sources* 169:85–91.
153. Tan, J., et al. 2008. Degradation characteristics of elastomeric gasket materials in a simulated PEM fuel cell environment. *J. Mater. Eng. Perform.* 17:785–792.

3

Cathode Contamination

Hui Li, Jianlu Zhang, Zheng Shi, Datong Song, and Jiujun Zhang

CONTENTS

3.1 Introduction

The membrane electrode assembly (MEA) in a proton exchange membrane (PEM) fuel cell has been identified as the key component that is probably most affected by the contamination process [1]. An MEA consists of anode and cathode catalyst layers (CLs), gas diffusion layers (GDLs), as well as a proton exchange membrane, among which the CLs present the most important challenges due to their complexity and heterogeneity. The CL is several micrometers thick and either covers the surface of the carbon base layer of the GDL or is coated on the surface of the membrane. The CL consists of: (1) an ionic conductor (ionomer) to provide a passage for proton transport;

(2) Pt (platinum) catalysts supported on a conductive matrix, such as carbon, to provide electron conduction; and (3) a hydrophilic agent, such as polytetrafluoroethylene (PTFE) to provide sufficient porosity and adjust the hydrophobicity/hydrophilicity of the CL for gaseous reactants to be transferred to active sites [2,3]. With each of those elements optimized to provide the best overall performance, the CL functions as a place for electrochemical reactions. The processes occurring in a CL include mass transport of the gaseous reactants, interfacial reactions of the reactants (e.g., H_2 at anode and O_2 at cathode) at the electrochemically active sites, proton transport in the electrolyte phase, and electron conduction in the electronic phase. When contaminants are present in the reactant streams, one or more of the above processes can be adversely affected, causing degradation in fuel cell performance or even fuel cell failure.

In general, PEM fuel cell contamination effects are classified into three major categories: (1) kinetic effect (poisoning of the catalyst sites or decreased catalyst activity); (2) ohmic effect (increases in the membrane and ionomer resistances, caused by alteration of the proton transport path); and (3) mass transfer effect (mass transport problems caused by changes in the structure of CLs and GDLs, and in the ratio between their hydrophilicity and hydrophobicity). Of these, the kinetic effect of the electrocatalysts on both anode and cathode sides is the most significant.

The sluggish kinetics of the cathode oxygen reduction reaction (ORR), the rate of which is four to six orders of magnitude lower than that of the hydrogen oxidation reaction (HOR) [4], has drawn most of the research attention in the study of PEM fuel cell catalysis. However, cathode contamination has not received as much attention as anode contamination, due to (1) the popularity of H_2 production through a reformation process that inevitably produces traces of CO, CO_2, NH_3, and sulfur compounds (including H_2S) in hydrogen-rich reformate gas [5–8], and (2) the catastrophic poisoning of the Pt anode catalyst by CO, even at a level of 50 ppm, or more severally by H_2S, even at a level of 1 ppm. In PEM fuel cell operation, air containing O_2, has been exclusively accepted as the most practical and economical oxidant to feed fuel cell stacks. Unless the air is supplied to the fuel cells through a contained source (e.g., bottled air), which may not be practical for large-scale commercialization, any pollutants in the immediate atmosphere can be potential contaminants. Therefore, studying the effect of atmospheric impurities on fuel cell durability contributes toward overcoming fuel cell contamination problems and developing mitigation strategies to improve the environmental adaptability of the fuel cell system.

Common air pollutants, including NO_x (NO and NO_2), SO_x (SO_2 and SO_3), NH_3, O_3, and some volatile organic compounds (VOCs, such as benzoic compounds), derive mainly from automotive exhausts, industrial emissions, and agricultural activities. Unusual air pollutants may also be a concern in some special atmospheres where fuel cells would be operating, such as battlefields that contain chemical warfare or even normal warfare contaminants. Battlefield contaminants include sarin, sulfur mustard, cyanogens, chloride, and hydrogen cyanide, all of which could cause significant irreversible fuel cell performance degradation [9].

In this chapter, we will focus on contamination by major air pollutants NO_x and SO_x, looking at mechanisms, experimental results, and mitigation. Contamination by other pollutants also will be briefly discussed.

3.2 Impacts of SO_x

Sulfur oxides (SO_x), especially SO_2, are emitted from the combustion of coal and oil, which contain variable proportions of sulfur (0.5 to 10% for coal and 0.5 to 3% for oil). They can also be found in urban areas with heavy traffic, and areas with active agricultural activities. SO_x (SO_2 and SO_3), acidic in nature, can cause a decrease in pH inside the MEA, resulting in free acids in the MEA and causing potential shifts. SO_x can also adsorb on the Pt surface, blocking the active catalyst sites that would otherwise be used for oxygen adsorption and reduction, and leading to fuel cell performance degradation.

The contamination impact of SO_2 was studied by Fenning et al. [10] through exposing a fuel cell to 1 ppm SO_2/air for 100 hours at 70°C with a constant-current discharge at 0.5 A cm^{-2}. From an initial value of 0.68 V (Figure 3.1), the cell performance fell to 0.44 V, a 35% decrease. Clearly, even a trace amount of SO_2 can cause a significant degradation in PEM fuel cell performance.

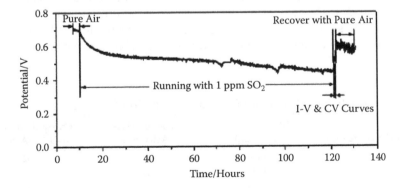

FIGURE 3.1
Constant-current discharging curve of a PEM fuel cell during operation with 1 ppm SO_2/air for 100 h at 70°C. Current density: 0.5 A cm^{-2}. (From Fenning, J. et al. 2007. *J. Power Sources.* 166:172–6. With permission.)

The contamination effect is strongly dependent on SO_2 concentration in the air. For example, a fuel cell performance degradation of 78% was observed after exposure to 5 ppm SO_2/air for 23 h, but only 53% was observed after exposure to 2.5 ppm SO_2/air for 46 h (i.e., the same applied dosage) [11]. No fuel cell performance degradation was observed if the fuel cell was exposed to 500 ppb SO_2/air [9]. The adverse effect of SO_2 is also related to operating cell voltage. As with other contaminants, lower cell voltage promotes the poisoning effect of SO_2 [12].

Several groups have studied the mechanism of SO_x poisoning. A general belief is that the S-containing species adsorbs on the active sites of a catalyst surface, occupying the polyatomic sites and thereby preventing the reactant oxygen from adsorbing at the catalyst surface. Contractor et al. [13] and Loucka et al. [14] reported that SO_2 adsorbs on the Pt surface to produce linearly and bridged adsorbed S species. These two forms of chemisorbed S species on the Pt surface were responsible for catalyst poisoning. Contractor et al. [15] also suggested that on a Pt electrode surface SO_2 could be electrochemically reduced to S, producing SO intermediates that could lead to difficulties in oxygen reduction:

$$Pt + SO_2 + 2H^+ + 2e^- \rightarrow Pt - SO_{ads} + H_2O \tag{3.1}$$

$$Pt - SO_{ads} + 2H^+ + 2e^- \rightarrow Pt - S_{ads} + H_2O \tag{3.2}$$

Garsany et al. [16,17] studied the kinetics of the ORR under the influence of adsorbed sulfur-containing species, using a rotating ring disk electrode (RRDE) technique. They studied the adsorption of sulfur-containing species on both Vulcan carbon-supported Pt (Pt/VC) [16] and Pt_3Co (Pt_3Co/VC) catalysts [17] by submersing the catalyst-coated electrodes in a SO_2-containing solution, and then evaluating the electrodes' activity toward the ORR. Table 3.1 shows the Pt mass activity of the Pt/VC electrode at 0.9 V as a function of sulfur coverage. As can be seen from Table 3.1, when sulfur coverage

TABLE 3.1

Pt/VC Mass Activities at 0.9 V as a Function of $\theta_{S,i}$ (Sulfur Coverage)

Sulfur Coverage ($\theta_{S,i}$)	Mass Activity (A/ mg_{Pt})	Loss of Mass Activity %	Overpotential /V (j=1.5 mA cm^{-2})
0	0.16	100	0.92
0.012	0.11	69	0.90
0.14	8.01×10^{-3}	5	0.81
0.37	1.71×10^{-4}	0.11	0.69
1	8.76×10^{-4}	0.56	0.54

Source: Garsany, Y. et al. 2007. *J. Electrochem. Soc.* 154:B670–5. With permission.

on the catalyst surface was increased from 1.2 to 14%, the loss in Pt mass activity rose from 33 to 95%. They also found that when sulfur coverage on the Pt surface reached 37% for the Pt/VC electrode and 96% for the Pt₃Co/ VC electrode, the reaction pathway of the ORR changed from a 4-electron to a 2-electron process.

The recoverability of fuel cell performance after SO_x contamination has also been studied extensively. Techniques include operating the fuel cell under open circuit voltage (OCV) conditions [12,18], replacing the SO_x-containing air with neat air [10,11], and using cyclic voltammetry (CV) scanning [10,11]. Figure 3.2 shows the fuel cell performance recovery behavior after using neat air to replace 5 ppm of SO_2/air and 2.5 ppm of SO_2/air, respectively [11]. In both cases, only partial recovery was observed after applying neat air for 24 h, indicating that the active surface area of the electrocatalyst occupied by the adsorbed SO_2 was partially reactivated. Compared to running with neat air, the use of CV scanning led to better fuel cell performance recovery, as shown in Figure 3.3; this performance recovery can be explained by Figure 3.4, which presents two cycles of CV scanning [10]. The smaller hydrogen desorption peak in the first cycle suggests that the adsorbed SO_2 on the electrocatalyst occupies the active positions, while the increased hydrogen desorption peak in the second cycle indicates a recovery in cell performance.

The cell performance recovery could be attributed to the electrochemical oxidation of adsorbed SO_2 at high potential. Mangun et al. [19] suggested that the adsorbed SO_2 might have electrochemically reacted with chemisorbed

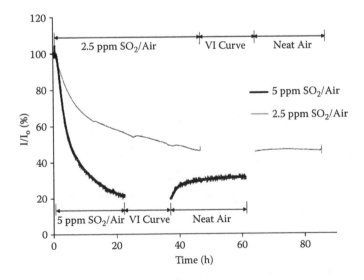

FIGURE 3.2
Fuel cell performance recovery after replacing 5 and 2.5 ppm SO_2/air with neat air. (From Mohtadi, R. et al. 2004. *J. Power Sources* 138:216–25. With permission.)

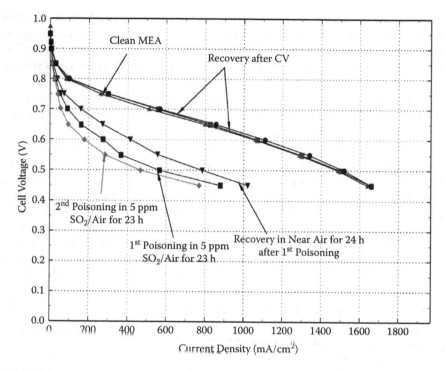

FIGURE 3.3
Polarization curves showing the contamination effect of 5 ppm SO₂/air, the recovery of running with neat air, and the recovery of applying CV scanning. (From Mohtadi, R. et al. 2004. *J. Power Sources* 138:216–25. With permission.)

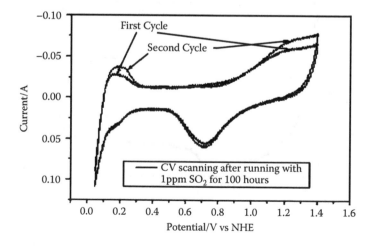

FIGURE 3.4
The CV scanning curve after running with 1 ppm SO₂/air for 100 h, scanning rate: 20 mV s⁻¹. (From Fenning, J. et al. 2007. *J. Power Sources*. 166:172–6. With permission.)

oxygen to form SO_3, which then reacted with water to form H_2SO_4, thus releasing the active sites of the electrocatalyst that were occupied by adsorbed SO_2. Those reactions can be described by the following equations:

$$Pt + SO_2 \rightarrow Pt - SO_{2(ads)} \tag{3.3}$$

$$2Pt - SO_{2(ads)} + O_2 + 4e^- \rightarrow 2Pt - SO_{3(ads)} \tag{3.4}$$

$$Pt - SO_{3(ads)} + H_2O \rightarrow H_2SO_4 + Pt \tag{3.5}$$

With respect to the recovery of the catalyst surface from a sulfur-bound surface (Pt–S_{ads}, from equation (3.2)) to a free Pt surface, Loucka [14] suggested that the adsorbed sulfur could be oxidized (1) by six electrons to SO_3, which then reacted with water to form H_2SO_4, or (2) by two electrons to SO_4^{2-}, according to the following two equations:

$$Pt - S_{ads} + 3H_2O \rightarrow SO_3 + 6H^+ + 6e^- + Pt \tag{3.6}$$

$$Pt - S_{ads} + 4H_2O \rightarrow SO_4^{2-} + 4H^+ + 2e^- + Pt \tag{3.7}$$

However, neither neat air operation nor CV scanning led to complete fuel cell performance recovery because the adsorbed sulfur species could not be completely desorbed and, therefore, still partially occupied the active sites of the Pt electrocatalysts.

3.3 Impacts of NO_x

Although the U.S. Environmental Protection Agency (EPA) has set air quality standards (i.e., concentrations of SO_2 and $NO_2 < 0.05$ ppm, $O_3 < 0.20$ ppm, and $H_2S < 1$ ppb) [20], the actual local concentrations of these impurities can be much higher. For example, the concentration of NO_x in the exhaust of internal combustion vehicles can exceed 5 ppm, which could cause cell performance to decline by more than 50% [11,21].

Even a trace amount of NO_x in the air can cause significant cell performance loss. Knight et al. [22] reported that 115 ppb of NO_x in the air stream caused a cell performance loss of about 25 mV at 0.175 A cm^{-2}. Many research groups have studied the effect of NO_x concentration on cell performance [10,11,21–23], and the general finding is that cell performance degradation is

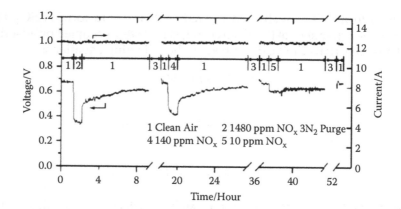

FIGURE 3.5
Effect of three NO_x concentrations on cell voltage at 0.5 A cm^{-2}. (From Yang, D. et al. 2006. *Electrochim. Acta* 51:4039–44. With permission.)

strongly dependent on NO_x concentration in the air stream. Figure 3.5 shows cell performances under the effect of three NO_x concentrations: 10, 140, and 1480 ppm in the air [21]. With the introduction of 1480 ppm NO_x into the cathode, the cell voltage rapidly dropped to below 0.37 V within 5 min, before leveling off at about 0.34 V (approximately 50% of the initial 0.67 V voltage). When the NO_x-containing air was replaced with neat air, the cell voltage recovered to 0.48 V (70% of the initial voltage) within 3 min, and then slowly reached ~0.60 V (about 90% of the initial voltage).

Recoverability from NO_x contamination has also been studied by several other groups [10,11,22,23]. It is generally agreed that NO_x contamination is fully recoverable by replacing the NO_x-containing air with neat air. However, if the PEM fuel cell is continuously exposed to NO_x for too long (e.g., 500 h), the cell performance may only be partially recoverable [23].

Since only a few groups have studied NO_x contamination mechanisms, there is as yet no consensus about these processes. Thus far, two primary mechanisms for NO_x contamination have been proposed: the NO_x reduction mechanism and the NO_x oxidation mechanism. Mohtadi et al. [11] proposed the reduction mechanism, suggesting that NO_2 was electrochemically reduced and thereby competed with O_2 for Pt sites. The product of NO_2 reduction was NH_4^+ (via equation (3.8)). NH_4^+ has been reported to be an ionomer poisoning species that might affect the ionomer and/or catalyst–ionomer interface.

$$NO_2(g) + 8H^+ + 7e^- \rightarrow NH_4^+ + 2H_2O \tag{3.8}$$

Fenning et al. [10] proposed the oxidation mechanism, and studied NO_2 contamination using CV techniques. They suggested that cell performance degradation was related to adsorption of NO_2 on the catalyst surface, which was then electrochemically oxidized to produce NO_3^-, according to equation (3.9):

$$NO_2 + H_2O \rightarrow NO_3^- + 2H^+ + e^- \tag{3.9}$$

Daijun et al. [21] hypothesized that NO could easily be chemically oxidized by O_2 to form NO_2 through the catalysis of Pt, and NO_2 would then react with water at the cathode to form nitric acid:

$$2NO + O_2 \rightarrow 2NO_2 \tag{3.10}$$

$$2NO_2 + H_2O \rightarrow HNO_3 + HNO_2 \tag{3.11}$$

Both HNO_3 and HNO_2 could easily dissociate in wet conditions, and the formation of HNO_3, HNO_2, NO_3^-, and NO_2^- might change the interface property of the cathode, thus causing cell performance degradation.

3.4 Impacts of NH_3 and H_2S

NH_3 and H_2S, especially the latter, have been most extensively studied with respect to their effects as fuel-side contaminants, due to their unavoidable presence in hydrogen-rich reformate gases. However, they are also present as impurities in the atmosphere as a result of automotive vehicle exhaust and industrial manufacturing processes.

No experimental data have been reported in the literature regarding the effect of NH_3 contamination on the cathode side of PEM fuel cells, although Halseid et al. [24] mentioned NH_3's potential poisoning effect on the cathode, suggesting that the electrochemical oxidation of NH_3 might produce species that could poison the catalytic sites. However, we believe that this is very unlikely because NH_3 should be stable at the cathode potential region. The most probable reason for NH_3 poisoning may be the reaction of NH_3 with protons in the membrane (equation (3.12) and equation (3.13)) [24,25] and/or with ionomer in the catalyst layer, either of which could reduce the proton conductivity of the membrane and the catalyst layer (the conductivity of the proton form of Nafion® is about 3.8 to 4.2 times higher than the ammonium form [25,26]). Halseid et al. [27] used a rotating disk electrode to study the ORR in the presence of NH_3; they found that the kinetic current density at 0.7 V (versus RHE) was reduced by a factor greater than 10 on Pt electrode when comparing the ORR in pure sulfuric acid to that in 0.18 M H_2SO_4+0.1 M NH_3 solution. The Tafel slope at low current density was about –90 mV/decade in the presence of NH_3, compared to about –74 mV/decade in pure acidic solution; the exchange current density was also lower in the presence of NH_3. Although the direct origin of the ORR activity loss was not identified, Halseid et al. suspected that adsorption and oxidation of NH_3, including intermediate species as well as nitrogen–oxygen species formed at high potential, might be involved.

$$NH_3(g) \rightarrow NH_3(membrane) \tag{3.12}$$

$$NH_3(membrane) + H^+ \rightarrow NH_4^+ \tag{3.13}$$

Just as H_2S is a notorious poisoning species for the anode side of the PEM fuel cell, it can also severely degrade fuel cell performance at the cathode side. Mohtadi et al. [11] studied the durability of cathodes when exposed to common air impurities, including H_2S. Figure 3.6 shows the degradation in cell performance during cathode exposure to 200 ppm H_2S for 10.5 h. The cell experienced complete deterioration during this time and at this high concentration. After the H_2S-containing air was replaced with neat air for 70.5 h, cell performance was partially recovered to 170 mA cm^{-2}, as compared to the original current density of about 480 mA cm^{-2}. Mohtadi et al. attributed the poisoning effect to strong H_2S adsorption on the Pt surface. This adsorption then blocked the active catalyst sites that

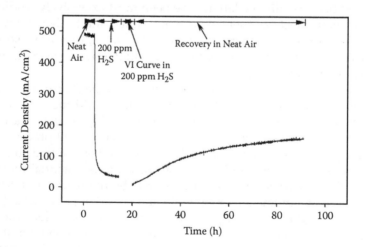

FIGURE 3.6
Transient response of fuel cell performance when the cathode was exposed to 200 ppm H_2S at a cell voltage of 0.68 V. (From Mohtadi, R. et al. 2004. *J. Power Sources* 138:216–25. With permission.)

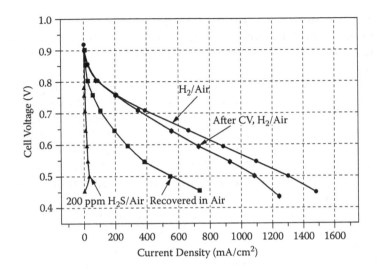

FIGURE 3.7
Polarization curves collected at the end of contamination with 200 ppm H$_2$S, and after recovery with neat air and CV scan, in comparison with the baseline. (From Mohtadi, R. et al. 2004. *J. Power Sources* 138:216–25. With permission.)

would otherwise have been used for oxygen, according to the following mechanism proposed by Mathieu et al. [28]:

$$Pt + H_2S \rightarrow Pt - S_{ads} + H_2 \tag{3.14}$$

In addition to using neat air to recover cell performance, Mohtadi et al. [11] also used CV scanning. As shown in Figure 3.7, the cell performance recovery after CV scanning was significantly better than after neat air. They ascribed the recovery to the oxidation of two sulfur species occurring during CV scanning.

3.5 Impacts of VOCs

VOCs are present in the atmosphere as a result of (1) decorative materials and dry cleaning solvents in indoor environments, and/or (2) industrial, agricultural, and biological processes in outdoor environments. VOCs are also present in battlefield situations. Moore et al. [9] briefly reported the contamination effect of benzene present in a battlefield environment [9]; Hui et al. systematically studied the effect of toluene in the air stream, including contamination testing [29–31]; Zheng et al. [32] conducted both empirical and theoretical model development; and Khalid et al. [33] studied the adsorption and electrochemical reaction of toluene on a Pt electrode.

Benzene has a significant contamination effect on fuel cell performance. For example, with 50 ppm benzene in the air, cell voltage can drop by 160 mV within 15 min at 0.2 A cm^{-2} [9]. The severity of benzene contamination was found to be a function of operating current density. When a fuel cell was fed with 50 ppm benzene in the air for 30 min, rather than pure air feed, the power output dropped 5% at 50 mA cm^{-2}, and 28% at 200 mA cm^{-2}. In addition, cell performance did not fully recover until after the benzene-containing air was replaced with neat air.

Hui et al. [29] conducted toluene contamination tests at different toluene concentrations. They also studied the effects of different operational conditions on toluene contamination, including the effects of fuel cell relative humidity (RH), of Pt loading in the cathode catalyst layer, of back pressure, and of air stoichiometry [30]. Figure 3.8 shows a set of representative results of contamination tests at 1.0 A cm^{-2} with various levels of toluene concentration in the air. It can be seen that the cell voltage starts to decline immediately after the introduction of toluene, and then reaches a plateau (steady state). These plateau voltages indicate the saturated nature of the toluene contamination. For example, the cell voltage drops from 0.645 V to 0.522 V at 1.0 A cm^{-2} within 30 min of the cathode

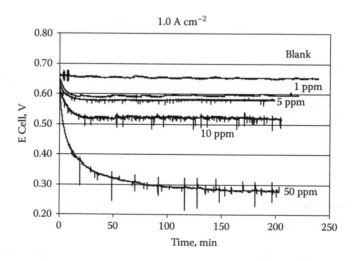

FIGURE 3.8
Cell voltage versus time at 1.0 A cm^{-2} with various levels of toluene concentration. Operating conditions: stoichiometries of air/H$_2$: 3.0/1.5; cell temperature: 80°C; RH of the fuel cell: 80%; back pressure: 30 psig. The MEA was based on Nafion® 211 membrane with Pt loading of 0.4 mg cm^{-2} on the anode and cathode sides. (From Li, H. et al. 2009. *J. Electrochem. Soc.* 156:B252–B257. With permission.)

being continuously exposed to 10 ppm toluene, and then starts to level off. Figure 3.9 illustrates the steady-state polarization curves affected by contamination from various levels of toluene concentration. It is observable that the magnitude of the negative impact of toluene on cell performance increases when both toluene concentration and current density increase. As reported in references [29–31], increases in RH, operating pressure, and cathode Pt loading led to less severe performance degradation, while an increase in air stoichiometry resulted in more severe performance degradation. In addition, toluene contamination was only partially recoverable.

In Hui et al.'s work, during each round of contamination testing, diagnostic investigation of fuel cell performance degradation using both alternating current (AC) impedance and CV techniques was also performed to gain insight into the toluene contamination mechanism. AC impedance measurement revealed that toluene contamination could result in an increase in both kinetic and mass transfer resistances, but the increase in kinetic resistance was the more dominant contributor to the drop in cell performance. Based on these observations and electrochemical studies of toluene adsorption on solid Pt electrode [33], they proposed a toluene contamination mechanism based on three reactions occurring at the catalyst surface, as shown below:

$$C_7H_8 + nPt \rightarrow Pt_n(C_7H_8)_{ads} \qquad (3.15)$$

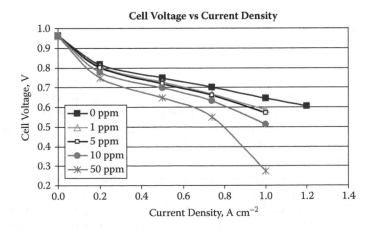

FIGURE 3.9
Steady-state polarization curves of a fuel cell contaminated by toluene in the air stream. Operating conditions: stoichiometry: 1.5/3.0 for H_2/air; RH: 80%; cell temperature: 80°C; back pressure: 30 psig. MEA: anode/cathode; Pt loading: 0.4 mg cm^{-2}. (From Li, H. et al. 2008. *J. Power Sources* 185:272–9. With permission.)

TABLE 3.2

Contamination Effects of Warfare Gases on Fuel Cell Power Output

Warfare Gas	Concentration ppm	Loss of Power Output %	Recoverability
HCN	1780	87	Partial
Sarin	170	70	Nonrecoverable
Sulfur mustard	15	87	Nonrecoverable

Source: Moore, J.M. et al. 2000. *J. Power Sources* 85: 254–60. With permission.

$$Pt_n(C_7H_8)_{ads} + 9O_2 \rightarrow 7CO_2 + 4H_2O + nPt \tag{3.16}$$

$$Pt_n(C_7H_8)_{ads} + mH_2O \rightarrow 7CO_2 + nH^+ + ne^- \tag{3.17}$$

Reaction (3.15) is the adsorption of toluene on the Pt surface, which can increase the overpotential of the ORR by blocking Pt active sites, thus degrading fuel cell performance. Under PEM fuel cell operating conditions with excess oxygen, the adsorbed toluene may experience deep chemical oxidation through reaction (3.16) and produce CO_2 [18], or electrochemical oxidation through reaction (3.17), thus releasing some of the Pt sites for the ORR.

Other VOCs in air include warfare gases, as listed in Table 3.2. These gases are for the most part disastrous for PEM fuel cell cathodes [9].

3.6 Impacts of Multiple Contaminants

In real-world cases, the ambient air always contains a mixture of multiple contaminants. For example, an air sample, according to the Ambient Air Quality Standard of the PRC [34], contains 1 ppm SO_2, 0.8 ppm NO_2, and 0.2 ppm NO. Therefore, it is important to study the impacts of multicontaminants on PEM fuel cell performance. Unfortunately, only two studies have been published, to date [10,22].

Fenning et al. [10] studied the contamination effect by feeding the fuel cell cathode with four types of gases for comparison: (1) pure air, (2) 1 ppm NO_2/air, (3) 1 ppm SO_2/air, and (4) a mixture of 1 ppm NO_2 and 1 ppm SO_2 balanced with air. Figure 3.10 shows the four performance curves of the fuel cell running for 100 h. It can be seen that the contamination effect of the gas mixture on fuel cell performance was between that of 1 ppm NO_2/air and that of 1 ppm SO_2/air, indicating that the effect of the mixed

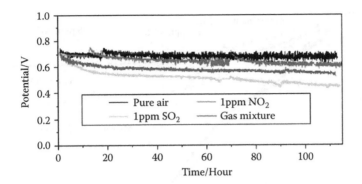

FIGURE 3.10
The constant-current discharge curves of the PEM fuel cell during running for 100 h with pure air, 1 ppm NO_2/air, 1 ppm SO_2/air, and air mixture containing 1 ppm NO_2 + 1 ppm SO_2. Current density: 500 mA cm^{-2}. (From Fenning, J. et al. 2007. *J. Power Sources.* 166:172–6. With permission.)

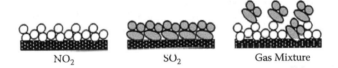

FIGURE 3.11
Schematic of air contaminants adsorbing on the catalyst layer. (From Fenning, J. et al. 2007. *J. Power Sources.* 166:172–6. With permission.)

contaminants was not the sum of the effects of the individual contaminants. This phenomenon was explained by competitive adsorption between NO_2 and SO_2, as shown in Figure 3.11 [10]. The presence of NO_2 suppressed the adsorption of SO_2. Therefore, the adsorption of NO_2 on the catalyst layer resulted in less area remaining for the adsorption of SO_2, thereby reducing the quantity of SO_2 that came into direct contact with the electrocatalyst. Thus, the contribution of SO_2 to the degradation of fuel cell performance in the gas mixture was lower than that of the same concentration of SO_2 alone in air.

Knight et al. [22] studied the additive effects of anode and cathode contaminations on fuel cell performance by feeding the cathode with air containing a mixture of NO_2 and SO_2, and feeding the anode with hydrogen containing H_2S [22]. Figure 3.12 shows the effects of each individual contaminant, and the additive effect of the anode and cathode contaminants. Contrary to what was observed by Fenning et al. [10], the effect of the multiple contaminants in both the anode and cathode sides was the sum of their individual effects.

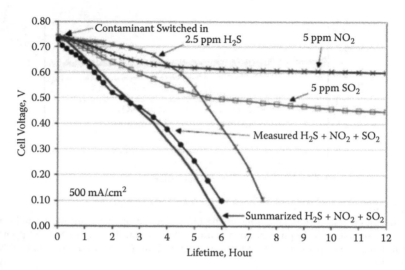

FIGURE 3.12
Additive effects of multiple contaminants: 2.5 ppm H_2S in fuel, 5 ppm SO_2, and 5 ppm NO_2 in air. Current density: 500 mA/cm²; cell temperature: 55°C; stoich: H_2 1.25/O_2 2.0; RH: 100%. (From Knight, S. et al. 2005. Fuel cell reactant supply—Effects of reactant contamination. *Fuel Cell Seminar 2005. Fuel Cell Progress, Challenges and Markets,* Palm Springs, CA, 121–5. With permission.)

3.7 Mitigation Strategies for Airside Contamination

Research on mitigating PEM fuel cell contamination has mainly focused on the anode side, driven primarily by the intent to use reformed hydrogen from fossil fuels containing impurities poisonous to sensitive Pt catalysts. However, the cathode side on which the sluggish ORR occurs requires twice as much Pt as the anode side and, therefore, is potentially more sensitive to contamination [35]. As reported above, air impurities cause degradation in fuel cell performance, and, in most cases, the performance is not fully recoverable. Therefore, it is critical to develop mitigation strategies to protect cathodes from contamination when fuel cells are operated in real environments.

There are two methods to deal with air contamination of fuel cells: one is to improve the catalyst's tolerance to contaminants, while the other is to filter contaminants from the air. The latter is always incorporated with chemical adsorption [36]. Recently, Bouwman et al. [37] developed a Pt-Fe phosphate catalyst (Pt-FePO), which had strong tolerance to CO and SO_2 in the air stream. However, due to the wide variety of contaminants in air, developing impurity-tolerant catalysts is a difficult, time-consuming, and possibly expensive alternative. Therefore, the more effective and feasible way to operate fuel cells in a contaminated environment is adsorptive filtration of contaminants from the ambient air. A wide variety of filtration technologies are available, but the filter design for fuel cell applications must be optimized

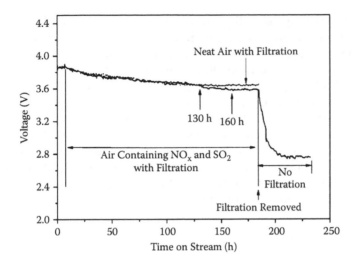

FIGURE 3.13
Cell voltage versus time of 250 W fuel cell stack with NO_x and SO_x filtration (air pressure = 1.5 bar; bed depth = 3 cm; air flow rate = 30 L min⁻¹; current density = 280 mA cm⁻²). (From Ma, X. et al. 2008. *J. Power Sources* 175:383–9. With permission.)

with consideration for air properties and fuel cell attributes. Kennedy et al. [35] created a cathode air filter design methodology that connected properties of the cathode air stream, filter design options, and the filter footprint to a set of adsorptive filter parameters, in order to optimize fuel cell operation.

One of the most important factors for adsorptive filtration is the development of effective adsorbents to remove NO_x, SO_x, and so forth. Xiaowei et al. [36] investigated activated carbon as the adsorbent for cathode air filtration, and found it very effective in protecting the stack from NO_x and SO_x contamination, as shown in Figure 3.13.

3.8 Example Analysis: Development of Cell Performance Predictor under the Influence of Toluene in the Air Stream [38]

Theoretical models may be comprehensive in terms of encompassing all the phenomenological processes involved in a PEM fuel cell operation, but are often difficult to solve due to an abundance of parameters that are often not easy to define. On the other hand, empirical or semiempirical models can identify the key processes, make reasonable assumptions, and utilize experimental data to assist in solving equations to obtain the key parameters [39]. From an engineering point of view, an empirical or semiempirical model with fewer and straightforward parameters may make more sense and be easier to use in predicting fuel cell performance in the presence of contaminants. In this section, we present a semiempirical model as an example of how to develop a performance predictor for practical applications. This semiempirical model (performance predictor) is an excellent fit for the experimental data on fuel cell performance in the presence of toluene in the air stream. To validate the model and make it more useful for practical applications, we carried out toluene contamination testing using a fuel cell assembled with commercially available catalyst-coated membrane (CCM) and GDLs. The toluene concentrations studied in both the experiments and the model were in the range of real-life levels. The importance of this model is its ability to predict cell performance under the influence of toluene in a concentration range of 0 to 50 ppm at various current densities.

3.8.1 Experimental

The MEA used for the contamination tests had an active area of 50 cm². The CCMs were manufactured by Ion Power, using a Nafion 211 membrane and a Pt loading of 0.4 mg cm⁻² on both cathode and anode sides. The GDL, purchased from SGL Carbon Group (Wiesbaden, Germany), was a PTFE (20 wt%) and carbon black impregnated carbon paper. The schematic configuration of the contaminant-mixing system and details on the fuel cell components, including flow field and hardware, are those described in our previous publications [29,30].

The contamination tests were conducted in a constant-current pattern in the presence of toluene, and the data of cell voltage versus time were collected using FC Power software to evaluate the degradation in cell voltage caused by toluene contamination. The following experimental conditions were used for all tests: 80% fuel cell RH, 80°C cell temperature, 30 psig back pressure, and 1.5/3.0 stoichiometries for H₂/air. In our published work [29,30], we reported contamination testing results at concentration levels of toluene ranging from 1 to 50 ppm, with the primary purpose of screening out a "typical" concentration level for subsequent tests on operating conditions, and gaining a fundamental understanding of toluene contamination mechanisms. However, we knew from the VOC concentrations in indoor and outdoor air (provided by Environment Canada, Gatineau, Quebec) that the true level of toluene falls in the ppb range. In the present work, therefore, we extended the concentration level of toluene down to the ppb level in order to develop and validate a semiempirical model.

3.8.1.1 Description of Semiempirical Models

3.8.1.1.1 Semiempirical Model for Fuel Cell Performance in the Absence of Contaminants

Fuel cell performance (cell voltage versus current density) in the absence of contaminants has been described by both theoretical and empirical expressions, among which the semiempirical equation (3.18) is the most simplified [40,41].

$$E_{cell} = E^o - b\log(i) - R_o i - m\exp(ni) \tag{3.18}$$

and

$$E^o = E_r + b\log(i_o) \tag{3.19}$$

where E^o is the open circuit voltage, E_r is the reversible cell voltage, i is the current density, b is the Tafel slope, and i_o is the exchange current density for the ORR. R_o represents the cell resistance, which is mainly due to the ohmic resistances of the PEM and the ionomer in the catalyst layer, and the electronic resistance of the fuel cell end plates and flow field plates. The term $m\exp(ni)$ in equation (3.18) primarily represents the contribution of oxygen mass transport to the fuel cell overpotential, where m and n are parameters associated with features of the catalyst layer structure.

Performance was established in the absence of toluene, and this served as a baseline to evaluate the performance degradation caused by contamination, as well as to simulate the parameters of E^o, b, R_o, m, and n in equation (3.18), which are considered to be independent of the contamination process. Figure. 3.14 shows the baseline performance (cell voltage versus current density) obtained with multiple measurements of different MEAs. The variation bars define the acceptable performance variation of a freshly assembled MEA. Within the current density range in which the fuel cell was operated (below 1.2 A cm^{-2}), the polarization/performance curve does not show a sign of mass transfer control at high current densities (i.e., a curve bending downwards). This indicates that the fuel cell has been optimized for water management [42] and, therefore, the mass transfer term in equation (3.18) can be neglected to obtain equation (3.20) [40,41].

$$E_{cell} = E^o - b\log(i) - R_o i \tag{3.20}$$

Simulation of Figure 3.14 using equation (3.20) yielded the parameter values shown in Table 3.3.

3.8.1.1.2 Semiempirical Model for Fuel Cell Performance in the Presence of Toluene

In the presence of toluene in the air stream, the fuel cell performance degraded. Figure 3.15 illustrates two sets of representative results of toluene contamination tests, conducted with various levels of toluene concentration at current densities of 0.75 and 1.0 A cm^{-2}, respectively. The cell voltage experienced a transient period (nonsteady state) immediately after the introduction of toluene, then reached a plateau (steady state). The duration of the transient period and the magnitude of the cell voltage drop to the plateau were strongly dependent on toluene concentration and current density.

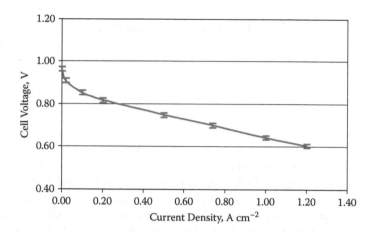

FIGURE 3.14

Baseline performance (cell voltage versus current density) in the absence of toluene. Operating conditions: cell temperature: 80°C; fuel cell RH: 80% on both anode and cathode sides; back pressure: 30 psig; air and hydrogen stoichiometries: 3.0 and 1.5, respectively.

TABLE 3.3

Simulated Values of Electrode Kinetic Parameters from the Baseline Polarization Curve, In the Absence of Toluene, Using Equation (3.20)

Parameter	Units	Value of Parameter
E^o	V	0.977
b	V dec⁻¹	0.0527
R_o	Ω cm²	0.174

In our published work [29], we concluded from AC impedance measurements and resistance data analysis that toluene caused fuel cell performance degradation mainly through a kinetic effect, i.e., the blocking of the active Pt sites via toluene adsorption. Therefore, an empirical model taking into account only the effect of toluene on the kinetics was constructed to describe the cell voltage under the influence of toluene, as shown in equation (3.21):

$$E_{cell} = E^o - (b + K_{CK}C) \times \log(i) - R_o i + K_{CK}C \times \log(i) \times \exp(-K_t t) \quad (3.21)$$

where E_{cell} is the fuel cell voltage, C is the toluene concentration in the air stream, and t is the contamination time. In addition to parameters E^o, b, and R_o, which were previously obtained from the baseline performance

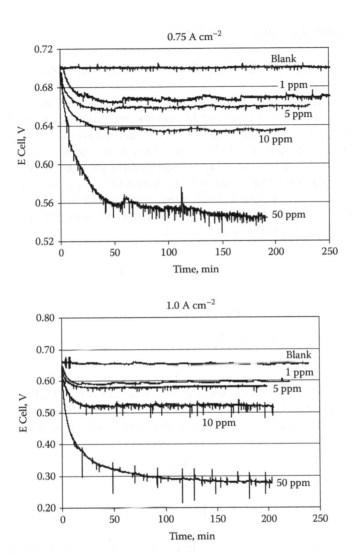

FIGURE 3.15
Cell voltage versus time at 0.75 and 1.0 A cm⁻², with 1, 5, 10, and 50 ppm toluene in the air stream. Toluene was introduced at zero minute. Operating conditions: stoichiometries of air/ H_2: 3.0/1.5; cell temperature: 80°C; RH of the fuel cell: 80%; back pressure: 30 psig.

curve in the absence of toluene (shown in Table 3.1), two new parameters are introduced in equation (3.21) as a result of toluene contamination. K_{CK} accounts for the steady-state performance behavior in the presence of toluene, and K_t represents the kinetic degradation rate that determines the transient behavior of the contamination process. Both K_{CK} and K_t are functions of current density, i, and toluene concentration, C.

Rearranging equation (3.21) leads to:

$$E_{cell} = E^o - [b + K_{CK}C(1 + \exp(-K_t t))] \times \log(i) - R_o i \tag{3.22}$$

where the term $K_{CK}C(1 + \exp(-K_t t))$ reflects the change in Tafel slope caused by the effect of toluene contamination on the activation overpotential of the ORR. The presence of contamination time, t, in the exponential term reflects the pattern according to which the cell performance degrades, i.e., cell voltage decays exponentially in the early contamination process (transient period) due to accumulation of the adsorbed toluene on the catalyst surface, and then levels off as $t \to \infty$, reaching a steady-state plateau. The fuel cell performance at the plateau of the contamination process can be obtained from simplifying equation (3.22) to the following equation:

$$E_{cell} = E^o - (b + K_{CK}C) \times \log(i) - Ri \tag{3.23}$$

3.8.2 Results and Discussion

3.8.2.1 Simulation of K_{CK} and K_t

First, values for K_{CK} and K_t were found through equation (3.22) using contamination testing data obtained with 50 ppm, 10 ppm, 5 ppm, 1ppm, 500 ppb, 250 ppb, and 50 ppb toluene under current densities of 0.2, 0.5, 0.75, and 1.0 A cm^{-2}. Then K_{CK} and K_t were correlated as functions of current density (i) and toluene concentration (C), as shown in equation (3.24) and equation (3.25):

$$K_{CK} = Ai^3 + Bi^2 + Ci + D \tag{3.24}$$

$$K_t = ai^3 + bi^2 + ci + d \tag{3.25}$$

where the constants A, B, C, D, a, b, c, and d are all a function of toluene concentration. Thus, the fuel cell performance can be simulated for any current density and toluene concentration within the range of the model. Figure 3.16 presents a comparison between the simulated and measured cell performances under the influence of various levels of toluene concentration at four current densities. The model simulation (solid curves) agrees well with the experimental data (dotted curves).

3.8.2.2 Model Prediction

3.8.2.2.1 Prediction of Cell Performances in the Presence of Toluene

To validate the model, contamination tests were also conducted at levels of 100 and 20 ppb toluene in the air stream; these tests were not used in the model

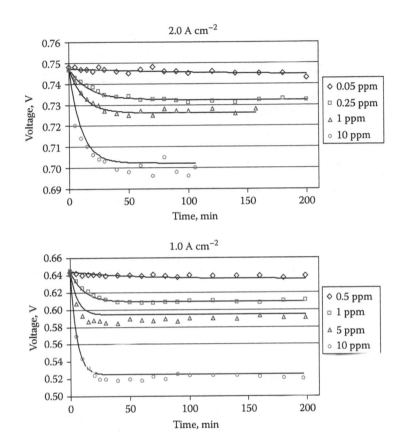

FIGURE 3.16
Measured (dotted curves) and model-simulated (solid curves) cell voltage versus time at 0.2 and 1.0 A cm^{-2} with various levels of toluene concentration. Operating conditions: the same as described for Figure 3.15.

simulation. Figure 3.17 presents a comparison between the experimentally tested and model-predicted cell performances. The comparison suggests that the empirical model is able to provide a reasonably good prediction for cell performance decay in the presence of toluene with concentration levels as low as 20 ppb.

Therefore, the validated toluene concentration range for the empirical model is 0 to 50 ppm in a current density range of 0.2 to 1.0 A cm^{-2}.

3.8.2.2.2 *Prediction of the Steady-State Polarization Curves*

Figure 3.18 shows the predicted and experimentally tested polarization curves for the test cell across the full range of toluene concentrations encompassed by the model. The top curve (solid black) is the baseline polarization curve (i.e., no toluene contamination). Curves below that indicate decreasing performance as the toluene concentration increases.

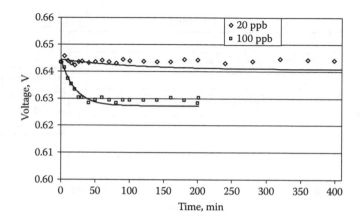

FIGURE 3.17
Experimentally tested (dotted curves) and model-predicted (solid curves) cell voltage versus time at 1.0 A cm^{-2} with 20 and 100 ppb toluene in the air stream. Operating conditions: the same as described for Figure 3.15.

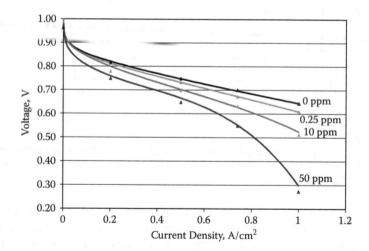

FIGURE 3.18
Model-predicted (solid) and tested (dotted) polarization curves at various concentrations of air side toluene. Operating conditions: stoichiometries: H$_2$/air: 1.5/3.0; cell temperature: 80°C; RH of the fuel cell: 80%; back pressure: 30 psig.

3.8.2.2.3 *Prediction of the Threshold Toluene Concentration*

As discussed earlier, the effect of toluene contamination on PEM fuel cell performance features a transient decay followed by a plateau in the cell performance. The performance plateau represents the maximum performance depression (MPD %) that toluene can cause to the cell performance at a given current density and toluene concentration. Based on the difference between equation (3.20) and equation (3.23), the maximum performance depression at a given current density and toluene concentration can be calculated from equation (3.26):

$$MPD\% = \frac{K_{CK}C \times \log(i)}{(E^\circ - b \times \log(i) - Ri}$$

(3.26)

Equation (3.26) can also be used to estimate the threshold toluene concentration in the air stream for a given performance requirement (MPD %) at a given current density. For example, if 10% MPD is the acceptable performance threshold for a specific PEM fuel cell application, the threshold toluene concentration would be 48.1 ppm at 0.5 A cm^{-2} and 4.5 ppm at 1.0 A cm^{-2}.

3.8.2.2.4 Prediction of the Degradation Rate

The rate of cell performance degradation is another interesting measure of the impact of toluene contamination on cell performance. This can be estimated by differentiating equation (3.21) with respect to t to obtain dE_{cell}/dt, as shown by equation (3.27):

$$\frac{dE_{cell}}{dt} = -K_{CK}K_t C \times \log(i) \times \exp(-K_t t)$$

(3.27)

It can be seen that the degradation rate, which is the rate at which the cell performance decays to the steady-state plateau, is a function of toluene concentration, current density, and toluene exposure time. Figure 3.19

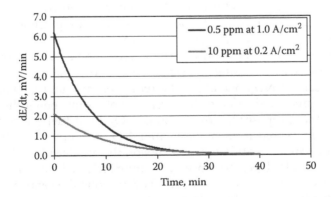

FIGURE 3.19
Degradation rate versus time at 1.0 A cm^{-2} with 0.5 ppm toluene and 0.2 A cm^{-2} with 10 ppm toluene. Operating conditions: stoichiometries: H$_2$/air: 1.5/3.0; cell temperature: 80°C; RH of the fuel cell: 80%; back pressure: 30 psig.

shows the model-predicted degradation rates over time of two different cases: 0.5 ppm toluene at 1.0 A cm^{-2} and 10 ppm at 0.2 A cm^{-2}.

3.8.3 Conclusions

A semiempirical model that considered only the effect of toluene on kinetics was constructed to describe cell voltage as a function of contamination time and current density. The parameters independent of the contamination process, i.e., open circuit voltage ($E°$), cell resistance (R_o), and Tafel slope (b), were first estimated based on experimental data in the absence of toluene (giving a baseline). Then these parameters were used to empirically obtain the expressions of two other parameters, K_t and K_{CK}, which accounted for the effect of toluene contamination on transient and steady-state cell performance, using experimental data at various levels of toluene concentration under four current densities. The model was validated by comparing the contamination testing results with model-predicted results. Several other definitions were also presented, based on the model, such as the threshold toluene concentration and the degradation rate.

3.9 Summary

Common air contaminants, such as SO_x, NO_x, H_2S, NH_3, and VOCs, can impose significant contamination effects on PEM fuel cells, causing fuel cell performance degradation. The degree of performance degradation depends on the type of air contaminants, their concentration, and the operating conditions. Generally, higher concentrations lead to a more severe contamination effect. The contamination effect of SO_x is due to adsorbed sulfur on the Pt surface, produced from SO_x reduction, which not only poisons the catalyst, but also changes the ORR mechanism. In addition, the fuel cell performance degradation caused by SO_x is only partially recoverable. With respect to NO_x contamination, two opposing hypotheses exist to explain the contamination mechanism of NO_x, one being the reduction mechanism that suggests the electrochemical reduction of NO_2, and the other one being the oxidation mechanism that proposes the oxidation of NO_2. PEM fuel cell performance can be fully recovered in most cases after NO_x are replaced with neat air. Similar to the contamination mechanisms on the anode side of the fuel cell, NH_3 degrades fuel cell performance by forming NH_3^+, which reduces the proton conductivity of the membrane, and H_2S poisons the catalyst surface by strongly adsorbing on the Pt surface and blocking the active catalyst sites for the ORR. VOC contamination of the cathode catalyst layer is not severe, and is mostly recoverable. There are several ways of mitigating cathode contamination, but adsorptive filtration is the most effective and feasible measure.

Acknowledgments

For the development of this semiempirical model, the authors gratefully acknowledge financial support from the Institute for Fuel Cell Innovation, National Research Council of Canada (NRC-IFCI), Ballard Power Systems Inc., Hydrogenics Corp., and Angstrom Power Incorporated.

References

1. Cheng, X., Shi, Z., Glass, N., Zhang, L., Zhang, J., Song, D., Liu, Z-S., Wang, H., Shen, J. 2007. A review of PEM hydrogen fuel cell contamination: Impacts, mechanisms, and mitigation. *J. Power Sources* 165:739–56.
2. Ticianelli, E.A., Derouin, C.R., Redondo, A., Srinivasan, S. 1988. Methods to advance technology of proton exchange membrane fuel cells. *J. Electrochem. Soc.* 135:2209–14.
3. Wilson, M.S., Gottesfeld, S. 1992. High performance catalyzed membranes of ultra-low Pt loadings for polymer electrolyte fuel cells. *J. Electrochem. Soc.* 139:L28–30.
4. Song, C., Tang, Y., Zhang, J-L., Zhang, J., Wang, H., Shen, J., McDermid, S., Li, J., Kozak, P. 2007. PEM fuel cell reaction kinetics in the temperature range of 23–120°C. *Electrochim. Acta* 52:2552–61.
5. Dicks, A.L. 1996. Hydrogen generation from natural gas for the fuel cell systems of tomorrow. *J. Power Sources* 61:113–24.
6. Hohlein, B., Boe, M., Bogild-Hansen, J., Brockerhoff, P., Colsman, G., Emonts, B., Menzer, R., Riedel, E. 1996. Hydrogen from methanol for fuel cells in mobile systems: Development of a compact reformer. *J. Power Sources* 61:143–47.
7. Larminie, J., Dicks, A. 2003. *Fuel cell systems explained*, 2nd ed., chap. 4. London: John Wiley & Sons.
8. Xianguo, L. 2006. *Principles of fuel cells*, chap. 7. New York: Taylor & Francis Group.
9. Moore, J.M., Pal, L.A., Barry, J.L., Gary, O.M. 2000. The effects of battlefield contaminants on PEMFC performance. *J. Power Sources* 85: 254–60.
10. Fenning, J., Ming, H., Weiyu, S., Jie, F., Hongmei, Y., Pingwen, M., Baolian, Y. 2007. The effect of ambient contamination on PEMFC performance. *J. Power Sources*. 166:172–6.
11. Mohtadi, R., Lee, W.K., John, W.Z. 2004. Assessing durability of cathodes exposed to common air impurities. *J. Power Sources* 138:216–25.
12. Garzon, F., Brosha, E. Pivovar, B., Rockward, T., Uribe, F., Urdampolleta, I. 2006. U.S. DOE Hydrogen Program–FY 2006 *Annual Progress Report*, Washington, D.C. 905–9.
13. Contractor, A.Q., Lal, H. 1979. Two forms of chemisorbed sulfur on platinum and related studies. *J. Electroanal. Chem.*, 96: 175–81.
14. Loucka, T. 1971. Adsorption and oxidation of sulphur and of sulphur dioxide at the platinum electrode. *J. Electroanal. Chem.* 31:319–32.

15. Contractor, A.Q., Lal, H. 1978. The nature of species adsorbed on platinum from SO₂ solution. *J. Electroanal. Chem.* 93:99–107.
16. Garsany, Y., Baturina, O.A., Swider-Lyons, K.E. 2007. Impact of sulphur dioxide on the oxygen reduction reaction at Pt/Vulcan carbon electrocatalysts. *J. Electrochem. Soc.* 154:B670–5.
17. Garsany, Y., Baturina, O.A., Swider-Lyons, K.E. 2007. Impact of SO₂ on the kinetics of Pt₃CO/Vulcan carbon electrocatalysts for oxygen reduction. *ECS Transactions*, 11:863–75.
18. Brosha, E., Garzon, F., Pivovar, B., Rockward, T., Springer, T., Uribe, F. 2006. *Annual U.S. DOE Fuel Cell Program Review*. Arlington, VA, (May) 16–19.
19. Mangun, C.L., DeBarr, J.A., Economy, J. 2001. Adsorption of sulphur dioxide on ammonia-treated activated carbon fibers. *Carbon* 39:1696–89.
20. U.S. Environmental Protection Agency. *Air Trends*, 2004. http://www.epa.gov/airtrends/sixpoll.html
21. Yang, D., Ma, J., Xu, L., Wu, M., Wang, H. 2006. The effect of nitrogen oxides in air on the performance of proton exchange membrane fuel cell. *Electrochim. Acta* 51:4039–44.
22. Knight, S., Jia, N., Chuy, C., Zhang, J. 2005. Fuel cell reactant supply—Effects of reactant contamination. *Fuel Cell Seminar 2005: Fuel Cell Progress, Challenges and Markets*. Palm Springs, CA, 121–5.
23. Uribe, F., Smith, W., Wilson, M., Valerio, J., Rockward, T., Garzon, F., Brosha, E., Saab, A., Bender, G., Adcock, P., Xie, J., Norman, K., Havrilla, G. *FY 2003 Report*, Los Alamos National Lab (LANL). http://www.eere.energy.gov/hydrogenandfuelcells/pdfs/ive12_uribe.pdf
24. Halseid, R., Preben, J.S.V., Tunold, R. 2006. Effect of ammonia on the performance of polymer electrolyte membrane fuel cells. *J. Power Sources* 154:343–50.
25. Uribe, F., Gottesfeld, S. Jr., Zawodzinski, T.A. 2002. Effect of ammonia as potential fuel impurity on proton exchange membrane fuel cell performance. *J. Electrochem. Soc.* 149:A293–6.
26. Halseid, R., Preben, J.S.V., Tunold, R. 2004. Influence of ammonia on conductivity and water content of Nafion 117 membranes. *J. Electrochem. Soc.* 151:A381–8.
27. Halseid, R., Bystron, T., Tunold, R. 2006. Oxygen reduction on platinum in aqueous sulphuric acid in the presence of ammonium. *Electrochim. Acta* 51:2737–42.
28. Mathieu, M., Primet, M. 1984. Sulfurization and regeneration of platinum. *Appl. Catal.* 9:361–70.
29. Li, H., Zhang, J., Fatih, K., Wang, Z., Tang, Y., Shi, Z., Wu, S., Song, D., Zhang, J., Jia, N., Wessel, S., Abouatallah, R., Joos, N. 2008. PEMFC contamination: Testing and diagnosis of toluene-induced cathode degradation. *J. Power Sources* 185:272–9.
30. Li, H., Zhang, J., Shi, Z., Song, D., Fatih, K., Wu, S., Wang, H., Zhang, J., Jia, N., Wessel, S., Abouatallah, R., Joos, N. 2009. PEM fuel cell contamination: Effects of operating conditions on toluene-induced cathode degradation. *J. Electrochem. Soc.* 156:B252–7.
31. Li, H., Zhang, J.L., Fatih, K., Shi, Z., Wu, S.H., Song, D.T., Zhang, J.J., Jia, N., Wessel, S., Abouatallah, R., Joos, N. 2008. Durability of PEMFC cathodes exposed to toluene-contaminated air. *ECS Transactions* 16:1059–67.
32. Shi, Z., Song, D., Li, H., Fatih, K., Tang, Y., Zhang, J., Wang, Z., Wu, S., Liu, Z-S., Wang, H., Zhang, J. 2009. A general model for air-side proton exchange membrane fuel cell contamination. *J. Power Sources* 186:435–45.

33. Fatih, K., Shi, Z., Song, D., Li, H., Zhang, J., Wu, S., Wang, H., Liu, Z-S., Zhang, J.J. 2009. Toluene adsorption on Pt electrode: Effect on oxygen reduction reaction and implication for PEM fuel cell contamination. Forthcoming. *Electrochim. Acta.*
34. Ambient Air Quality Standard, National Standard of PRC, GB 3095-1996, ICS 13.040.20 Z50, 1996.
35. Kennedy, D.M., Donald, R.C., Wenhua, H.Z., Kenneth, C.W., Mark, R.N., Bruce, J.T. 2007. Fuel cell cathode air filters: Methodologies for design and optimization. *J. Power Sources* 168:391–9.
36. Ma, X., Yang, D., Zhou, W., Zhang, C., Pan, X., Lin, X., Minzhong, W., Ma, J. 2008. Evaluation of activated carbon adsorbent for fuel cell cathode air filtration. *J. Power Sources* 175:383–9.
37. Bouwman, P.J., Teliska, M.E., Swider Lyons, K. 2006. Increased poisoning tolerance of Pt-FePo oxygen reduction catalysts. *Proton Conducting Membrane Fuel Cell IV: Proceedings of the International Symposium.* 222–6.
38. Li, H., Zhang, J., Song, D., Shi, Z., Fatih, K., Wu, S., Zhang, J., Jia, N., Wessel, S., Abouatallah, R., Joos, N., Schrooten, J. 2009. Unpublished results.
39. Rama, R., Chen, R., Thring, R. 2005. A polymer electrolyte membrane fuel cell model with multi-species input. *Proceedings of the Institution of Mechanical Engineering*, Part A. *J. Power Energy* 219:255–71.
40. Laurencelle, F., Chahine, R., Hamelin, J., Agbossou, K., Fournier, M., Bose, T.K., Laperriere, A. 2001. Characterization of a Ballard MK5-E proton exchange membrane fuel cell stack. *Fuel Cells* 1:66–71.
41. Kim, J., Lee, S.M., Srinivasan, S., Chamberlin, C.E. 1995. Modeling of proton exchange membrane fuel cell performance with an empirical model. *J. Electrochem. Soc.* 142:2670–4.
42. Li, H., Tang, Y., Wang, Z., Shi, Z., Wu, S., Song, D., Zhang, J., Fatih, K., Zhang, J., Wang, H., Liu, Z., Abouatallah, R., Mazza, A. 2008. A review of water flooding issues in the proton exchange membrane fuel cell. *J. Power Sources* 178:103–17.

4

Anode Contamination

Daijun Yang, Jianxin Ma, and Jinli Qiao

CONTENT

4.1 Introduction

Driven by the limited fossil fuel resources and the implicit political dependencies to secure and diversify energy supplies, the development of alternative and renewable energy has been and is adopted by more and more countries worldwide.

Proton exchange membrane fuel cell (PEMFC) is a promising device for energy saving and environment protection, and is also one of the main tools for the realization of "hydrogen economy." For this reason, it is just common sense that PEMFC can be used as a propulsion of electric vehicles, i.e., fuel cell vehicle (FCV). At the beginning stage of a hybrid FCV, which is powered by both PEMFC stack and other power systems, such as super capacitor or battery, commercialization does seem possible. In any case, hydrogen is used exclusively for the fuel in PEMFCs. Therefore, there is no doubt that the application of PEMFCs in transportation will cut down the dependency on petroleum oil because hydrogen can be retrieved from various feedstocks, thus reducing "carbon emissions." Although the emission of the main greenhouse gas (CO_2) cannot really be eliminated, as some exaggerated advocators have depicted, when hydrocarbons are used as the feedstocks of hydrogen, centralized production of hydrogen provides a possible pathway for collection and sequestration.

As an energy carrier, hydrogen plays a key role on the way to hydrogen energy. However, the production, storage, transportation, as well as the fueling infrastructure of hydrogen, is still in its infancy. So far, except for the field of space, there is almost no real market application for hydrogen energy. The niche market based on the governmental demonstration projects or small-scaled commercialization programs for the FCV may be shaped in the next 10 to 15 years. Nevertheless, as a chemical product, hydrogen was widely used more than 100 years ago in various industries, such as petrochemical, food, electronics, fertilizer, metallurgical, and others. The production, storage, and transportation of hydrogen have been rather mature, which makes its utilization as an energy carrier highly attractive. On the other hand, a number of technical challenges have to be overcome to make hydrogen a commercially available, large-scale player in the energy market. For instance, a hydrogen refueling infrastructure, regulation framework should first be built up, as well as the public's understanding and acceptance, to facilitate the safe introduction of FCVs and hydrogen energy. In this chapter, we will start with discussing the quality of hydrogen.

The PEMFC is an electrochemical reactor, which converts the chemical energy of hydrogen and oxygen into electric power in an effective and silent way, where catalysts for both anode and cathode are necessary and important. With the highest activity, precious metal platinum (Pt) dominates the anode catalysis. After the invention of scattering and supporting Pt on cheap carbon black, e.g., Vulcan® XC-72R from Cabot, its consumption on the anode has been substantially reduced, even to 0.1 mg/cm². The low loading of Pt is advantageous for the mass production of PEMFCs. Unfortunately, Pt is also a catalyst that is susceptive to the poisoning of carbon monoxide (CO) and sulfides. In fact, hydrogen today primarily comes from the reforming of hydrocarbons, of which the feedstock is either natural gas or coal. For these feedstocks, sulfur is a commonly existing element other than H and C, and, after the reforming processes, it will be present in various forms, such as hydrogen sulfide (H_2S), sulfide dioxide (SO_2), carbyl sulfide (COS), etc. Besides, N_2, O_2, CO, CO_2, H_2O, HCO, HCOOH, NH_3, and other substances will also be produced as by-products with hydrogen. Except for H_2, all the other end products, intermediates, or substances from air are referred to as impurities here. Getting rid of all the impurities and operating PEMFCs on ultra high purity (UHP) hydrogen would be ideal, but tremendous cost accompanies this process. For cost efficiency, it is almost impossible to run PEMFC with UHP H_2. Tradeoffs between hydrogen clean-up, fuel cell cost, and durability need to be determined. How to control the quality of H_2 and find a solution for the mitigation of anode contamination makes it necessary as well as important to investigate the influences of all the impurities.

This chapter attempts to summarize recent significant progress in the current knowledge of anode contamination with the help of the results reported in literature and obtained in the authors' laboratory. Less attention is given to an exhaustive review on the topic of anode contamination, since these subjects have been covered in recent publications [1,2]. In order to accomplish our objective, the relationship between hydrogen purity and its source is charted; the anode contamination effects of PEMFC from major impurities (CO, CO_2, H_2S, NH_3, and organic compounds) will be presented and summarized as well as the contamination mechanisms and mitigation strategies.

4.2 Hydrogen Production Processes

Hydrogen is the most abundant element in the universe, but binds into molecular compounds, existing first in water and next in hydrocarbons. Hydrogen production technologies convert a hydrogen-containing material, such as water, natural gas (NG), gasoline, methanol, etc., into hydrogen or a hydrogen-rich stream. With respect to the energy required, it is easier to produce hydrogen from compounds that are at a higher energy state, i.e., reforming from fossil fuels. However, it takes more energy to split hydrogen from water, done normally through electrolysis, because excess energy has to be input into the splitting process.

Currently, reforming is the principal and economical way and dominates the area of hydrogen manufacture. Up to now, ca. 48% of the world's hydrogen is produced through the reforming of natural gas, about 30% through the partial oxidation of liquid hydrocarbons, and about 18% through the gasification of coal [3]. Only 4% of the hydrogen product comes from the process of electrolysis.

4.2.1 Reforming of Hydrogen-Containing Fossil Fuel

Natural gas is believed to be the fourth generation of primary energy sources after firewood, coal, and petroleum oil, as well as the most-used feedstock for hydrogen production. There are some NG-reforming-hydrogen manufacturers worldwide with high yields. Natural gas drawn from underground is composed of mostly methane (75 to 85%), together with other hydrocarbons, N_2, CO_x (CO and CO_2), and other substances with less amounts, such as H_2S (organo-sulfur) and H_2O. The reforming of NG is normally comprised of steaming methane reforming (SMR), adiabatic prereforming, partial oxidation (POX), and autothermal reforming (ATR). Hydrogen-rich (typically from 40 to 70%) products obtained through these processes are called "reformates." However, the reformates cannot be directly applied to PEMFCs, because some impurities in the feedstock, along with some derivative substances and intermediates from the chemical reaction, will inevitably exist. On the other hand, reformates can also be produced through other processes including the gasification of coal and reforming of small molecular organic matters, like methanol [4], ethanol [5], and dimethyl ether (DME) [6]. Similar to natural gas reforming, three primary techniques, namely steam reforming, POX, and ATR, are used. Figure 4.1 shows the general process for the production of hydrogen from hydrocarbon fuels.

For the purpose of removing the large amounts of CO produced during the reforming, as shown in Figure 4.2, one or more water gas shift (WGS) reactors—typically a high temperature reactor and low temperature reactor—are used. The high temperature (>350°C) reactor has fast kinetics, but

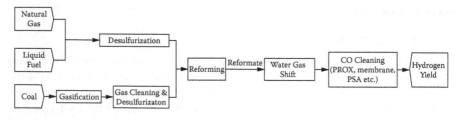

FIGURE 4.1
Reforming processes for hydrogen production.

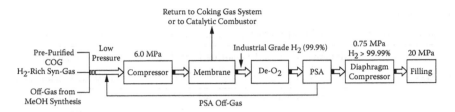

FIGURE 4.2
Schematic of by-product hydrogen purifying processes.

is limited by thermodynamics to the amount of carbon monoxide that can be shifted. Therefore, a lower temperature reactor (210 to 330°C) is used to convert the carbon monoxide to a lower level [7]. The reforming, WGS, and oxidation reactions can be generalized as follows for hydrocarbon and methanol fuels [7].

Steam reforming:

$$C_mH_n + mH_2O \rightarrow mCO + (m + 1/2n) + H_2 \tag{4.1}$$

$$CH_3OH + H_2O \rightarrow CO_2 + 3H_2 \tag{4.2}$$

Partial oxidation:

$$C_mH_n + m/2\ O_2 \rightarrow mCO + n/2\ H_2 \tag{4.3}$$

$$CH_3OH + 1/2\ O_2 \rightarrow CO_2 + 2H_2 \tag{4.4}$$

Autothermal reforming:

$$C_mH_n + m/2\ H_2O + m/4\ O_2 \rightarrow mCO + (m/2 + n/2)H_2 \tag{4.5}$$

$$4CH_3OH + 3H_2O + 1/2\ O_2 \rightarrow 4CO_2 + 11H_2 \tag{4.6}$$

Carbon (coke) formation:

$$C_mH_n \rightarrow xC + C_{m-x}H_{n-2x} + xH_2 \tag{4.7}$$

$$2CO \rightarrow C + CO_2 \tag{4.8}$$

$$CO + H_2 \rightarrow C + H_2O \tag{4.9}$$

Water gas shift:

$$CO + H_2O \rightarrow CO_2 + H_2 \tag{4.10}$$

CO oxidation:

$$CO + O_2 \rightarrow CO_2 \tag{4.11}$$

TABLE 4.1

Typical Hydrogen Yields Compositions by Fuel Reforming (before Purification)

Source	Methane, Methanol, Gasoline/Diesel [3,8]		Coal [9]
Process	SMR	ATR	Dry Coal Gasification
H_2	75.7–94.3%	47.3–93.2%	25.3–26.5%
N_2	0.2–1.9%	0.7–33.5%	0.6–4.3%
CO_2	2.5–19.9%	1.7–16.7%	2.6–10.6%
CO	0.1%	1.4%	42.2–63.6%
Ar	0*	0.6%	–
Source Fuel	2.5–2.9%	2.4–2.5%	~0.09%
H_2O	<250	<250	4.6–20.1%
H_2S	–	–	0.2–0.3%

Note: * = in ppm if no unit follows; – = no data.

$$H_2 + 1/2\ O_2 \rightarrow H_2O \tag{4.12}$$

Although different transformation processes take place while different feedstocks are used, similar types of impurities are produced. The concentrations of impurities depend on the feedstocks and the reformation process.

Some typical processes for hydrogen manufacture and the compositions of hydrogen-rich gases are compared and listed in Table 4.1.

4.2.2 Electrolysis

Compared with reforming of fossil fuels, producing hydrogen through electrolysis is a cleaner way. Theoretically, the feedstock of electrolysis is pure water, and the splitting reaction is very simple—only hydrogen and oxygen are produced. However, to obtain good electrolyte conductivity and energy efficiency, other salts, like sodium chloride and potassium hydroxide, must be added into water as electrolytes. If sodium chloride is used, as in the case of chlor-alkali sector, >99.9% H_2 can be produced [10]. For fuel cell application, where chlorine is not needed, potassium hydroxide must be used as an electrolyte. Normally 99.5 to 99.8% purity can be achieved, and the main impurities are oxygen, water, and nitrogen. However, as water is a substance in a very low energy state, external energy, in the form of electricity or heat, has to be added to its splitting process. In fact, to reach a considerable generation rate of H_2 and reduce the capital cost, higher current density must be used, which in turn results in increased voltage and lowers the energy efficiency. With state-of-the-art technologies, 4 to 5 kW h electric power is consumed for the production of 1 m³ hydrogen and the efficiency of ca. 82% is achieved [11]. Evidently, hydrogen produced by this way loses it advantage. Considering the production of the electricity, which is mainly generated from the combustion of fossil fuels, a cleaner pathway has to be taken into account. For the demonstration of truly clean hydrogen energy, some hydrogen refueling stations combined with solar electricity and water electrolyzing have been built.

4.2.3 Purification of By-Product Hydrogen

Hydrogen is produced as a by-product of some conventional industrial processes, such as chlorine and polyvinylchloride production, sodium hydroxide production, light gases production in crude oil refineries, coal coking, and chemical dehydrogenation processes. As a by-product, hydrogen is commonly emitted or burned. Due to their complex components and insufficient capacity, to recycle hydrogen from these by-products is costly; therefore, only a small part of these by-product sources is reused. Extra cost must be paid in the purification process to obtain a clean hydrogen product. However, in the infancy of "hydrogen economy," recycling hydrogen from these by-product sources will provide an easy pathway to supply hydrogen to demonstration FCV fleets, since these FCV fleets are commonly located in big cities normally equipped with chemical industry zones. In the Shanghai metropolitan area (China) as an example, a FCV demonstration fleet is using purified by-product hydrogen as fuel. Figure 4.2 describes the main membrane/PSA processes of coking oven gas (COG) as well as other hydrogen-rich feed gases. The qualities of the product hydrogen have been tested in our laboratory and example specifications are shown in Table 4.2.

4.3 Sources of Key Impurities

Many concerns arose about the quality of hydrogen rich gas because, during the hydrogen production process, impurities are also produced or introduced from feedstock or air and may adversely affect the performance of PEMFCs.

For industrially produced hydrogen, all the impurities are controlled in trace amounts. Using mature and reliable technologies, such as pressure swing adsorption (PSA), or membrane and cryogenic separations, hydrogen with >90% purity can be produced [12,13]. PSA is a process able to provide hydrogen with a purity ranging from 99 to 99.999% [8]. Of course, multilevel PSA processes must be employed and higher cost will be incurred. However, in the case of a fuel cell integrated with a reformer, where the hydrogen-rich gas cannot be purified to the same extent as by industrial scale process, the concentrations of impurities may be in the order of tens of percents. Pt or Pt-alloy membrane is commonly adopted to remove CO and CO_2 to achieve the purity needed for PEMFCs, with which a compact integration can be made. Figure 4.3 shows a sample for this kind of system, which uses logistic fuel JP-8 as fuel, to be integrated with a 1 kW PEMFC [14].

TABLE 4.2

An Example: Components of By-Product Sources and Purified H_2 Product

Feed Gas Component					Content					
	H_2	CO	CO_2	C_nH_m	O_2+Ar	N_2	CH_3OH	S	NH_3
Unit	%	%	%	%	%	%	%	ppm	ppm
Feed gas 1	85.63	13.21	/	0.01	0.60	0.56	/	0.42	/
Feed gas 1	68.17	27.58	3.42	0.25	0.16	0.41	0.012	0.1	/
Industry H_2	99.9	1 ppm	<0.1	1 ppm	<0.1	<0.1	/	0.14	/
Unit	%	ppm	ppm	ppm	ppm	ppm	ppm	ppb	ppb
Product	>99.99	<1	<1	<1	<1	<50	/	<10	<18

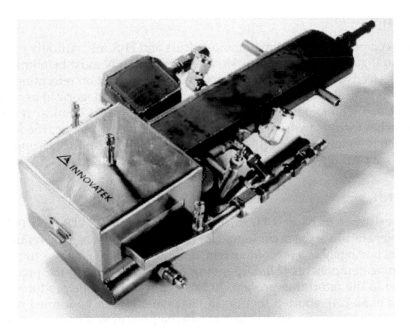

FIGURE 4.3
InnovaGen JP-8 fuel processor for integration with PEM fuel cell. (From Irving, P.M., Pickles, J.S. 2007. *ECS Transactions* 5: 665–71. With permission.)

4.3.1 CO_x

CO_2 is present in natural gas and other gaseous feedstocks. Besides, CO_2 can be produced in various reforming processes, which always are compose of the oxidation or splitting of C-containing substances. Although CO does not exist commonly in feedstock, it is always present in all kinds of hydrogen-rich reforming gases. The reformation of methanol results in a gas mixture of about 74% hydrogen, 25% carbon dioxide, and 1 to 2% CO [15]. Using a selective oxidation process, the CO concentration can be reduced further to about 2 to 100 ppm. Therefore, even after clean-up processes, a small amount of CO may not be avoided. On the other hand, in the presence of CO_2, CO will be formed through reverse water gas shift (RWGS) over high-temperature catalysts at temperatures higher than 300°C or low-temperature catalysts supporting the reaction to about 150°C:

$$CO_2 + H_2 \rightarrow CO + H_2O \qquad (4.13)$$

Thermodynamic calculations show that at a $H_2:CO_2$ ratio of 3:1, which is representative for a methanol-based reformate, CO produced by the RWGS, which is proceeded in the fuel cell itself, could be in the range of 20 to 50 ppm [16]. Therefore, RWGS is the reason for the "poisoning" effects of CO_2 and was added into a CO poisoning model [17].

4.3.2 H$_2$S

Sulfur compounds, including organo-sulfurs and H$_2$S, are naturally present in natural gas and coal feedstocks. These compounds must be eliminated from the feedstocks due to their serious poisoning effect on reforming catalysts. The typical approaches to desulfurization can be categorized as chemical reaction technologies and adsorptive technologies. Chemical reaction approaches include hydrodesulfurization (HDS) and alkylation, the former of which is widely adopted [18–20]. In the HDS process, sulfur-bearing molecules are partially or completely hydrogenated to H$_2$S [21].

4.3.3 NH$_3$

The coexistence of N$_2$ and H$_2$ in the processes of SMR, ATR, and POX at high temperature is the main source for NH$_3$, especially when the reforming involves homogeneous precombustion, or if the fuel itself contains nitrogen-containing components [22]. Depending on the catalyst and the processor adopted in the reforming process, 30 to 90 ppm NH$_3$ can be produced [23]. Halseid et al. [24] reported that up to 150 ppm NH$_3$ can be formed during the reforming process. Ammonia may also be present as a trace component if ammonia is used as feedstock for hydrogen production by thermal splitting [25]. Besides, in the NG distribution system, NH$_3$ is normally input on purpose as the tracer for the convenience of detection [1].

Understanding the tolerance of anode catalysts to all the impurities is no doubt helpful for hydrogen quality control. The tolerance of an anode to certain H$_2$ impurity is usually defined as its ability to electro-oxidize H$_2$ at an acceptable polarization loss in the presence of the impurity. Typically, 20 to 100 mV polarization loss is adopted for quantifying the anode tolerance at certain current density (typically 500 mA/cm^2), where polarization loss is referenced to the performance with pure H$_2$. Guided by this scheme, the maximum tolerable content in H$_2$ for certain impurity can be determined, but it is very difficult to obtain data among different groups. This can be attributed to the differences in electrode structure, operating conditions, electrolytes, and electrocatalysts. Even if a common understanding about the impurity tolerance, which in turn decides the H$_2$ quality, is achieved by all of us, the balance between H$_2$ quality and its cost must be carefully considered. Prior to the balance, the impacts of main hydrogen impurities as well as the mechanisms of how the impacts occur must be investigated.

4.4 Impact of CO$_x$

As the most ubiquitous fuel impurities, CO and CO$_2$ have attracted great interests as early as the 1980s, accompanying the reinflamed interests in fuel cell. Since 1988, Gottesfeld et al. [26] and Wilson et al. [27] have reported serious efficiency and power output reduction resulting from CO poisoning. Until now, some excellent reviews about the mechanisms, effects, and mitigation of CO_x on PEMFCs can be found in Baschuk et al. {28], Iwasita [29], Bellows et al. [30], Divisek et al. [15], and Hirschenhofer et al. [31].

4.4.1 Preferential Adsorption of CO on Platinum

It is well known that CO is a poisoning substance to hydrogen oxidation reaction (HOR). The poisoning effect of CO is due to strong and preferential CO adsorption on the platinum surfaces, at normal operation temperature (60 to 100°C). Such strong binding has been explained by Blyholder [32]. As shown in Figure 4.4, 5σ CO orbital donates an electron to the metal, while two d electrons transfer from the metal atomic orbits to the antibonding $2\pi^*$CO orbital, which is know as "back-donation" [33]. The co-existence of bonding energies strengthens the interaction between CO and the catalyst metal.

In addition, studies indicate that both linear- and bridge-bonded adsorbed CO species can be found on the catalyst [34]. This is confirmed by the fact that there is no 1:1 correspondence between the loss of an adsorbed H site and the coverage of CO on Pt surfaces since one CO molecule can block more than one H site, as in the case of bridge bonded CO [35].

The oxidation of hydrogen in an acid environment has been extensively studied. The mechanism for electrochemical HOR over smooth platinum in

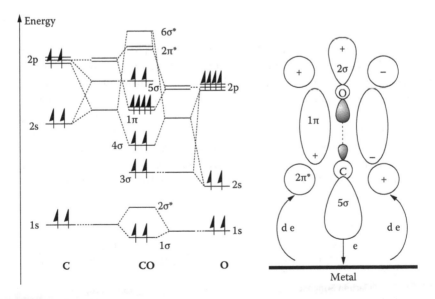

FIGURE 4.4
Illustration of the energy level of carbon monoxide molecules, and the formation of metal carbon monoxide bonding. (From Grgur, B.N. et al. 2001. *J. Serb. Chem. Soc.* 66: 785–97. With permission.)

an acid electrolyte is the slow dissociation of adsorbed hydrogen molecules to hydrogen atoms, known as the Tafel reaction, followed by the fast electrochemical oxidation of the adsorbed hydrogen atoms to form protons and electrons, known as the Volmer reaction [36]. This can be illustrated below:

$$2Pt + H_2 \rightarrow 2Pt\text{-}H \qquad (4.14)$$

$$Pt\text{-}H \rightarrow Pt + H^+ + e^- \qquad (4.15)$$

The mechanism of the poisoning effect of CO has been recognized in literature, especially those aiming at the exploring of CO-tolerant anode catalysts [37–40]. The poisoning process can be described as:

$$Pt + CO \rightarrow Pt\text{-}CO \qquad (4.16)$$

$$2CO + 2Pt\text{-}H \rightarrow 2Pt\text{-}CO + H_2 \qquad (4.17)$$

It can be seen that CO molecules cannot only be adsorbed onto the "naked" platinum atoms, but also replace the hydrogen atoms that have already been adsorbed, i.e., H in Pt-H. This is shown in the schematic diagram (see Figure 4.5).

Because of the preferential adsorption of CO, the current density, or reaction rate of HOR in the presence of CO is reduced and can be written as [30]:

$$i_{H2/CO} = i_{H2}(1-\theta_{CO})^2 \qquad (4.18)$$

where θ_{CO} is the coverage of CO (CO_{ad}) and dictated by the CO adsorption isotherm. It is the most important parameter in the discussion and modeling of the performance of an anode under the influence of CO [41]. This equation is based on the speculation that the limiting reaction requires two adjacent sites for equation (4.2) and the probability of finding two adjacent sites is reduced by the square of the surface fraction not covered by CO. To understand the nature of the polarization loss due to the CO coverage on Pt, a half-cell experiment with H_3PO_4 as the electrolyte has been done by Dhar et al. [41]. θ_{CO} was found to bear a linear relationship with $\ln[CO]/[H_2]$:

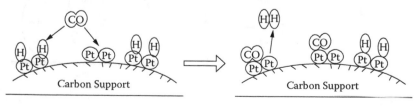

FIGURE 4.5
Schematic of PEMFC anode CO poisoning.

$$\theta_{CO} = -\Delta G_0^0/r - RT/r \cdot \ln H + RT/r \cdot \ln[CO]/[H_2] \qquad (4.19)$$

where ΔG_0^0 is the standard free energy of adsorption, r is the interaction parameter, and H is the Henry's law constant for CO solubility in the unit of atm/(mol/liter).

4.4.2 Reduction of CO_2 to CO

In most chemical engineering processes, CO_2 is commonly the end product and relatively stable. However, even if it does not adsorb onto platinum and compete with H_2 in HOR, CO_2 cannot be treated as inert gas, like N_2 and Ar. According to studies by Cheng [1] and Baschuk [28], the influence of CO_2 is too great to be simply explained by reactant dilution effect. The excessive impact of CO_2 can be accounted for through its reduction to CO. With higher concentration, a well-known reverse RWGS reaction (see equation (4.2)) takes place under fuel cell operation temperature and through the catalysis of Pt. A common concentration level, such as "25%" in Baschuk [28], is the basic requirement for RWGS reaction. The RWGS reaction is the prevalent explanation for the poisoning effect of CO_2 on anode catalysts. Besides, there is another pathway for CO_2 reduction which is the electrochemical reduction of CO_2 by Pt hydride at low potentials [42]:

$$CO_2 + 2Pt\text{-}H^+ + 2e \rightarrow Pt\text{-}CO + H_2O + Pt \qquad (4.20)$$

The formation of adsorbed CO on Pt has been testified experimentally by Gu et al. [43] (see Figure 4.6).

The shaded area in the stripping graph of Figure 4.6(a) stands for the oxidation of Pt-CO, which is produced during the 10 h exposure to 50%CO_2/50%H_2. It is the direct proof for the sentence of RWGS reaction. The Pt-CO formed by electro-reduction of CO_2 polarizes HOR in the same manner as Pt-CO formed by direct CO adsorption. Therefore, in the following section, the impact of CO_x will be focused on the impact of CO.

4.4.3 Deleterious Effect of CO

The severe impact of CO is the best known and well documented of hydrogen impurities, since large numbers of works involving CO have been done. Cheng [1] provided a comprehensive review on the deleterious effect of CO, which was categorized into CO concentration, exposure time, operation temperature, pressure, anode catalyst type, etc. Below, we will emphasize some fundamental aspects of the deleterious effects of CO.

4.4.3.1 Severe Poisoning Impact of Trace Amounts of CO

PEMFCs using pure Pt as an anode catalyst faces great problems when fuel gas contains CO, even at the level of a few parts per million (ppm). It has

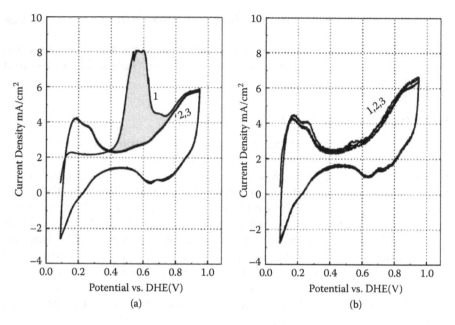

FIGURE 4.6
(a) CO stripping graph of Pt electrode, which was exposed to 100 sccm 50% CO_2/50% H_2 for 10 h before flushing with N_2 for 5 min. Scan rate = 10 mV/s; T_{cell} = 70°C; RH(A/C) = 100%/100%. (b) Same electrode exposed only to H_2. (From Gu, T. et al. 2005. *Appl. Catal. B: Environ.* 56: 43–9. With permission.)

been reported that 5 ppm CO could lead to a pronounced cell performance drop [44]. As mentioned above, the degradation of the cell performance is associated with preferential CO adsorption on Pt, thus blocking the active sites for HOR. Therefore, the degree of the performance drop of the cell will be highly correlated with the concentration of CO. The results from Oetjen et al. [45] clearly show this fact in Figure 4.7.

The steady-state polarization curves were obtained after the fuel cell was exposed to various concentrations of CO for a period of time. In all measurements, an open circuit voltage (OCV) of 1.05 V was observed in H_2/O_2 and $H_2(CO)/O_2$, indicating that the OCV is independent of the CO concentration in the hydrogen. This suggests that the open-circuit potential of the anode is predominantly determined by the $H_2/2H^+$ equilibrium. It was found that in the presence of 25 ppm CO, the polarization curve is linear. As expected, the slope of the curve is negative compared to that measured in pure H_2. This can be explained by a higher anode overpotential caused by CO poisoning. In Figure 4.8, two distinct slopes could be observed for CO concentrations greater than 100 ppm. The smaller slopes at higher current densities were explained by CO oxidation kinetics. At increasing current densities, the potential of the anode increased to values at which adsorbed CO could be oxidized to CO_2, thus leading to higher reaction rates for hydrogen adsorption and oxidation [28,45].

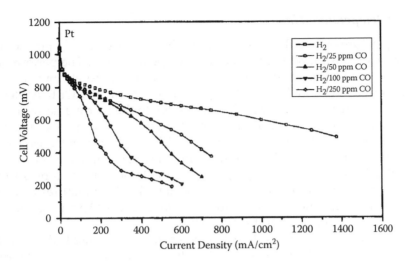

FIGURE 4.7
Steady-state polarization plots of a PEM single cell using H_2/CO (25 to 250 ppm); anode and cathode: 30 w/o Pt on Vulcan XC 72; $T_{cell} = 80°C$; $p(H_2) = 0.22$ MPa, $p(O_2) = 0.24$ MPa. (From Oetjen, H.F. et al. 1996. *J. Electrochem. Soc.* 143: 3838–42. With permission.)

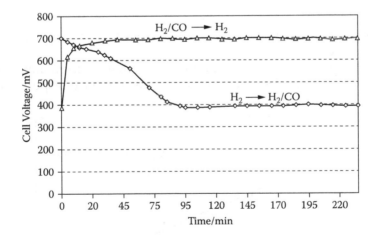

FIGURE 4.8
Cell voltage after changing the fuel gas at $j = 400$ mA/cm^{-2}; anode catalyst: $Pt_{0.5}Ru_{0.5}$; fuel gases: pure H_2 and $H_2/100$ ppm CO; T = 80°C. (From Divisek, J. et al. 1998. *Electrochim. Acta* 43: 3811–5. With permission.)

Fortunately, the severe impact of CO can be recovered by purging pure H_2 through the anode, as the consequence of the equilibrium between the coverage of CO on Pt surface and its concentrations in H_2. This recoverability and the strong dependence of cell performance on the CO concentration are shown in Figure 4.7 [15].

4.4.3.2 Temperature Effect

At temperatures lower than ≤80°C and in the presence of 250 ppm CO, cell performance was found to increase with temperature [15]. The performance increased with temperature for two reasons. One can be attributed to the kinetic activities of both HOR and oxygen reduction reaction (ORR) excited by the increased temperature; the other one is due to the thermal desorption of CO. With the increase of temperature, the standard free energy of CO adsorption on Pt was found to decrease linearly, which can be seen in Figure 4.9 [41].

In Figure 4.10, a plot is shown illustrating the equilibrium coverage of CO on Pt at CO concentrations ranging from 1 to 100 ppm with a H_2 pressure of 0.5 bar. It can be seen clearly that operation at higher temperatures increases the tolerance of the fuel cell anode to small amounts of CO by decreasing the coverage of CO on the catalyst surface [46].

But the problem is that it is almost impossible. It is nonsense to decrease the CO coverage through elevated operation temperature of a normal PEMFC, in which perfluorosulfonic acid (PFSA) membrane is most widely used as a proton conductor. PSFA membrane can conduct protons only with the participation of liquid water and its conductivity depends strongly on the relative humidity. At higher operating temperatures, membrane dehydration and the subsequent decrease in proton conductivity is a significant issue. For this reason, Inorganic compounds, such as SiO_2 [47] and TiO_2 [48], were incorporated into the hydrophilic domains of PFSA in order to improve their mechanical strength and thermal stability, in particular, water retention at elevated temperatures.

Another approach to achieving proton conductivity in membranes at high temperatures is to replace water with another proton transport-assisting

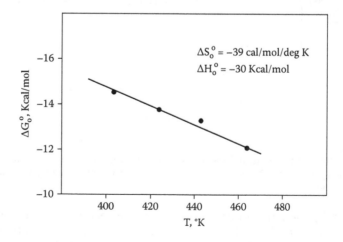

FIGURE 4.9
Standard free energy of CO adsorption versus temperature. (From Dhar, H. et al. 1987. *J. Electrochem. Soc.* 134: 3021–6. With permission.)

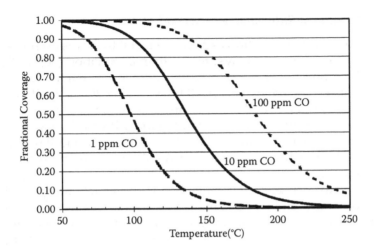

FIGURE 4.10
CO coverage on platinum as a function of temperature and CO concentration. The partial pressure of H_2 is 0.5 bar. (From Yang, C. et al. 2001. *J. Power Sources* 103: 1–9. With permission.)

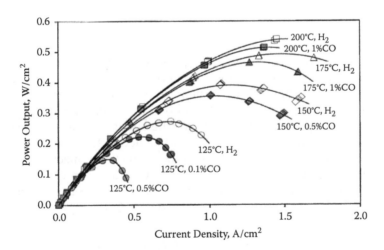

FIGURE 4.11
Power output of a PBI-based PEMFC with pure hydrogen and hydrogen containing CO. Temperature and CO contents are indicated in the figure. (From Li, Q.F. et al. 2003. *J. Electrochem. Soc.* 150: A1599–605. With permission.)

solvent that possesses a higher boiling point. Phosphoric acid-doped poly-benzimidazole (PBI) membrane is the most developed one and can operate at temperatures up to 200°C. As shown in Figure 4.11, the poisoning effect of CO in a concentration as high as 1%, can be sufficiently suppressed at elevated temperatures [49].

4.4.3.3 Addition of Ruthenium as a Second Metal

The addition of ruthenium (Ru) as a second metal to Pt was originated from the exploration of electrocatalyst for methanol oxidation [50]. Pt-Ru alloy catalysts have proved to be the most active ones at low temperatures (~80°C) in both the direct methanol fuel cells (DMFC) and H_2/O_2 fuel cells to oxidize adsorbed CO. Detailed studies on catalysts with defined Pt to Ru surface ratios have shown that $Pt_{0.5}Ru_{0.5}$ (atomic ratio 1:1) has the lowest overpotential in comparison to pure Pt [15]. The function and mechanism for the addition of Ru have been concluded by Ye [51]. In this work, we just have a brief review of it.

There are two debated viewpoints, one is the widely accepted "bifunctional" mechanism, and the other one is the direct mechanism enabled by the electronic effect. For the bifunctional mechanism, the Pt activates the fuel to dehydrogenation and the Ru sites provide the necessary oxygenated species, i.e., OH for CO elimination, which is covered on the Pt surface by oxidation. Platinum is, therefore, liberated after the oxidative removal and by of activity for HOR. A schematic representation of such a bifunctional catalyst is given in Figure 4.12

With the catalysis of Ru, the oxidation of the CO_{ad} occurs through the chemisorbed -OH at potential as low as 0.25 V, which is 0.2 V less than that required by Pt electrocatalysis [52].

However, the bifunctional mechanism neither takes into account a possible change in the CO binding energy on Pt by Ru, nor does it directly describe the effect of CO and OH competitive adsorption on the Ru. The direct mechanism proposes that ruthenium modifies the electronic properties of platinum, weakening the Pt-CO bond and thereby lowering the CO electro-oxidation potential. However, for the enhancement of CO oxidation through the addition of Ru, only 20% is attributed to the direct mechanism, while 80% of the enhancement is due to the bifunctional mechanism [53].

As the most famous commercial Pt-Ru/C catalyst provider, Johnson Matthey (London) has published data comparing the cell performance of Pt-Ru/C with Pt, in the presence of 10 to 100 ppm CO in H_2 [54].

$$Pt\text{-}CO_{ads} + M\text{-}OH = Pt + CO_2 + H^+ + M + e$$

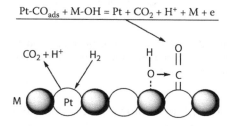

FIGURE 4.12
Bifunctional catalyst (schematic representation) is shown above. (From Ye, S. 2008. CO-tolerant catalyst. In *PEM fuel cell electrocatalysts and catalyst layers*, ed. Zhang J., 771–4. London: Springer-Verlag. With permission)

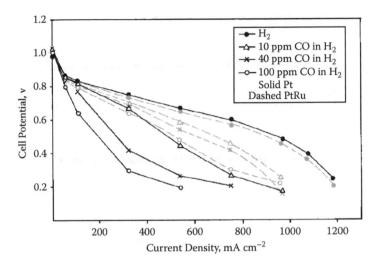

FIGURE 4.13
Progressive poisoning from 10 to 100 ppm CO on pure Pt and $Pt_{0.5}Ru_{0.5}$ alloy anodes. The anodes are prepared from 20 wt% Pt/C or 20 wt% Pt, 10 wt% Ru/C at a loading of 0.25 mg Pt/cm². The cathode uses 40 wt% Pt/Vulcan XC72R at a loading of 0.6 mg Pt/cm²; T = 80°C, 308/308 kPa, 1.3/2 stoichiometry with full internal membrane humidification. (From Ralph, T.R., Hogarth, M.P. 2002. *Platinum Metals Rev.* 46: 117–35. With permission.)

In Figure 4.13, significantly increased CO tolerance is shown by the $Pt_{0.5}Ru_{0.5}$ alloy anodes, especially when the concentration of CO is as low as 10 ppm.

4.4.3.4 Impact of Anode CO on Cathode

A surprising phenomenon was found by Qi et al. [55,56]: that CO could poison not only the anode but also the cathode, and, sometimes, the cathode potential drop is even larger than that of the anode as a result of CO crossover from anode to cathode. However, they did not test the cathode directly introducing CO to the cathode. This is controversial when we consider the mitigation method, "air/O_2 bleed" (see section 4.7). Based on the same mechanism, CO crossed over from anode can be easily oxidized to CO_2 in the presence of large amounts of O_2 and the cathode catalyst Pt. In this scenario no poisoning effect will be found in the cathode. To verify this prediction, we conducted single cell tests using a fuel cell that has 50 cm² active area, first with H_2 and clean air, and then with air containing 79 to 1500 ppm CO (see Figure 4.14).

It can be seen in Figure 4.14 that the cathode has shown good tolerance to CO that was directly introduced from the air stream, with concentrations even up to 1500 ppm, a value much higher than a "crossover" process can reach.

In literature, the other effects involving CO on PEMFC anode have also been investigated extensively, including the effects of H_2 dilution by N_2 [57],

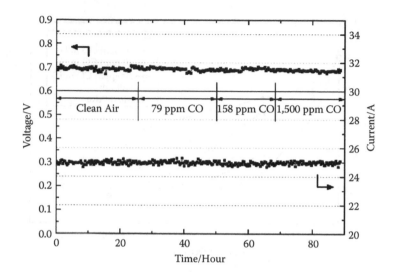

FIGURE 4.14
Cathode durability to CO with different concentrations in air stream. Catalyst loading: pure Pt, 0.3 mg Pt/cm² at each electrode; Nafion® 212; $T_{cell} = 70 \pm 5°C$; $P_{air} = P_{H2} = 1.0 \pm 0.1$Bar; $RH_{H2} = 95 \pm 5\%$, dry air.

the anode flow rate and cathode O_2 pressure [58], anode catalyst loading [59], binary and ternary catalysts (see section 4.7), etc.

4.4.4 Impact of CO_2

While it has been known for many years that CO poisons Pt-based electrocatalysts, it has only recently been discovered that CO_2 also acts as a poisoning species in the PEMFC. To certain degree, CO_2 itself is recounted inert on platinum. Therefore, deviations of cell voltage from Nernst behavior is expected with CO_2 as the fuel diluter [60]. For dilution effect, the impact of CO_2 is limited. We have tested the polarization curves for a PEMFC single cell exposed to H_2 containing various levels of CO_2 (see Figure 4.15).

Figure 4.15 shows the polarization curves for a cell exposed to CO_2 in the concentrations from 0.2 to 20%. No significant impact was observed. Compared with the results in Figure 4.13, even up to 20% CO_2 in the air stream (air as oxidant) causes less polarization loss than 10 ppm CO in the hydrogen feed (O_2 as oxidant). Similar results can be found in Wilson et al. [27]. However, at CO_2 concentrations in excess of 25%, the voltage losses can no longer be accounted for only through the dilution effect [28]. Either by RWGS or electroreduction, up to 200 ppm CO will be produced through CO_2 at such high CO_2 concentrations [30]. Unfortunately no detailed deduction or calculation has been made in this literature. A kinetic model assuming the formation of CO on Pt through RWGS reaction and its experimental validation through CO-stripping experiments have been published by Gu et al.

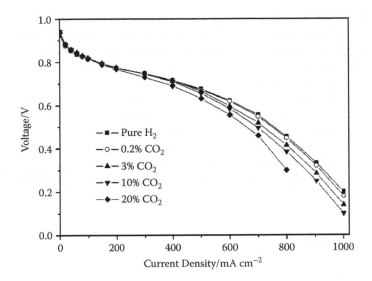

FIGURE 4.15
Polarization curves of a 50 cm² PEM fuel cell fueled with H_2 containing CO_2. Pt loading 0.3 mg Pt/cm² at each side; Nafion® 212; $T_{cell} = 70 \pm 5°C$; $P_{air} = P_{H2} = 1.0 \pm 0.1Bar$; $RH_{H2} = 95 \pm 5\%$, dry air.

[43]. The calculation result shows that even at a low CO_2 level of 1%, enough CO could be produced through RWGS reaction to poison more than 50% of total sites on platinum electrode at 101 kPa. It means 1% CO_2 in H_2 may cause pronounced anodic overpotential. Additionally, through the model calculation of Yan et al. [61], 24.69% cell performance decrease was observed with the poisoning effect of only 10% CO_2 in H_2. The controversy and disagreement between experimental data and model can be accounted for through the understanding of the fact that the CO_2 reduction process is several orders of magnitude slower than that for CO adsorption [43].

An interesting phenomenon has been found by Bruijn et al. [16] through cyclic voltammetry (CV) stripping of Pt/C and Pt-Ru/C catalysts. After exposure to CO_2 the CO-tolerant alloy catalyst, Pt-Ru/C, can suppress the RWGS reaction as well. The greatly improved CO_2 tolerance obtained with only modest CO-tolerant catalyst is also reported by Bellows et al. [30]. This conclusion shed light on the strategic advantage of PEMFCs, as which can tolerate significant levels of CO_2, over alkaline fuel cells (AFC).

4.4.5 Combined Impacts of CO and CO_2

As described above, the impacts of CO and CO_2 have been well studied in literature, but separately in general. Little investigation is carried out on the combined impacts of CO and CO_2. Yan et al. [61] investigated the unsteady behaviors of hydrogen concentration, CO concentration, Pt surface coverage, and current density distributions after anode catalyst when poisoned by CO and CO_2. The modeling results show that at low CO concentration (10 ppm), the H_2 coverage and cell performance decrease drastically when the rate constant of RWGS is not zero. Nevertheless, for high CO concentration conditions, the CO_2 poisoning effect has little effect on H_2 coverage and current density. This is due to the fact that the platinum catalyst layer surface would be occupied by CO for high CO concentration condition, and not enough Pt-H left for CO_2 to conduct RWGS reaction. According to this calculation, in practice, CO_2 poisoning should always be taken into consideration as low concentration of CO in H_2 product is often the case.

Due to the poisoning effect of CO_2 and the alterable fact, complete removal of CO from hydrogen product does not guarantee PEMFC performance. The tolerable concentration for both CO or/and CO_2 need to be determined.

4.5 Impact of H_2S

It is well known that, as a hydrogen impurity, H_2S has a severely detrimental impact on PEMFCs. The impact of H_2S is a result of the blockage of the Pt active sites by various adsorbed sulfur species. The adsorption and transformation of H_2S on Pt have been studied in both the aqueous [62,65] and gaseous phase [66].

4.5.1 Adsorption Behavior of H_2S

In aqueous solutions containing H_2S, strong H_2S adsorption on Pt occurs according to the following mechanism proposed by Mathieu et al. [65] and Ramasubramanian [67]:

$$Pt + H_2S \rightarrow Pt\text{-}H_2S \tag{4.21}$$

$$Pt + HS\text{-} \rightarrow Pt\text{-}HS\text{-} \tag{4.22}$$

$$Pt + HS\text{-} + H+ \rightarrow Pt\text{-}H_2S \tag{4.23}$$

$$Pt + H+ \rightarrow Pt+\text{-} H \tag{4.24}$$

$$Pt + H_2S \rightarrow Pt\text{-}H + Pt\text{-}SH \tag{4.25}$$

$$Pt\text{-}SH_n \rightarrow Pt\text{-}S + nH^+ + ne \tag{4.26}$$

where Pt^+ represents an equivalent positive charge on the Pt surface, and equation (4.26) involves the dissociation of H_2S, which produces a sulfur block on the Pt surface. According to the experimental results of Yu et al. [68], the value of the dissociation potential is a function of temperature, and it is about 0.4V at 90°C, 0.5V at 60°C, and about 0.6V at 30°C. The adsorbed H_2S and SH are highly unstable on Pt, while the adsorbed S and H are the most stable intermediates on Pt [69]. The formation of Pt-S on the catalyst surface makes it impossible for the fuel cell to recover.

4.5.2 Impact of H_2S

In most of the literature, relatively high H_2S concentrations were used to study the impact of this impurity. For example, the concentrations in the level of 1 to 2 ppm were used in [60,70,71], even higher concentrations (5 ppm to 50 ppm) were used in [68, 72,73]. In all of these studies, severe and irreversible impacts from H_2S were reported. Studies with H_2S concentrations less than 1 ppm were only reported in Besancon et al. [8], but no detailed data were provided. We believe that this low level of H_2S is of greater importance for practical cases where the purity of H_2 product is controlled. Therefore, to get a reasonable H_2 specification, we investigated the impacts of H_2S with concentrations ranging from 0.25 to 5 ppm in the hydrogen stream (see Figure 4.16) [74].

As shown in Figure 4.16, H_2S with concentration as low as 0.25 ppm still affects the cell performance significantly. This is very different from what was reported in Besancon et al. [8], in which no significant voltage loss was observed after 100 hours of operation at the same H_2S concentration. In addition, it can be seen clearly from Figure 4.16 that the voltage loss is

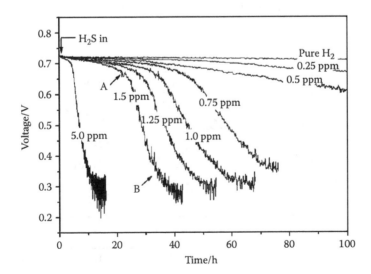

FIGURE 4.16
Durability to different concentrations of H_2S in the anode of a single cell. MEA: 50 cm²; Pt loading: 0.3 mg/cm² on each side; H_2/air utilization ratio: 70/30%; P = 1.0 bar; T cell = 70°C; current density: 500 mA/cm². (From Shen, M. et al. 2008. *J. Xi'an Jiaotong University* 42: 1054–8. With permission.)

highly correlated with the concentration of H_2S, and at higher concentrations (≥ 0.75 ppm) the voltage dropping rate is evidently slowing down with time. It seems that a saturated absorption has been reached, and the dependence of the saturated adsorption upon temperature has been established by Mohtadi et al. [72].

Using high concentrations, Shi et al. [68] investigated the effect of H_2S at different operating conditions. It was found that the higher the H_2S concentration and current density, the higher the poisoning effect, but this decreased with increasing temperatures. On the other hand, no relationship between anode humidification and the poisoning rate was observed.

4.5.3 Cyclic Voltammetry Study

Cyclic potential sweeping was frequently employed in literature to explore the adsorbed species of H_2S, as well as to recover the cell performance, because purging the anode with pure H_2 made almost no sense [8,71,73]. Shown in Figure 4.17 is a typical result obtained from a CV measurement after the poisoning effect of H_2S.

Two distinct oxidation peaks at potentials over 0.8 V can be found in Figure 4.17, especially in the first scan. It was believed that during the potential swoop, the adsorbed sulfur (through equation (4.23)) is oxidized to SO_3 and SO_4^{2-}, according to the following reactions [62].

$$Pt\text{-}S + 3H_2O \rightarrow SO_3 + 6H^+ + 6e^- + Pt \qquad (4.27)$$

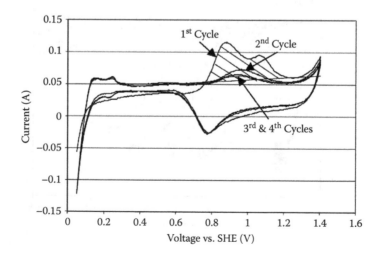

FIGURE 4.17

Evaluation of the charge of sulfur species oxidation during a CV at 70°C. The CV was measured while flowing N_2 on the anode (working electrode) and neat H_2 was flowing on the cathode; scanning rate: 5 mV/s. (From Mohtadi, R. 2005. *Appl. Catal. B: Environ.* 56: 37–42. With permission.)

$$Pt\text{-}S + 4H_2O \rightarrow SO_4^{2-} + 8H^+ + 6e^- + Pt \qquad (4.28)$$

It was pointed out by Mohtadi et al. [72] that the two oxidation peaks shifted negatively with the increasing temperature adopted during the adsorption of H_2S. It indicates that sulfur adsorbs more strongly at lower cell operating temperatures.

4.5.4 Open Circuit Voltage Operation

Not like in the case of CV, open circuit voltage (OCV) would not have been expected to have a strong electrochemical effect on the anode because anode polarization drops. However, Imamura et al. [71] reported an interesting result that OCV is of beneficial effect on H_2S tolerance (see Figure 4.18).

It can be seen in Figure 4.18 that after being poisoned by 2.0 ppm H_2S, the cell voltage suffered a great loss, even after being purged with pure H_2 for 1 h. However, after holding at OCV for another hour, while supplying the anode with pure hydrogen, the cell voltage greatly recovered to about the same extent as before the supply of H_2S.

Similar phenomena and reasonable explanations about "OCV relief" have been given by Urdampilleta et al. [70]. The authors stated that the cathode was poisoned by H_2S crossed over from the anode. Therefore, OCV relief effect could be explained in terms of a reactivation of the cathode by partial oxidative desorption of sulfur at high potentials.

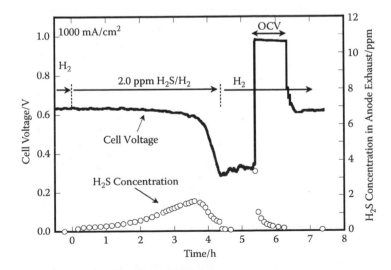

FIGURE 4.18
Effect of hydrogen sulfide on fuel cell performance and the results of anode exhaust gas analysis. (From Imamura, D., Hashimasa, Y. 2007. *ECS Transactions* 11: 853–62. With permission.)

4.6 Impact of NH₃

The presence of traces of NH_3 in hydrogen has deleterious effects on PEMFC performance. Different from CO and H_2S, as a simple molecular either, NH_3 has complex poisoning mechanism, which is not completely understood as of yet.

4.6.1 Deleterious Effects of NH₃

Although few publications have reported the deleterious effects of NH_3 on anode, an expanded range of NH_3 concentrations has been employed. Halseid et al. [24] and Besancon et al. [8] tested the effect of 1 ppm NH_3/H_2 for 7 d and 115 h, respectively. However, results reported by the two groups are significantly different. While Halseid et al. observed a significant contamination effect on PEMFC performance, results from Besancon et al. showed no noticeable effect on cell performance. For higher concentrations, consistent results were given [23]. Typical polarization plots under the influence of 30 and 130 ppm NH_3/H_2 are shown in Figure 4.19.

The extent of performance deterioration and recovery depends on NH_3 impurity level and the time of exposure to NH_3. With short times of exposure (1 to 3 h) the original performance can be fully recovered. Longer times of exposure (e.g., 15 h) decrease the cell performance to impractical levels, and the cell does not recover even after several days of operation with pure H_2 [23].

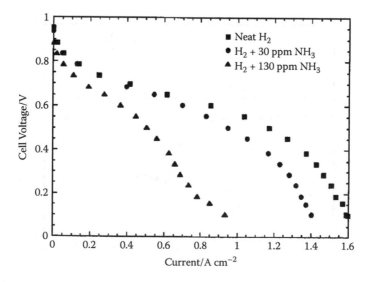

FIGURE 4.19
Polarization curves showing the effect of two levels of NH_3 impurity on H_2-air fuel cell performance at 80°C. (From Uribe, F.A. et al. 2002. *J. Electrochem. Soc.* 149: A293–6. With permission.)

4.6.2 Mechanism for the Impact of NH_3

Borup et al. [60] ascribed the harmful effect of NH_3 to the conductivity loss, which was due to the reaction between NH_3 and H^+ to form ammonium (NH_4^+), and consequently, lowering the protonic conductivity of both the membrane and catalyst layer (CL). However, only part of the performance loss can be attributed to this process because the conductivity of the membrane and CL is affected after longer times of exposure, whereas cell performance starts dropping immediately after NH_3 is brought into the cell, as shown in all the literature [8,23,24].

To understand the adsorption of NH_3, cyclic voltammetry measurements were done. However the voltammogram after NH_3 exposure was the same as that without any NH_3 exposure [23,75]. The absence of oxidation peak during a CV sweep implies that almost no catalyst poisoning occurred. Therefore, there must be other reasons besides that of conductivity decrease and electrocatalyst poisoning.

Halseid et al. [24] have tested the ohmic resistance of the cell while introducing 1 and 10 ppm NH_3 into the anode stream. It was found that the increase in cell resistance after the cell was exposed to NH_3 was about 20% in most cases, and made just 5% (1 ppm) and 15% (10 ppm) contribution to the performance losses. "Ion exchange" may provide a reasonable explanation for the quick and severe impact of NH_3, i.e., NH_3 would react with protons in the membrane, thus staying in the membrane and causing the decrease in the protonic conductivity of the membrane. In addition, the water content in the membrane phase decreases linearly with increasing NH_4^+ fraction [22]. For PFSA membrane, dehydration will cause catastrophic consequences to the fuel cell performance.

4.7 Impact of Organic Compounds

Even though not key impurities, organic compounds exist in H_2 in various forms. Smaller molecular hydrocarbons, such as methane (CH_4), ethane (C_2H_6), ethylene (C_2H_4), etc., can always be found in H_2. Also, oxygen-containing organic compounds, such as HCHO, HCOOH, etc., may be found according to the fuel-processing pathway adopted. With the intention to reconstruct the hydrogen quality standard, Japanese Automotive Research Institute (JARI) has investigated the impact of various hydrocarbons and oxygen-containing compounds on the performance of PEMFC [76]. It was recommended that the allowable concentration limits for CH_4, C_2H_6, and C_2H_4 are all 5%. Obviously, after purification, the concentrations of the hydrocarbons will be far lower than 5%. Therefore, their influences can be ignored. However, the limits for oxygen-containing compounds are recommended to be as low as the ppm level. That means oxygen-containing compounds are a poison to PEMFCs.

The results reported by Narusawa et al. [77] and Dorn et al [78] have supported the fruit of JARI; even so, the adverse effects of HCHO and HCOOH are only 0.1 and 0.004 times that of CO, respectively [77].

4.8 Mitigation Methods for Anode Contamination

Anode catalyst poisoning, as well as cathode catalyst poisoning, is the main electrocatalytic problem for the efficiency and long-term stability of traditional low temperature PEMFCs. CO and H_2S can be strongly adsorbed onto the surface of Pt, thus blocking part of Pt active sites. The conductivities of the membrane, CL, and gas diffusion layer (GDL) can be reduced through the invasion of NH_4^+ converted formed from NH_3, and other foreign cations, such as Fe^{n+}, Cu^{n+}, Si^{n+}, from fuel cell components. To protect the anode from contamination or recover the anode after contamination, great efforts have been made and some useful schemes are presented below.

4.8.1 Impurity-Tolerant Catalyst

The development of Pt-based binary or ternary metallic electrocatalysts and Pt-free electrocatalysts is an intuitional idea to mitigate the problem caused by the strong adsorption of impurities on Pt. Therefore, numerous efforts have been made in this area. However, almost all of these works exclusively focused on CO.

It is well established that binary catalysts, with Pt as one of the central components, can exhibit a substantial resistance to CO. It has been found that the use of a second element, such as Ru [79,80], Fe [81], Ni [82], Mo [83], etc., in the form of alloy or co-deposit yields significant improvement in the CO-tolerance relative to pure Pt [84]. The superior CO tolerance is mainly attributed to the so-called "bifunctional" mechanism, i.e., the lower CO oxidation potential due to the addition of a second element. Among these various Pt-based binary catalysts, the most commonly used and the most promising catalyst is Pt-Ru/C, with the best atomic ratio being 1:1. Nevertheless, Pt-Ru/C does not provide tolerance against H_2S poisoning of the anode [85].

Another approach for the preparation of CO-tolerant catalyst is to combine a highly active element for HOR with a second metal to produce the surface (ideally), which does not adsorb CO under the fuel cell operating conditions. Along this line, PdAu-black alloys have already been proposed. Indeed, much lower CO adsorption energies on different poly- and single-crystalline PdAu surfaces were found in ultrahigh vacuum (UHV) studies compared with pure Pd or pure Pt [86]. In the oxidation measurement of Schmidt [87], which was based on an earlier study by Fishman [88], the Pd-Au/C catalyst showed superior HOR activity at fuel cell relevant potential (0.50 to 0.1 V) in the presence of 250 and 1000 ppm CO, compared with the Pt-Ru/C catalyst.

Attempts have been made to improve the CO tolerance through a ternary catalyst, typically based on a Pt-Ru/C alloy. The addition of W [89], WO_x [90], Sn [89], or Pd [91] to Pt-Ru provides improved CO tolerance.

Removing CO by virtue of a catalyst outside the CL [92] or fuel cell itself, for example, adopting a CO oxidation reactor, to remove CO is practical and may bring good result [93,94]. He et al. [95] studied the possibility of integrating a CO removal reactor into a fuel cell stack, for which six typical catalysts were screened under fuel cell operation conditions. Nearly 100% CO conversion was achieved with an $Ir/CoO_x-Al_2O_3$/carbon catalyst with good selective oxidation ($O_2/CO = 1.5$) at 76°C and 100% relative humidity.

4.8.2 Elevated Temperature Operation

Increasing the temperature (>80°C) is one of the main strategies for improving the contamination tolerance of PEMFCs, which is also one of the driving forces for the large efforts in developing suitable, more temperature resistant membranes with sufficient proton conductivity [49,96]. Removal of catalyst poisoning by thermal desorption plays an important role in the mitigation of CO contamination as well as other weakly adsorbed poisons. On the other hand, at temperatures below 80°C, the CO poisoning can be mitigated by negatively shifted oxidation potential with increased temperature, rather than by lower steady-state CO_{ad} coverage due to the thermal desorption [97].

4.8.3 Chemical Oxidant Bleed

Introducing a certain amount air or O_2 into the anode to chemically react with the preadsorbed poisons is called *air or O_2 bleed operation* [98]. This air or O_2 bleed method is electrochemically promoted and is catalyzed by Pt [26]. The performance of the fuel cell can be completely restored by the bleed of O_2, normally in the level of 2 to 5% [26,99] in H_2. Although this scheme is reported useless for the removal of H_2S [70], Shen et al. [74] purged the anode, which has been poisoned by 10 ppm H_2S for 4 h, for 30 min with clean air, and ~95% performance recovery was obtained.

However, one important side effect must be taken into account during the air/O_2 bleed operation, which is the formation of H_2O_2. This is the well-known intermediate of cathode oxygen reduction. Degradation of proton exchange membrane fuel cells due to CO and CO_2 poisoning on reaction (ORR), may have devastating consequences for the long-term stability of the membrane [100–103].

Regarding the safety problem of mixing H_2 with O_2 in the case of air/O_2 bleed, "H_2O_2 bleed," which involves adding H_2O_2 into the humidification water, is of great advantages. H_2O_2 is relatively safe since the explosive vapor mixtures of H_2O_2 and H_2 are formed only when the H_2O_2 concentration in the vapor phase exceeds 26% [104]. Reported by Schmidt et al. [105] and Divisek et al. [15], the performance of a Pt catalyst anode was completely recovered by the addition of 5% H_2O_2 in the humidification water at 80°C. The significant improvement was reported to be associated with the heterogeneous decomposition of H_2O_2 on Pt or Pt-alloy surface.

4.8.4 Electrochemical Removal

This scheme is originated from the CV measurement, during which adsorbed impurities, such as CO and H_2S, on the electrocatalyst can be stripped. Derived methods include cycling the current and raising the anode overpotential [106], starvation of fuel and electrical shorting of the cell [107], etc. They can all be used to enhance the electrochemical oxidation of CO [99].

Operating the fuel cell at OCV to recover the catalyst from adsorbed H_2S, as tested in Urdampilleta et al. [70], is another embodiment of the electrochemical scheme.

Limitation for this scheme is obvious. Equipment for electrochemical oxidation of the impurities may not be easily achieved onsite, although it can be simply realized on a test station. In addition, dangers may occur to either fuel cell stack or operating person if electrical shorting is adopted.

4.9 Summary

As an energy carrier, hydrogen is exclusively used as the fuel for PEMFCs, the power train of the energy-efficient, environmentally friendly, and the most promising vehicles, FCVs. Reforming of hydrocarbons is the most economic and dominant pathway for H_2 production. However the existence of trace amounts of impurities in hydrogen products is inevitable. Some impurities, such as CO, CO_2, H_2S, and NH_3, are detrimental to the PEMFC anode, which normally operates at a low temperature and uses Pt as the catalyst.

The severe effect of CO, even at the level of few ppm, occurs by blocking the active sites of Pt for H_2 oxidation. The poisoning effect of CO_2 can be explained by its reduction to CO. Comparing with CO_x, the adsorption of H_2S on Pt is even stronger, which results in the permanent poisoning of the anode. NH_3 has relatively slight impact on the anode because no poisoning adsorption of NH_3 onto the Pt has been found. Although the answer to the instant effect of NH_3 on PEMFC is still open, the formation of NH_4^+ can be the reason. With the stabilization of NH_4^+ in the PFSA membrane, the protonic conductivity as well as the water content in the membrane phase decrease. Compared with these key impurities, hydrocarbons and oxygen-containing compounds in H_2 stream have only a slight influence on the anode.

Several mitigation methods for anode contamination are briefly introduced, mainly centered on CO mitigation. The addition of Ru to Pt is the widely adopted scheme to generate excellent CO tolerance. Elevating the operation temperature (>80°C) may improve the tolerance to impurities that poison the catalyst through adsorption, but the adoption of a novel proton exchange membrane is necessary. Chemical oxidants, like H_2O_2 or just O_2, can be bled into the anode stream to react with CO and excellent results can be expected. Other than that, electrochemical methods, such as current or potential pulse, cyclic voltammetry, and OCV operation, have also been explored for the removal of CO and H_2S.

Acknowledgments

The authors gratefully acknowledge the support from the Ministry of Science and Technology of China (2007DFC61690) and the Henkel Professorship.

References

1. Chen, X., Shi, Z., Glass, N., et al. 2007. A review of PEM hydrogen fuel cell contamination: impacts, mechanisms, and mitigation. *J. Power Sources* 165: 739–56.
2. Li, H., Song, C., Zhang, J., et al. 2008. Catalyst contamination in PEM fuel cells. In *PEM fuel cell electrocatalysts and catalyst layers*, ed. Zhang J., 331–54. London: Springer-Verlag.
3. Lanz, A., Heffel, J., Messer, C. 2001. *Hydrogen fuel cell engines and related technologies*. Palm Desert, CA: College of the Desert.
4. Palo, D.R., Dagle, R.A., Holladay, J.D. 2007. Methanol steam reforming for hydrogen production. *Chem. Rev.* 107: 3992–4021.
5. Yang, Y., Ma, J., X., Wu, F. 2006. Production of hydrogen by steam reforming of ethanol over a Ni/ZnO catalyst. *Int. J. Hydrogen Energy* 31: 877–82.
6. Takeishi, K., Suzuki, H. 2004. Steam reforming of dimethyl ether. *Appl. Catal.B: General.* 260: 111–7.
7. Holladay, J.D., Hu, J., King, D.L., et al. 2009. An overview of hydrogen production technologies. *Catal. Today* 139: 244–260.
8. Besancon, B.M., Hasanov, V., Imbault-Lastapis, R., et al. 2009. Hydrogen quality from decarbonized fossil fuels to fuel cells. *Int. J. Hydrogen Energy* 34: 2350–60.
9. Cormos, C.C., Starr, F., Tzimas, E., et al. 2008. Innovative concepts for hydrogen production processes based on coal gasification with CO_2 capture. *Int. J. Hydrogen Energy* 33: 1286–94.

10. Uhde. 2009. Chlor-alkali electrolysis plants, superior membrane process. http://www.uhde.eu/cgi-bin/byteserver.pl/archive/upload/uhde_brochures_pdf_en_10.00.pdf (accessed ???)
11. Düttel, A. 2007. Introduction. In *Hydrogen as a Future Energy Carrier*, ed. Düttel, A., Borgschulte, A., Schlapbach, L. 1–6. Weinheim, Germany: WILEY-VCH Verlag.
12. Ribeiro, A.M., Grande, C.A., Lopes, F.V.S., et al. 2008. A parametric study of layered bed PSA for hydrogen purification. *Chem. Eng. Sci.* 63: 5258–73.
13. Tagliabue, M., Delnero, G. 2008. Optimization of a hydrogen purification system. *Int. J. Hydrogen Energy* 33: 3496–98.
14. Irving, P.M., Pickles, J.S. 2007. Operational requirements for a multi-fuel processor that generates hydrogen from bio- and petroleum-based fuels for both SOFC and PEM fuel cells. *ECS Transactions*, 5: 665–71.
15. Divisek, J., Oetjen, H.F., Peinecke, V., et al. 1998. Components for PEM fuel cell systems using hydrogen and CO containing fuels. *Electrochim. Acta* 43: 3811–5.
16. Bruijn, F.A. de, Papageorgopoulos, D.C., Sitters, E.F. 2002. The influence of carbon dioxide on PEM fuel cell anodes. *J. Power Sources* 110: 117–24.
17. Janssen, G.J.M. 2004. Modelling study of CO_2 poisoning on PEMFC anodes. *J. Power Sources* 136: 45–54.
18. Babich, I.V., Moulijn, J.A. 2003. Science and technology of novel processes for deep desulfurization of oil refinery streams a review. *Fuel* 82: 607–31.
19. Song, C., Ma, X. 2003. New design approaches to ultra-clean diesel fuels by deep desulfurization and deep dearomatization. *Appl. Catal. B: Environmental* 41: 207–38.
20. Song, C. 2003. An overview of new approaches to deep desulfurization for ultra-clean gasoline, diesel fuel and jet fuel. *Catal. Today* 86: 211–63.
21. Breysse, M., Djega-Mariadassou, G., Pessayre, S., et al. 2003. Deep desulfurization: Reactions, catalysts and technological challenges. *Catal. Today* 84: 129–38.
22. Halseid, R., Vie, P.J.S., Tunold, R. 2004. Influence of ammonium on conductivity and water content of Nafion 117 membranes. *J. Electrochem. Soc.* 151: A381–8.
23. Uribe, F.A., Gottesfeld, S., Zawodzinski, T. Jr. 2002. Effect of ammonia as potential fuel impurity on proton exchange membrane fuel cell performance. *J. Electrochem. Soc.* 149: A293–6.
24. Halseid, R., Vie, P.J.S., Tunold, R. 2006. Effect of ammonia on the performance of polymer electrolyte membrane fuel cells. *J. Power Sources* 154: 343–50.
25. Chellappa, A.S., Fischer, C.M., Thomson, W.J. 2002. Ammonia decomposition kinetics over Ni-Pt/Al_2O_3 for PEM fuel cell applications *Appl. Catal. A: General* 227: 231–40.
26. Gottesfeld, S., Pafford, J. 1988. A new approach to the problem of carbon monoxide poisoning in fuel cells operating at low temperatures. *J. Electrochem. Soc.* 135: 2651–2.
27. Wilson, M.S., Derouin, C., Valerio, J., et al. 1993. Electrocatalysis issues in polymer electrolyte fuel cells. *Proc. 28th IECEC*, 1.1203–8.
28. Baschuk, J.J., Li, X. 2001. Carbon monoxide poisoning of proton exchange membrane fuel cells. *Int. J. Energy Res.* 25: 695–713.
29. Iwasita, T. 2003. Methanol and CO electrooxidation. In *Handbook of Fuel Cells-Fundamentals, Technology and Applications*, Vol. 2: *Electrocatalysis*, Ed. 3. Vielstich, W. V., Lamm, A., and Gasteiger, H.A., 603–10, Chichester, U.K.: John Wiley & Sons.

30. Bellows, R.J., Marucchi-Soos, E.P., Buckley, D.T. 1996. Analysis of reaction kinetics for carbon monoxide and carbon dioxide on polycrystalline platinum relative to fuel cell operation. *Ind. Eng. Chem. Res.* 35: 1235–42.
31. Hirschenhofer, J.H., Stauffer, D.B., Engleman, R.R., et al. 1998. *Fuel Cells: A Handbook*, Ed. 4. 6.1–6.12, Philadelphia: Parsons Corporation Reading.
32. Blyholder, G. 1964. Molecular orbital view of chemisorbed carbon monoxide. *J. Phys. Chem.* 68: 2772–7.
33. Grgur, B.N., Marković, N.M., Lucas, C.A., et al. 2001. Electrochemical oxidation of carbon monoxide: From platinum single crystals to low temperature fuel cells catalysis. Part I: carbon monoxide oxidation onto low index platinum single crystals. *J. Serb. Chem. Soc.* 66: 785–97.
34. Camara, G.A., Ticianelli, E.A., Mukerjee, S. 2002. The CO poisoning mechanism of the hydrogen oxidation reaction in proton exchange membrane fuel cells. *J. Electrochem. Soc.* 149: A748–53.
35. Wang, K., Gasteiger, H.A., Markovic, N.M., et al. 1996. On the reaction pathway for methanol and carbon monoxide electrooxidation on Pt-Sn alloy versus Pt-Ru alloy surfaces. *Electrochim. Acta* 41: 2587–93.
36. Stonehart, P., Ross, P.N. 1975. The commonality of surface processes in electrocatalysis and gas phase heterogeneous catalysis. *Catal. Rev.-Sci. Eng.* 12: 1–35.
37. Gustavo, L., Pereira, S., Santos, F.R. 2006. CO tolerance effects of tungsten-based PEMFC anodes. *Electrochim. Acta.* 51: 4061–6.
38. Garcia, G., Silva-Chong, J.A., Guillen-Villafuerte, O. 2006. CO tolerant catalysts for PEM fuel cells spectroelectrochemical studies. *Catal. Today* 116: 415–21.
39. Wee, J.H., Lee, K.Y. 2006. Overview of the development of CO-tolerant anode electro- catalysts for proton exchange membrane fuel cells. *J. Power Sources* 157: 128–35.
40. Marques, P., Ribeiro, N.F.P., Schmal, M. 2006. Selective CO oxidation in the presence of H$_2$ over Pt and Pt-Sn catalysts supported on niobia. *J. Power Sources* 158: 504–8.
41. Dhar, H., Christner, L., Kush, A. 1987. Nature of CO adsorption during H$_2$ oxidation in relation to modeling of CO poisoning of fuel cell anode. *J. Electrochem. Soc.* 134: 3021–6.
42. Giner, J. 1963. Electrochemical reduction of CO$_2$ on platinum electrodes in acid solution. *Electrochim. Acta* 8: 857–65.
43. Gu, T., Lee, W.K., Van Zee, J.W. 2005. Quantifying the "reverse water gas shift" reaction inside a PEM fuel cell. *Appl. Catal. B: Environmental* 56: 43–9.
44. Wilson, M.S., Derouin, C., Valerio, J., et al. 1993. Electrocatalysis issues in polymer electrolyte fuel cells. *Proc. 28th IECEC*, 1: 1203–8.
45. Oetjen, H.F., Schmidt, V.M., Stimming, U., et al. 1996. Performance data of a proton exchange membrane fuel cell using H$_2$/CO as fuel gas. *J. Electrochem. Soc.* 143: 3838–42.
46. Yang, C., Costamagna, P., Srinivasan, S., et al. 2001. Approaches and technical challenges to high temperature operation of proton exchange membrane fuel cells. *J. Power Sources* 103: 1–9.
47. Wang, H.T., Holmberg, B.A., Huang, L., et al. 2002. Nafion-bifunctional silica composite proton conductive membranes. *J. Mater. Chem.* 12: 834–7.
48. Shao, Z.G., Xu, H., Li, M., et al. 2006. Hybrid Nafion–inorganic oxides membrane doped with heteropolyacids for high temperature operation of proton exchange membrane fuel cell. *Solid State Ionics* 177: 779–85.

49. Li, Q.F., He, R.H., Gao, J.A., et al. 2003. The CO poisoning effect in PEMFCs operational at temperatures up to 200°C. *J. Electrochem. Soc.* 150: A1599–605.
50. Wasmus, S., Vielstich, W.J. 1993. Methanol oxidation at carbon supported Pt and Pt-Ru electrodes: An online MS study using technical electrodes. *Appl. Electrochem.* 23: 120–4.
51. Ye, S. 2008. CO-tolerant catalyst. In *PEM fuel cell electrocatalysts and catalyst layers*, ed. Zhang J., 771–4. London: Springer-Verlag.
52. Venkataraman, R., Kunz, H.R., Fenton, J.M. 2003. Development of new CO tolerant ternary anode catalysts for proton exchange membrane fuel cells. *J. Electrochem. Soc.* A278–84.
53. Lu, C., Masel, R.I. 2001. The effect of ruthenium on the binding of CO, H_2, and H_2O on Pt(110). *J. Phys. Chem. B* 105: 9793–7.
54. Ralph, T.R., Hogarth, M.P. 2002. Catalysis for low temperature fuel cells. *Platinum Metals Rev.* 46: 117–35.
55. Qi, Z.G., He, C.Z., Kaufman, A. 2001. Poisoning of proton exchange membrane fuel cell cathode by CO in the anode fuel. *Electrochem. Solid-State Lett.* 4: A204–5.
56. Qi, Z., He, C., Kaufman, A. 2002. Effect of CO in the anode fuel on the performance of PEM fuel cell cathode. *J. Power Sources* 111: 239–47.
57. Gu, T., Lee, W.K., Xan-Zee, J.W., et al. 2004. Effect of reformate components on PEMFC performance. *J. Electrochem. Soc.* 151: A2100–5.
58. Zhang, J.X., Thampan, T., Datta, R. 2002. Influence of anode flow rate and cathode oxygen pressure on CO poisoning of proton exchange membrane fuel cells. *J. Electrochem. Soc.* 149: A765–72.
59. Kim, J.D., Park, Y.I., Kobayashi, K., et al. 2001. Effect of CO gas and anode-metal loading on H_2 oxidation in proton exchange membrane fuel cell. *J. Power Sources.* 103: 127–33.
60. Borup, R., Meyers, J., Pivovar, B., et al. 2007. Scientific aspects of polymer electrolyte fuel cell durability and degradation. *Chem. Rev.* 107: 3904–51.
61. Yan, W.M., Chub, H.S., Lub, M.X. 2009. Degradation of proton exchange membrane fuel cells due to CO and CO_2 poisoning. *J. Power Sources* 188: 141–7.
62. Loučka, T. 1971. Adsorption and oxidation of sulphur and of sulphur dioxide at the platinum electrode. *J. Electroanal. Chem.* 31: 319–32.
63. Najdeker, E., Bishop, E. 1973. The formation and behaviour of platinum sulphide on platinum electrodes. *J. Electroanal. Chem.* 41: 79–87.
64. Contractor, A.Q., Lal, H. 1979. Two forms of chemisorbed sulfur on platinum and related studies. *J. Electroanal. Chem.* 96: 175–81.
65. Mathieu, M.V., Primet, M. 1984. Sulfurization and regeneration of platinum. *Appl. Catal.* 9: 361–70.
66. Wang, Y., Yan, H., Wang, E. 2001. The electrochemical oxidation and the quantitative determination of hydrogen sulfide on a solid polymer electrolyte-based system. *J. Electroanal. Chem.* 497: 163–7.
67. Ramasubramanian, N. 1975. Anodic behavior of platinum electrodes in sulfide solutions, the formation of platinum sulfide. *J. Electroanal. Chem.* 64: 21–37.
68. Shi, W.Y., Yi, B.L., Hou, M., et al. 2007. The influence of hydrogen sulfide on proton exchange membrane fuel cell anodes. *J. Power Sources* 164: 272–277.
69. Michaelides, A., Hu, P. 2001. Hydrogenation of S to H_2S on Pt(111): A first-principles study. *J. Chem. Phys.* 115: 8570–4.
70. Urdampilleta, I.G., Uribe, F.A., Rockward, T., et al. 2007. PEMFC poisoning with H_2S: Dependence on operating conditions. *ECS Transactions* 11: 831–42.

71. Imamura, D., Hashimasa, Y. 2007. Effect of sulfur-containing compounds on fuel cell performance. *ECS Transactions* 11: 853–62.
72. Mohtadi, R., Lee, W.-K., Van Zee, J.W. 2005. The effect of temperature on the adsorption rate of hydrogen sulfide on Pt anodes in a PEMFC. *Appl. Catal. B: Environmental* 56: 37–42.
73. Shi, W.Y., Yi, B.L., Hou, M., et al. 2007. The effect of H_2S and CO mixtures on PEMFC performance. *Int. J. Hydrogen Energy* 32: 4412–7.
74. Shen, M., Yang, D.J., Wang, J.H., et al. 2008. Reasonable threshold of H_2S concentration in H_2 for PEMFC. *J. Xi'an Jiaotong University* 42: 1054–8.
75. Soto, H.J., Lee, W.K., Van Zee, J.W., et al. 2003. Effect of transient ammonia concentrations on PEMFC performance. *Electrochem. and Solid-State Lett.* 6: A133–5.
76. Watanabe, S., Tatsumi, M., Akai, M. 2004. Hydrogen quality standard for fuel cell vehicles. *Fuel Cell Seminar.*
77. Narusawa, K., Hayashida, M., Kamiy, Y., et al. 2003. Deterioration in fuel cell performance resulting from hydrogen fuel containing impurities: Poisoning effects by CO, CH_4, HCHO and HCOOH. *JSAE Review* 24: 41–6.
78. Dorn, S., Bender, G., Bethune, K. 2008. The impact of trace carbon monoxide/toluene mixtures on PEMFC performance. *ECS Transactions* 16: 659–67.
79. Janssen, M.M.P., Moolhuysen, J. 1976. Binary systems of platinum and a second metal as oxidation catalysts for methanol fuel cells. *Electrochim. Acta.* 21: 869–76.
80. Gasteiger, H.A., Markovic, N., Ross, P.N., et al. Carbon monoxide electrooxidation on well-characterized platinum-ruthenium alloys. *J. Phys. Chem.* 98: 617–25.
81. Maeda, N., Matsushima, T., Uchida, H., et al. 2008. Performance of Pt-Fe/mordenite monolithic catalysts for preferential oxidation of carbon monoxide in a reformate gas for PEFCs. *Appl. Catal. A: General* 341: 93–7.
82. Napporn, W.T., Léger, J.M., Lamy, C. 1996. Electrocatalytic oxidation of carbon monoxide at lower potentials on platinum-based alloys incorporated in polyaniline. *J. Electroanal. Chem.* 408: 141–7.
83. Grgur, B.N., Zhuang, G., Markovic, N.M., et al. 1997. Electrooxidation of H_2/CO mixtures on a well-characterized $Pt_{75}Mo_{25}$ alloy surface. *J. Phys. Chem. B* 101: 3910–3.
84. Wee, J.H., Lee, K.Y. 2006. Overview of the development of CO-tolerant anode electrocatalysts for proton-exchange membrane fuel cells. *J. Power Sources* 157: 128–35.
85. Mohtadi, R., Lee, W.K., Cowan, S., et al. 2003. Effect of hydrogen sulfide on the performance of a PEMFC. *Electrochem. Solid State Lett.* 6: A272–4.
86. Eley, D.D., Moore P.B. 1981. The adsorption and reaction of CO and O_2 on Pd–Au alloy wires. *Surf. Sci.* 111: 325–43.
87. Schmidt, T.J., Jusys, Z., Gasteiger, H.A., et al. 2001. On the CO tolerance of novel colloidal PdAu/carbon electrocatalysts. *J. Electroanal. Chem.* 501: 132–40.
88. Fishman, J.H. 1970. Method of generating electricity comprising contacting a Pd/Au alloy black anode with a fuel containing carbon monoxide. U.S. Patent 3,510,355.
89. Götz, M., Wendt, H. 1998. Binary and ternary anode catalyst formulations including the elements W, Sn and Mo for PEMFCs operated on methanol or reformate gas. *Electrochim. Acta* 43: 3637–44.
90. Chen, K.Y., Shen, P.K., Tseung, C.C. 1995. Anodic oxidation of impure H_2 on teflon-bonded Pt-Ru/WO_3/C electrodes. *J. Electrochem. Soc.* 142: L185 –7.

91. Stonehart, P., Watanabe, M., Yamamoto, N., et al. 1993. Electrocatalyst. U.S. Patent 5,208,207.
92. Shi, W.Y., Hou, M., Shao, Z.G., et al. 2007. A novel proton exchange membrane fuel cell anode for enhancing CO tolerance. *J. Power Sources* 174: 164–169.
93. Yan, J., Ma, J.X., Cao, P., et al. Preferential oxidation of CO in H_2-rich gases over Co-promoted Pt-gamma-Al_2O_3 catalyst. *Catal. Lett.* 93: 55–60.
94. Lee, S.H., Han, J.S., Lee, K.Y. 2002. Development of 10-kWe preferential oxidation system for fuel cell vehicles. *J. Power Sources* 109: 394–402.
95. He, C Z., Kunz, H.R., Fenton, J.M. 2001. Selective oxidation of CO in hydrogen under fuel cell operation conditions. *J. Electrochem. Soc.* 148: A116–24.
96. Zhang, J.L., Xie, Z., Zhang, J.J., et al. 2006. High temperature PEM fuel cells. *J. Power Sources* 160: 872–91.
97. Behm, R.J., Jusys, Z. 2006. The potential of model studies for the understanding of catalyst poisoning and temperature effects in polymer electrolyte fuel cell reactions. *J. Power Sources* 154: 327–42.
98. Du, B., Pollard, R., Elter, J.F. 2006. CO-air bleed interaction and performance degradation study on proton exchange membrane fuel cells. *ECS Transctions* 3: 705–13.
99. Murthy, M., Esayian, M., Hobson, Al. 2001. Performance of a polymer electrolyte membrane fuel cell exposed to transient CO concentrations. *J. Electrochem. Soc.* 148: A1141–7.
100. Marković, N.M., Schmidt, T.J., Stamenković, V., et al. 2001. Oxygen reduction reaction on Pt and Pt bimetallic surfaces: A selective review. *Fuel Cells* 1: 105–16.
101. Jusys, Z., Kaiser, J., Behm, R. J. 2003. Simulated "air bleed" oxidation of adsorbed CO on carbon supported Pt. Part I. A differential electrochemical mass spectrometry study. *J. Electroanal. Chem.* 554, 555: 427–37.
102. Jusys, Z., Behm, R.J. 2004. Simulated "air bleed" oxidation of adsorbed CO on carbon supported Pt. Part 2. Electrochemical measurements of hydrogen peroxide formation during O_2 reduction in a double-disk electrode dual thin-layer flow cell. *J. Phys. Chem. B.* 108: 7893–901.
103. Stamenkovic, V., Grgur, B.N., Ross, P.N., et al. 2005. Oxygen reduction reaction on Pt and Pt-bimetallic electrodes covered by CO. *J. Electrochem. Soc.* 152: A277–82.
104. Arpe, H.J. 1989. *Ullman's Encyclopedia of Industrial Chemistry*, 5th ed., Vol. A 13, 462., Weinheim, Germany: VCH Publishers.
105. Schmidt, V.M., Oetjen, H.F., Divisek, J. 1997. Performance improvement of a PEMFC CO by addition of oxygen-evolving compounds. *J. Electrochem. Soc.* 144: L237–8.
106. Carrette, L.P.L., Friedrich, K.A., Huber, M., et al. 2001. Improvement of CO tolerance of proton exchange membrane (PEM) fuel cells by a pulsing technique. *Phys. Chem. Chem. Phys.* 3: 320–4.
107. Wilkinson, D.P., Chow, C.Y., Allan, D.E., et al. 2000. U.S. Patent 6,096,448.

5

Membrane Electrode Assembly Contamination

S. R. Dhanushkodi, M. W. Fowler, A. G. Mazza, and M. D. Pritzker

CONTENTS

5.1 Introduction

Polymer electrolyte membranes (PEM) are semipermeable barriers sandwiched between two catalyzed electrodes in a polymer electrolyte membrane fuel cell (PEMFC) [1,2]. The membranes serve a number of roles within the fuel cell: proton conductor, electronic insulator, and gas separator. Perfluorosulfonic acid (PFSA) is the most widely used polymeric electrolyte membrane material in PEMs at this time [3]. In order to maintain the performance and life of fuel cells, the proton conductivity of the PEM should be close to 0.1 S cm^{-1}. At the same time, the properties of the membrane must be able to accommodate pure hydrogen and reformate hydrogen, which are the most commonly used fuels. Thus, the hydrogen permeability through this membrane should be less than 10^{-12} mol H$_2$ cm^{-1} s^{-1} kPa^{-1} to avoid gas crossover. Operating parameters, such as temperature, pressure, and relative humidity, have important effects on these properties and on determining the lifetime of fuel cells. Every engineering and scientific device has limitations associated with its operation, particularly over the long term, due to the effects of degradation and contamination. The PEM fuel cell is no exception in this regard. This chapter begins with an overview of membrane chemistry and different membrane degradation mechanisms in section 5.1. It follows in section 5.2 with a discussion of the various contamination sources and associated degradation mechanisms. Finally, the chapter closes with a discussion of strategies to mitigate PEM contamination.

5.2 Background

5.2.1 Membrane Chemistry

The membrane layer consists of a polymer capable of transporting H$^+$ ions with high conductivity. The membrane should have high proton conductivity, low gas permeability, and low temperature sensitivity [1,2]. PFSA polymers have been widely used in PEM fuel cells. The most widely studied and used membranes in PEM fuel cells are made from Nafion®, a sulfonated tetrafluoroethylene copolymer that was first manufactured by DuPont™ in the 1960s. Its chemical structure is shown in Figure 5.1. The details of perfluorocarbon sulphonic acid membranes produced by a variety of manufacturers are reported in the literature [4]. Generally, PFSA membranes are made up of a polytetrafluoroethylene (PTFE) backbone with sulfonic acid-containing side chains. The sulfonic acid group H$^+$SO$_3^-$ that enables H$^+$ transport is attached at the end of an arm containing CF$_2$-O-CF$_2$ groups. The hydrophilic sulfonic acid end groups in PSFA segregate into nano-sized clusters that cause the ionomer in the membrane to swell under humidified conditions [5–7]. These water clusters form interconnected acidic domains, which are thought to be the major path for proton transport. These membranes do not have sufficient conductivity under dry conditions and high temperatures (>90°C) and thus function best under fully humidified conditions. Improvements in membrane structure and conductivity have been achieved by producing

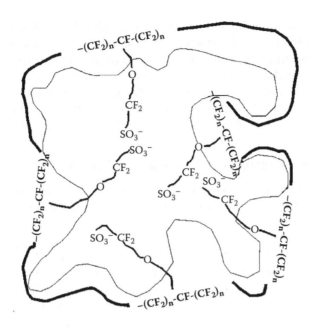

FIGURE 5.1
Structure of a common PSFA polymer.

composite membranes [8,9]. These composite membranes can be prepared by: (1) reinforcing the perfluorosulfonic membrane by PTFE components for greater mechanical strength; (2) impregnating a membrane with a solution or with a solid powder to decrease the permeability of the reactant gases; (3) dissolving the membrane into an appropriate solvent and mixing it with inorganic materials, such as silica, titania, or zirconia; (4) radiation grafting; or (5) plasma polymerization [10].

5.2.2 Types of MEA Membrane Degradation

The successful commercialization of PEM fuel cells depends primarily on their cost and durability. Consequently, there has been considerable effort to bring these costs down by reducing the Pt content of the catalysts and using thinner perfluorinated, partially fluorinated, or nonfluorinated membranes [11]. Limits to the durability and lifetime of membranes have had negative effects on their introduction into the automotive sector, which requires that the PEM fuel cells be able to withstand a wide range of conditions.

Membrane degradation can be classified according to whether it occurs primarily by chemical, mechanical, or thermal factors. Chemical degradation of the membrane mainly occurs due to the presence of contaminants and the formation of free radicals. Mechanical degradation is driven primarily by mechanical stresses. Inadequate compression of the cell, nonuniform torque at the end plate, and insufficient humidification are the main factors contributing to mechanical stress in PEM fuel cells. Thermal degradation can be defined as the molecular deterioration of membrane components due to the high temperature operation of the cell. At high temperatures, the components of the long chain backbone (Teflon) of PFSA can begin to separate due to molecular scission and react with one another to change the physical properties, such as conductivity. This degradation mechanism generally involves changes in the polymer molecular weight and chain length distribution.

5.2.3 Concept of "Reliability, Durability, and Stability"

To understand important aspects of membrane electrode assembly (MEA) integrity and the failure modes of PEM fuel cells, the following terms are useful:

- *Reliability* is the capability of a PEM fuel cell stack to achieve the required performance under a given environment for a period of time. Catastrophic failure and performance losses of the cell can be considered as examples of reliability failure modes [12].
- *Durability* is the capability of a PEM fuel cell stack to resist permanent change in performance over time. A durability failure may not cause catastrophic failure in the fuel cell. However, this mode of failure will decrease the performance of the fuel cell. It also can involve irreversible failures, such as electrochemical surface area reduction, carbon corrosion, etc., and losses mainly related to ageing [12].
- *Stability* is the capability to recover performance losses during continuous operation. Stability decay is always related to the response of a fuel cell to a given set of operating conditions (such as water management) and reversible material changes [12].

5.2.4 Perspectives on MEA Integrity

PFSA membranes are widely used due to their high ionic conductivity and chemical stability. They have been shown to remain relatively durable during most conditions [13]. Nevertheless, membrane integrity can be adversely affected by chemical, mechanical, and thermal degradation. A complication in assessing membrane integrity is that the solubility and crystalline structure of the Nafion membrane are often nonuniform, making it difficult to uniquely characterize their structure. The mechanisms for the changes to the morphology of membranes can be classified based on the distribution of the sulfonic acid and polymeric clusters. The membrane integrity can be gauged using the following criteria:

1. Sensitivity to chemical and mechanical degradation
2. Performance during relative humidity (RH) cycles

3. Sensitivity to localized material degradation

4. Performance loss due to contamination effects

5.2.5 Chemical and Mechanical Degradation

The longevity of the ionomer plays a major role in determining the useful lifetime of the fuel cell stack. The ability to prevent gas from crossing over from the anode to cathode will be reduced due to the failure or thinning of ionomer in the membrane. Again, this failure can be caused by mechanical, thermal, and chemical degradation.

5.2.5.1 Mechanisms of Membrane Failure by Chemical Attack

The chemical degradation of the membrane can occur due to the action of hydroxyl ($^\bullet$OH) and peroxyl ($^\bullet$OOH) radicals that are most commonly formed from reactions involving H_2O_2. H_2O_2 can be produced as an intermediate during oxygen reduction [14]:

$$O_2 + 2H^+ + 2e^- \rightarrow H_2O_2 \tag{5.1}$$

However, it also believed that H_2O_2 can also form due to reactions caused by the crossover of oxygen or air from the cathode side to the anode side and hydrogen in the opposite direction. The generation of hydroxyl radicals from H_2O_2 can occur in several ways. One route involves Fenton's reaction in which H_2O_2 combines with Fe(II) present due to corrosion or contamination as follows:

$$H_2O_2 + Fe^{2+} \rightarrow {}^\bullet OH^- + Fe^{3+} + OH^- \tag{5.2}$$

Also, the formation of these radicals due to reactions involving contaminants has been reported [15].

Studies revealed that attack by peroxide radicals of polymer end groups containing residual H-containing bonds is chiefly responsible for the severe chemical degradation of the membrane when the cell operates with gas-fed streams containing pure hydrogen and oxygen at lower humidity and/or higher temperatures [16]. This is believed to occur by the following steps. Hydrogen is first extracted from an acid end group by a hydroxyl radical to produce a perfluorocarbon radical (R - CF_2 $\bullet$), carbon dioxide, and water via reaction (5.3):

$$R\text{-}CF_2COOH + {}^\bullet OH \rightarrow R\text{-}CF_2\bullet + CO_2 + H_2O \tag{5.3}$$

The perfluorocarbon radical can then react with another hydroxyl radical to form an intermediate that ultimately leads to the formation of $R - COF$ and HF as follows:

$$R\text{-}CF_2 \bullet + \bullet OH \rightarrow R - CF_2OH \rightarrow R - COF + HF \qquad (5.4)$$

$$R\text{-}COF + H_2O \rightarrow R\text{-}COOH + HF \qquad (5.5)$$

Membrane thinning and fluoride ion detection are indicators of chemical degradation. The leakage of fluoride ions into the water product has been shown to be an effective measure for predicting the life of the membrane [17] Besides these factors, platinum particle migration and precipitation in the membrane have been observed during fuel cell operation at very high potential and low pH. This leads to the formation of platinum bands within the membrane. Hydrogen and oxygen can diffuse to these Pt bands and form H_2O_2 and localized hotspots. The oxidation of Pt(II) to Pt(IV) by hydrogen peroxide and other contaminants also can lead to the formation of more peroxide radicals. These additional radicals can severely damage the Nafion polymer at its chain ends. As a result, membrane degradation can be triggered by electrode degradation as well as by reactions involving hydrogen and oxygen [18].

5.2.5.2 Mechanism of Membrane Failure by Mechanical Attack

The material failure of the PEMFC can also be due to mechanical damage. Undue mechanical stress arises due to nonuniform pressure, inadequate humidification, or penetration of contaminants into the membrane. These stresses lead to delamination, flaws, and pinholes in the solid electrolyte. Many of these defects can form when the membrane is being fabricated. In addition, mechanical failure can be accelerated by chemical degradation [19]. An example of the potentially synergistic interaction between chemical and mechanical effects is pinhole formation due to radical attack. These pinholes act as initiation points for the development of membrane cracks due to residual stresses [20]. The distribution of these mechanical stresses is not uniform at the MEA surface. This mechanical failure can also arise from a reduction of membrane ductility combined with mechanical strain caused by constrained drying or hydration cycling [21]. A common cause of the sudden failure of the PEM fuel cell is the delayed mechanical failure of the membrane that occurs when the relative humidity is cycled under conditions conducive for mild chemical degradation. Hence, the slow strain-failure rate of the membrane can be the limiting criterion for the delayed mechanical failure. These failure areas have been termed *a priori* by Beuscher et al. [22]. The state of the membrane can be gauged from its failure stress, strain-to-failure, or tear resistance.

The temperature, relative humidity of the cell, and pressure also play roles in the stress-related failures in the membrane during the operation of the fuel cell. The delamination between the membrane and gas diffusion layer has been studied in detail by several researchers [21,23,24]. The precise mechanisms of membrane failure in this case are not well understood so far. However, studies support the possibility of mechanical failure due to hydration-thermal cycles (HTCs) even in the absence of an electric potential or a reactive gas [22,25]. Pinholes and the introduction of foreign materials or contaminants during the manufacturing stage can initiate and propagate cracks leading to irreversible failure [26]. The distribution of the initial gas pressure over the membrane surface can lead to membrane fatigue as well. Operating parameters, such as relative humidity, temperature, and the torque applied to the MEA, also can initiate membrane failure. The common practice of cycling the relative humidity during operation has been shown to cause higher stresses and thinning of Nafion (N111) membranes [27]. The development of thin membranes can potentially solve problems associated with deterioration of proton conductivity. However, attention would be necessary and steps taken, if necessary, during manufacturing to ensure that the mechanical properties (e.g., tensile and tear strength, puncture resistance) of these thin membranes meet the required levels.

5.3 Membrane Contamination

Membrane contaminants can be undesirable gas-phase, solid-phase, or liquid-phase materials that can adversely affect the performance of PEM fuel cells during operation. The presence of these foreign species leads to different modes of fuel cell degradation, as discussed in the previous sections. Contamination mainly affects the ionomer conductivity of the membrane and the catalytic activity of both anode and cathode electrodes. The contaminant effects on the membrane are the focus of this section, although it is difficult to ignore catalyst contamination here because of the confounding effects on PEM performance. The effects of contaminants originating from the air stream, fuel stream, and fuel cell components on PEM fuel cell performance, life, and durability can be significant and have been identified as critical research topics for further study. Common contamination sources and the affected fuel cell properties are listed in Table 5.1.

5.3.1 Contaminants in Fuel Gas Stream

Hydrogen used in the PEM fuel cell can be produced using fossil fuels via steam reforming, partial oxidation of natural gas, coal gasification, and electrolysis. Most of the hydrogen fuel is produced by natural gas

TABLE 5.1

Contamination Sources of the Membrane

Sources	Contaminant	Affected Property
Leaching of bi-polar plate and endplates	Fe^{3+}, Cu^{2+} Cr^{2+}, Ni^{2+}, Al^{3+}	Membrane conductivity
Fuel	CO, CO_2, NH_3, H_2S	Catalyst surface area and proton conductivity
Fuel	NO_x and SO_x	Cathode and cathode-membrane interface
Air mixed with emissions from other automotive exhaust	NO_x, SO_x, CO_x, NH_3, O_3	Membrane permeability and conductivity, catalyst-membrane interface, and flow fields
Compressors	oils	Porosity of the GDL, PEM, and catalyst layer
Sealing gasket	Si	PFSA contamination
Coolant, DI water	Na^+, Ca^+, Si, Al, S, K, Cu, Cl, V, Cr	Membrane proton conductivity, corrosion of components
Reformate hydrogen	CO_x, H_2S, NH_3, CH_4	MEA poisoning principally the catalyst

reforming, a situation that is likely to remain the dominant mode of production for a number of years [2]. Unfortunately, this process generates contaminants, such as NO_x, SO_x, CO_x, H_2S, NH_3, and CH_4, that can damage fuel cell components, particularly the membrane and electrocatalyst, and degrade cell performance [2]. Consequently, the quality of the hydrogen is very important [28,29]. Thus, it is important to study the reactions of these contaminants with the MEA and procedures to mitigate their effects [13]. Apart from CO, the mechanism by which these contaminants react with the MEA of these pollutants has not been studied in detail [30–32]. Although many details of the interactions between impurities in the hydrogen feed stream, water, and the fuel cell components are not completely understood, the need to control and manage their effects is well recognized and remains as a critical challenge for the success of fuel cell technology.

5.3.2 Contaminants in Air Stream

In addition to the fuel source, contamination can come from the environment or from other fuel cell components. The major sources of the contaminants in the air stream fed to fuel cells are determined by the general air quality standards. Thus, these feed streams will contain contaminants from vehicle emissions, such as NO_x, SO_x, CO_x, and specific chemical species. The impact of SO_x is particularly critical since its presence can cause fuel cell death depending on its concentration or dosage [33]. Contamination sources can also be from fuel cell components, such as the seals, lines, or fittings.

5.3.3 NO$_x$ Contamination

The NO$_x$ emitted into the atmosphere creates serious environmental issues, such as acid rain and excessive ground-level ozone with leads to photochemical smog, which is common in the urban environment [2,34]. The continual exposure of NO$_x$ to the PEM fuel cell, even at low concentration, can be detrimental to cell performance. It is estimated that the NO$_2$ concentration in a PEM fuel cell should remain below 0.05 ppm in order that its performance and lifetime not be unduly compromised [35,36]. Although the adsorption of NO on Pt (platinum) is weak, it can block sites on the platinum surface that would otherwise be available for oxygen reduction [37]. The following reaction mechanism for the detrimental effect of NO$_x$ has been proposed based on a study by [38]:

$$NO + \frac{1}{2}O_2 \rightarrow NO_2 \tag{5.6}$$

$$2NO_2 + H_2O \rightarrow HNO_3 + HNO_2 \tag{5.7}$$

$$4HNO_2 + 2O_2 \rightarrow 4HNO_3 \tag{5.8}$$

$$4H^+ + O_2 + 4e \rightarrow 2H_2O \tag{5.9}$$

The mechanism postulates the formation of nitrous acid (HNO$_2$) and nitric acid (HNO$_3$) by the reaction of NO with oxygen at the cathode under wet conditions. The simultaneous occurrence of the anodic and cathodic steps comprising reactions (5.6) and (5.8) above on the cathode create a mixed potential at the electrode that tends to reduce the cathode potential and retard the oxygen reduction reaction (5.9) [37,38]. The impact of NO$_x$ contamination is shown in Figure 5.2 and Figure 5.3.

Another deleterious effect of the presence of NO$_x$ is the electrochemical reduction of NO$_2$ at the cathode by the reaction below [39]:

$$NO + 6H^+ + 5e^- \rightarrow NH_4^+ + H_2O \tag{5.10}$$

NH$_4^+$ formed from the electrochemical reduction of NO$_x$ at the cathode has been reported to be a potential poison for the cell [37,40]. This reaction competes with the oxygen reduction reaction for catalytic active sites at the cathode electrode. In addition, NH$_4^+$ formed by this reaction can exchange with the protons in the membrane and can be a potential poison for the ionomer. By minimizing or preferably eliminating NO oxidation, one can recover the performance loss associated with its occurrence [41].

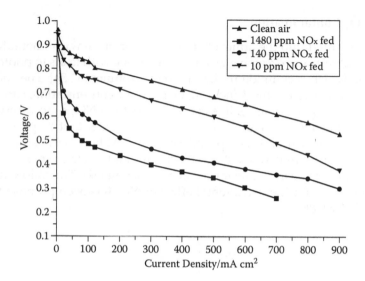

FIGURE 5.2
Performance loss of the fuel cell due to presence of NO_x. (From Yang, D. et al. 2006. *Electrochimica Acta*, 51, 4039–4044. With permission.)

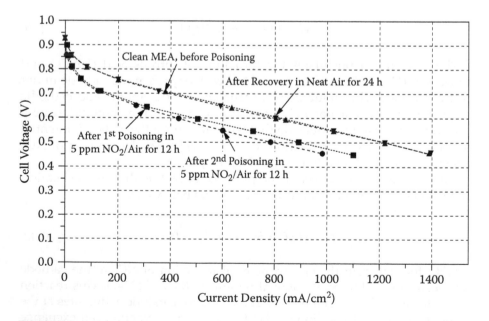

FIGURE 5.3
Effect of NO_2 on fuel cell polarization curve and recovery of the performance loss after sufficient operation with NO_2-free feed. (From Mohtadi, R. et al. 2004. *Journal of Power Sources*, 138, 216–225. With permission.)

The presence of NO reduces the efficiency of proton transfer through the PFSA membrane both throughout the membrane and at the catalyst–membrane interface. Figure 5.2 shows the effect of continuous exposure to NO on the resulting polarization curve of the PEM fuel cell. As shown, a noticeable deterioration of cell performance is observed when the feed stream contains as little as 10 ppm NO [38]. The degradation becomes progressively more severe as the NO concentration rises, although the incremental effect becomes weaker. NO oxidation to NO_2 on Pt catalyst active sites can have an irreversible effect if the rate of oxidation reaction and exposure time of NO increases. Some previous studies have shown evidence of performance loss because of competition between NO_2 and H^+ [33,35,38]. Figure 5.3 shows the effect on fuel cell performance due to 12 h of contact with an air feed stream containing 5 ppm NO_2 and the subsequent recovery after 24 h of contact with a pure air stream containing no NO_2 [37]. Contact with NO_2 causes a drop in current density of 200 to 300 mA/cm² at most cell potentials. However, the results show that this loss is completely recovered when NO_2 is removed and enough time is allowed for its effects to be flushed out.

5.3.4 SO$_x$ and Hydrogen Sulfide Contamination

Sulfur poisoning has a direct and dramatic effect on platinum-based catalysts in the fuel cell. Although SO_2 and H_2S contaminants are usually present at very low levels in both the fuel and air streams, they are strong poisons for the fuel cell. The presence of SO_2 influences both the anode and cathode catalysts by blocking the platinum active sites [37,42–44]. The performance of the cell is affected due to the reduction in the active area of the catalyst. Contamination by SO_2 in the PEM and at the electrode has been shown to cause sudden and complete shutdown of the fuel cell [33]. The mechanism of SO_2 poisoning has been studied in detail [45] and proposed to be given by the following reactions [43]:

$$Pt + SO_2 + 2H^+ + 2e^- \rightarrow Pt - SO + 2H_2O \tag{5.11}$$

$$Pt - SO + 2H^+ + 2e^- \rightarrow Pt - S + H_2O \tag{5.12}$$

These chemisorbed sulfur species on platinum surfaces are very stable and so are not easily detached. As a result, the presence of SO_2 causes permanent damage to the catalyst [46–48]. The introduction of sulfur-containing species in a continuous manner and/or at high concentration was shown to irreversibly damage the catalysts [49,50]. The rate at which a platinum catalyst is poisoned depends strongly on the SO_2 concentration in the bulk gas

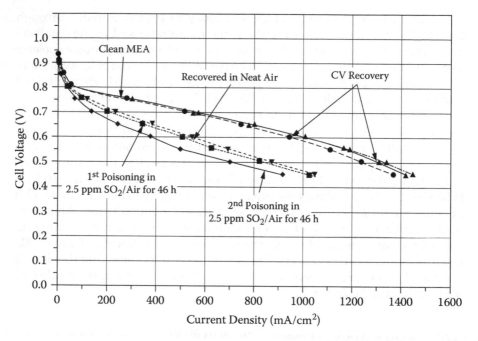

FIGURE 5.4
Comparison between recovery of the performance loss for 5 ppm of SO₂/air. (From Mohtadi, R. et al. 2004. *Journal of Power Sources*, 138, 216–225. With permission.)

phase. The formation of the adsorbed SO and S species by reaction (5.11) and reaction (5.12) above also directly competes with oxygen reduction reaction (5.9). The oxygen used for the (5.9) reactions can create the change in the potential at the cathode. This will adversely affect the oxygen reduction reaction (ORR) and reduce the overall cell potential [17].

Mohtadi et al. investigated the influence of different SO_2 concentrations in the air stream on cell performance degradation. The cell performance was reduced by 53% during 46 h exposure to 2.5 ppm SO_2 in air (Figure 5.4) [37,43]. Furthermore, the results also indicate that only partial recovery of the cell performance is possible after exposure of the cell to SO_2. To recover the poisoned Pt cathode, cyclic voltammetry was used to oxidize the sulfur adsorbed on the Pt at anode potentials of approximately 0.89 and 1.05 V normal hydrogen electrode (NHE). The cell regained its original performance after this cyclic voltammetry treatment.

In comparison to SO_2, hydrogen sulfide is reported to have an even more adverse effect on fuel cell catalysts [33,37,43,50]. H_2S can strongly adsorb onto Pt and affect the catalyst layer by disrupting its morphology [43]. By occupying polyatomic sites on the catalyst, it prevents the reactants from reaching reaction sites. A mechanism proposed for the interaction of hydrogen sulfide with a platinum surface involves the formation of adsorbed hydrogen sulfide, hydrogen, and sulfide complexes through the following sequence of steps [50]:

$$Pt + H_2S \rightarrow Pt - H_2S \tag{5.13}$$

$$Pt + HS^- + H^+ \rightarrow Pt + H_2S \tag{5.14}$$

$$Pt + H^+ \rightarrow Pt^+ - H \tag{5.15}$$

$$2Pt + H_2S \rightarrow Pt - HS + Pt - H_{(ads)} \tag{5.16}$$

$$Pt - SH_n \rightarrow Pt - S + nH^+ + ne \; (n = 1 \; or \; 2) \tag{5.17}$$

The dissociative adsorption of hydrogen sulfide has been shown to occur in the gas phase according to the above reactions at elevated temperatures [49]. Although the conductivity of the solid electrolyte membrane can increase due to oxidation of the adsorbed SH and H_2S (i.e., reaction (5.16)), the formation of a stable platinum sulfide film can lead to fuel cell failure [38].

Shi and Anson [80] studied the effect of H_2S on the Pt catalyst using cyclic voltammetry (CV) and obtained the results shown in Figure 5.5. In this

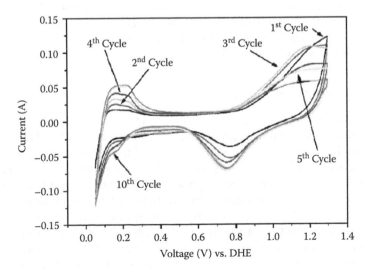

FIGURE 5.5
Cyclic voltammograms obtained after the cells were poisoned with 50 ppm H_2S for 1 h at 600 mA cm^{-2}. (From Shi, W. et al. 2007. *International Journal of Hydrogen Energy*, 32, 4412–4417. With permission.)

study, the cell was poisoned with 50 ppm H_2S during operation for 1 h at 600 mA cm^{-2} before beginning the CV scan. The common reaction used to describe the desorption of hydrogen from platinum surfaces in PEMFC electrodes is given below. The hydrogen desorption is expected to occur in the potential below 0.4 V (NHE) [51].

$$Pt - H_{ads} \Leftrightarrow Pt + H^+ + e^- \qquad (5.18)$$

One can infer that the absence of such a peak for hydrogen desorption (~0.15 V) in the first cycle arises due to preferential H_2S adsorption via the steps given above in reaction (5.13) to reaction (5.17). These steps yield a layer of adsorbed sulfur on the Pt surface. However, as the scan proceeds in the anodic direction and reaches a potential of about 1.2 V, oxidation of this adsorbed sulfur begins and leads to the partial restoration of Pt sites. However, not all of the adsorbed sulfur is likely removed in the first cycle since the peak associated with the formation of adsorbed H by the reduction of H^+ ions at potentials between 0.7 and 0.8 V when the scan is reversed in the cathodic direction is smaller than what it becomes in subsequent cycles. In the following cycles, the sulfur oxidation peaks become smaller and the expected electrode response of clean platinum is recovered as the poisoning effect of sulfur wanes. These results show that electrochemical scanning of the potential of the cell can be an effective method for reversing the effects of sulfur poisoning.

The severity of the PEM degradation depends on the amount of adsorption of hydrogen sulfide on the MEA. A recent study reported the potentially important finding that the introduction of a small quantity of Ru (ruthenium) or WO_3 into the carbon support can inhibit Pt-S formation and improve the tolerance of the MEA to H_2S [52][94].

5.3.5 CO_2 and CO Contamination

Hydrogen produced from natural gas is often used as a fuel [2]. Great efforts are made to purify this hydrogen stream, but amounts of CO and CO_2 inevitably remain. Furthermore, the concentrations of these contaminants will also build up inside a PEM fuel cell system when the anode stream is recycled [28,29,49]. Because the feed stream produced contains a significant amount of CO_2, attention has been paid to the influence of this contaminant on the behavior at the anode [53]. The poisoning effects of CO_2 mainly depend on the properties of the anode catalyst material. Evidence has shown that CO_2 poisoning may not affect the performance of the membrane directly. However, the conversion of CO_2 to CO via the reverse water–gas shift reaction (WGSR) can cause the displacement of adsorbed hydrogen from the anode [39,54]:

$$2Pt + H_2 \rightarrow 2Pt - H \qquad (5.19)$$

$$CO_2 + 2Pt - H \rightarrow Pt - CO + Pt + H_2O \tag{5.20}$$

Factors that increase the rate of the reverse WGSR and reduce the hydrogen dissociation rate enhance the CO_2 poisoning effects. This study also showed that operation of the cell at higher current densities tended to reverse the adverse effect of CO_2 on its performance. This finding was confirmed in another study that reported that complete recovery was possible by purging with neat hydrogen (Castellanos et al. [41]). Some research on elucidating the exact structure of the adsorbed species and the adsorption mechanism has been reported, although further work is necessary in order to develop procedures to mitigate the effect of CO_2 poisoning on Pt catalysts. However, the formation of the stable adsorbed layer on the catalyst surface is believed to involve linearly bonded CO (adsorption of CO only at a single Pt site) and bridge-bonded CO (adsorption on two adjacent Pt sites) produced from CO_2 [20,55,56].

By increasing the temperature and water content of the anode feed, the CO concentration can be reduced by the WGSR [45]. Another approach to mitigate the poisoning effect is to pretreat the electrode by electrochemically oxidizing it at high electrode potentials (≥ 0.7 V versus NHE) to convert strongly adsorbed CO to CO_2, which then desorbs [57]. The following reactions have been proposed for the mechanism by which adsorbed CO is converted to CO_2 by electrochemical oxidation at higher electrode potentials [58]:

$$Pt - CO + Pt - O \rightarrow 2Pt + CO_2 \tag{5.21}$$

$$Pt - CO + Pt - OH \rightarrow CO_2 + H^+ + 2Pt + e \tag{5.22}$$

In another study, the reaction of CO_2 via the reverse WGSR on a surface containing preadsorbed hydrogen has been shown to be suppressed at low temperatures when carried out on a Pt-Ru/C catalyst. This catalyst shows better tolerance to CO than Pt or Pt/WO catalysts [59].

Kinetic models for the WGSR on Pt catalysts suggest that the adverse effects on cell performance can be even stronger when the feed stream to the anode contains a mixture of both CO and CO_2 [60]. The presence of CO_2 and CO in the fuel stream results in monolayer coverage of the anode and appears to accelerate poisoning at the catalyst surface. Trace concentrations of CO (30 to 50 ppm) in 30% CO_2 can cause a remarkable decrease in cell performance [39]. Purging the anode with hydrogen at high flow rates or raising the operating temperature has been shown to break up this blanket layer and inhibit the adsorption of CO [61,62].

Since CO poisoning has a more adverse effect on the catalyst than CO_2, it is important to study its poisoning effect in detail. The adsorption of CO at the Pt electrode can directly block active surface sites otherwise used for H_2 electrochemical oxidation. A proposed mechanism for CO adsorption involves two pathways by which CO can occupy surface sites: adsorption onto empty Pt sites and displacement of adsorbed H intermediates formed in the first step of H_2 oxidation [49].

$$2Pt + H_2 \rightarrow 2Pt - H \tag{5.23}$$

$$Pt + CO \rightarrow Pt - CO \tag{5.24}$$

$$2CO + 2Pt - H \rightarrow 2Pt - CO + H_2 \tag{5.25}$$

$$2Pt - H \rightarrow 2Pt + 2H^+ + 2e^- \tag{5.26}$$

According to this mechanism, CO adsorption competes with that of H_2 for surface sites for several reasons and, thus, can reduce the H_2 oxidation rate and the cell voltage. As noted above, CO can adsorb onto empty Pt sites by reaction (5.24) and displace adsorbed H sites by reaction (5.26). The H_2 adsorption step in reaction (5.22) involves dissociation of hydrogen atoms on two adjacent active Pt sites. On the other hand, the displacement of adsorbed H by CO by reaction (5.25) occurs relatively fast due to the lower activation energy requirements arising from the higher bond energy between CO and Pt. The CO coverage on the electrode surface also has been found to depend on the electrode surface state (surface roughness) and partial pressure of CO [63].

The removal of adsorbed CO is described in the following steps:

$$Pt + H_2O \rightarrow Pt - OH + H^+ + e^- \tag{5.27}$$

$$Pt - OH + Pt - CO \rightarrow 2Pt + CO_2 + H^+ + e^- \tag{5.28}$$

Active unoccupied Pt sites tend to be inefficient for electro-catalytic oxidation of CO to eliminate its poisoning effect at the catalyst surface. The presence of water molecules at the Pt surface appears to inhibit the adsorption of CO onto Pt and promote the evolution of CO_2 from the surface. At the same time, the H_2 oxidation rate in the presence of CO depends on several other factors including the electrode structure, operating conditions, polymer electrolyte, and, most importantly, the nature of the electro-catalyst [64]. The rate of removal of adsorbed CO through reaction (5.27) and reaction (5.28) is increased by raising the cell temperature [65].

The mitigation of both external and internal contaminants is necessary to increase the life of the fuel cell. CO_x poisoning can be mitigated in PEM fuel cells as follows:

1. Promotion of CO oxidation by pretreating reformate
2. Introduction of low levels of an oxidant, such as hydrogen peroxide, into the anode fuel stream to oxidize CO to CO_2 or reduction of CO levels in reformate gas by the WGSR reaction
3. Development of CO-tolerant catalysts
4. Use of CO_2 scrubbers with subsequent methanation to reduce CO
5. Further development of Pt-Ru alloy catalysts that minimize contaminant effects

5.3.6 Ammonia Contamination

Ammonia contamination can arise when the hydrogen fuel has been produced from ammonia or reforming of fossil fuels that contain traces of nitrogenous compounds or its functional groups [66]. Ammonia should not be present when pure fuel components are produced. The amount of ammonia present in the fuel will depend on many factors, such as the type of fuel, operating parameters, and concentration. The degradation of the cell voltage due to ammonia contamination occurs due to its effect on the anode, cathode catalyst, and the ionomer [67,68]. Although ammonia in the feed stream may not adsorb on the carbon support, it can react with protons in the ionomer and membrane and form ammonium ions. Ammonia uptake and reaction within the membrane has been explained in terms of the relatively simple mechanism given below [39]:

$$NH_3(g) \rightarrow NH_3(membrane) \tag{5.29}$$

$$NH_3(membrane) + H^+ \rightarrow NH_4^+ \tag{5.30}$$

The presence of ammonium ions can decrease the proton conductivity and, therefore, degrade the cell performance, particularly when relatively high amounts of ammonia accumulate and exposure becomes long. Since NH_4^+ ions displace protons from the ionomer, another important effect of ammonia is on the membrane transport properties. The presence of ammonium ions can reduce the proton conductivity of the Nafion 117 membrane by a magnitude of 3.8- to 3.9-fold [69,70]. The reduction in the proton conductivity of a PFSA 1035 membrane by a factor of 4 to 4.2 has also been reported in the literature [35]. It has been hypothesised that the platinum catalyst can help to decompose ammonia and thereby reduce the poisoning effects on the membrane, particularly at the "three-phase contact" PFSA/platinum/carbon region. [68,69,71]. The basis for this idea comes from the effect of ammonia on Brønsted acid sites and the observation that it is stronger on the PFSA/carbon sites than on the three-phase region sites. Pulsed chemisorption experiments have shown that ammonia adsorbs at the platinum/carbon interface and subsequently exchanges with protons on sulfonic acid sites of the ionomer [68,69,71].

Soto et al. [72] studied the effect of ammonia concentration in the feed stream on cell performance. An increase in the ammonia concentration in the fuel feed stream from 200 ppm to 5000 ppm at 70°C was shown to reversibly drop the cell voltage from 0.67 to 0.28 V. Once ammonia was removed from the feed stream, the original cell voltage was recovered after another 10 h of operation.

The ability of ammonium to exchange with protons on ionomer sites creates problems because it enables ammonium to move through the membrane from the anode side to the cathode side and, therefore, degrade the performance at both electrodes [72].

A study on a platinum rotating disk electrode in aqueous sulfuric acid solutions has shown that the presence of ammonium retards the kinetics of the oxygen reduction reaction regardless of whether the electrode is in a preoxidized or prereduced state [73,74]. Although this effect has not yet been clearly explained, it is consistent with the idea that the adsorption of ammonium, an intermediate of ammonium oxidation and/or bisulphate ions (HSO_4) stabilized by the presence of ammonium, competes with the adsorption of oxygen for surface platinum sites [70]. More research has been conducted on the effect of ammonium on hydrogen oxidation and hydrogen evolution where the poisoning effect is expected to be more severe [35,70,72]. Cyclic voltammetry studies of the electrode reactions at a platinum electrode in acidic solutions show that the presence of ammonium ions has little effect on the portion of the anodic-going scan at potentials below about 0.25 V standard hydrogen electrode (SHE) associated with H_2 adsorption [70]. However, a combination of CV, differential electrochemical mass spectrometry, and electrochemical quartz crystal microbalance experiments suggests that ammonium ions have a number of significant effects at more positive potentials—adsorption at potentials of about 0.25 V followed by its oxidation to N_2 and nitrogen oxide species beginning at potentials of about 0.65 V [74]. The generation of N_2 involves several adsorbed nitrogen and/or nitrogen–hydrogen species that are quite stable and interfere with the normal formation of Pt oxides at more positive potentials. Because these adsorbed species are not easily removed when the potential is shifted in the negative direction, they also tend to interfere with the cathodic processes, such as the reduction of Pt oxides and evolution of H_2, that subsequently occur. The influence of the adsorbed ammonium and its subsequent oxidation products on both the anodic and cathodic processes on Pt has been found to be heightened when the electrolyte also contains sulfuric acid. Evidence has been presented that HSO_4 ions also adsorb onto platinum in such solutions [70]. This co-adsorption of HSO_4 is believed to enhance the stability of ammonium on the electrode surface.

The decrease in the proton conductivity of the membrane due to ammonia contamination causes an increase in ohmic resistance that accounts for approximately 5 to 20% of the voltage loss. Although complete recovery of membrane conductivity appears to be possible, only partial recovery of the overall cell performance was observed at low contaminant concentrations [70]. This observation suggests that other degradation mechanisms associated with ammonia contamination may also be operating. The diffusion rate of ammonia through the hydrogen-filled gas infusion layer (GDL) was estimated to figure out the possibility of mass transfer limitations for absorption of ammonia. The mass transfer coefficient reported is very high. The results predicted the increase in mass transfer rate during the adsorption of ammonia [70]. Although the mass transfer limitations on the GDL are not significant, the oxidation of ammonium ions to form nitrogen at the anode Pt catalyst sites may lead to performance gain.

Despite the ample evidence that the presence of ammonia is deleterious to PEM fuel cell performance, many details concerning the mechanism for the effect of ammonia on the membrane and electrocatalyst layer are uncertain. However, fuel cell performance can be recovered by frequent acid washing and purging with pure hydrogen for a long enough time to flush ammonia and ammonium ions from the cell [35,36]. The electrochemical oxidation of ammonium to nitrogen at relatively high potential, typically >700 mV RHE, may be another way by which ammonium is removed from the cell [73].

5.3.7 Metal Ion Contamination

Ionic impurities present in the PEM fuel cell can also accelerate membrane degradation. Also, although most of these contaminants do not participate in the fuel cell reactions, they will affect the transport properties of the membrane. Metal ions are generated from the corrosion of stack and system components, such as the flow field bipolar plates, seals, inlet/outlet manifolds, humidifier reservoirs, and cooling loops. Also, metal ions from the coolant streams may leak into the cell through cracks that develop in the channels. Metal ion content can also be concentrated in hydration systems, which employ water recycling. Impurities, such as Fe, Cr, Al, and Ni, have been identified near the endplate and flow plates [75,76]. The water that is transported from the anode to the cathode by electro-osmotic drag and the water generated on the cathode side, in particular, facilitate the transport of these impurities toward the membrane and electrode catalyst layers. Unfortunately, the removal of these cationic or anionic contaminants during cell operation is difficult.

The presence of these ions can have a significant effect on the ohmic losses in a cell [77]. In general, the cations with higher charge density and higher hydration enthalpy tend to carry more water molecules during transport and produce a high streaming potential. Since these highly charged cations tend to carry more water molecules, they tend to lower the water content in the membrane as they move. This eventually decreases the transference number (fraction of the total current carried in a membrane by a cation) of the cations and increases the ohmic resistance [25]. Alkali, alkaline earth, rare earth, and transition metal ions and ammonium derivatives originating from reactants or materials used in the cell can enter the membrane by displacing H$^+$ ions and decrease its ionic conductivity [78,79]. The formation of oxides of these metals on the surface of the membrane will decrease both its ionic and electronic conductance. This occurs mainly because of the mobility of the transition metal ions through the membrane and affects the proton conductivity. The detrimental effect of transition metals on Nafion 117 has been previously reported [80]. The displacement of protons from the ionomer by ionic contaminants through cationic exchange can also affect the structure and properties of the membrane [80]. These ionic species, particularly the cations, have high affinity for the sulfonic group in the PFSA membranes and can form sulfonate salts that can retract and squeeze out water from the membrane, thereby causing steric hindrance between similarly charged ions. The mobility of the contaminant ions, water dipoles, and sulfonated sites can lead to morphological changes in the membrane microstructure [81]. The cation hydrated radius, relative size of the ion and channel diameters, membrane microstructure, and the interactions between the mobile ions, water dipoles, and fixed sulfonated sites are the main factors influencing ionic contaminant transport. The transport properties of the various ionic contaminants have been reported in the literature [82,83]. Conductance and water uptake of different ionic contaminants have been estimated on the basis of absolute rate theory [84]. The gas flow field through the cell and the MEA components can be designed in such a way in order to help minimize the ionic contamination problems [80].

Because the ferric and ferrous iron contaminants are generated primarily from corrosion mainly on the anode side, their significance to fuel cell performance has been addressed in some detail. These ionic species move toward the cathode side and contaminate the membrane. This can accelerate the leaching of fluoride ions from the membrane itself depending on the concentration of the Fe^{2+} ions [75]. Multivalent cations, such as Fe^{3+} and Ni^{2+}, have a higher affinity for the sulfonic acid sites in the membrane than H$^+$ ions. The presence of these ions can cause shrinkage of the membrane [85]. Furthermore, if peroxide is generated during oxygen reduction at the cathode, it can combine with Fe^{2+} to generate undesired radicals and drastically reduce the lifetime of the PFSA membrane [86,87]. Thus, soluble iron can accelerate membrane degradation. The mechanism by which radicals are formed in this way is presented below [30–31]:

$$H_2O_2 + Fe^{2+} \rightarrow HO\bullet + OH^- + Fe^{3+} \tag{5.31}$$

$$Fe^{2+} + HO\bullet \rightarrow OH^- + Fe^{3+} \tag{5.32}$$

$$H_2O_2 + HO\bullet \rightarrow HO_2\bullet + H_2O \tag{5.33}$$

$$Fe^{2+} + HO_2\bullet \rightarrow HO_2^- + Fe^{3+} \tag{5.34}$$

$$Fe^{3+} + HO_2\bullet \rightarrow H^+ + O_2 + Fe^{2+} \tag{5.35}$$

A linear relationship between the soluble iron concentration and fluoride ion release within the cell has been observed [88]. In addition, higher levels of dissolved fluoride ions have been found on the anodic side where corrosion primarily occurs [75].

Apart from Fe^{2+}, other cations that have high affinity toward sulfonic acid groups can affect the proton conductivity and reduce the cell performance [39,89]. Even the presence of only a small amount of contaminants can displace the protons, significantly reducing the water flux through the membrane [79], and decreasing the membrane conductivity [16,90]. Salts of these contaminants can also precipitate on the carbon fibers and change their physical properties. This can enhance the carbon corrosion failure as well.

5.3.8 Si Contamination

Silicon gaskets can act as contaminant sources during continuous operation of a fuel cell [91]. These gaskets are firmly attached to the membranes and can affect the proton conductivity, pH at the ionomer/electrolyte, and reaction interfaces and catalyst activity if silicon dissolves from the gaskets and penetrates into the membrane [91]. During operation, dispersion of the platinum catalyst on both electrodes can change due to changes to the potential [92]. EPMA (electron probe microanalysis) of the MEA has detected significant amounts of silicon at the catalyst layers of both the anode and cathode as well as oxygen at the cathode in the form of platinum oxide [92]. On the basis of the distribution of the Si peak obtained in this study, it appears that Si dissolves into the PFSA membrane first and then migrates through the membrane, followed by penetration into the catalytic layer of the electrochemically stressed MEA. Thus, this EPMA analysis confirms that the catalysts in the cathode are degraded by the formation of platinum oxide and that the catalysts and PFSA membrane are contaminated by silicon dissolved from the gasket. This contamination can lead to chemical degradation.

5.3.9 Dilution Effects and Coolant Contamination

Reformate gas contains undesirable gases, such as CO, CO_2, and N_2 along with hydrogen. On the cathode side, the oxidant chiefly air contains 79% nitrogen. These gases not only affect the performance of fuel, but can also dilute the fuel. The dilution of the reactant species at both the cathode and anode significantly affects the performance of the fuel cell [93]. The diffusion of the diluted fuel at the anode side is slower in comparison to that of neat hydrogen. Higher concentrations of the inert gases with the hydrogen stream causes reversible losses in the performance of the cell [39]. Boillet et al. [93] studied the influence of nitrogen gases in either hydrogen or oxygen on the fuel cell impedance characteristics. Although the appearance of two impedance loops associated with hydrogen oxidation and oxygen reduction was not surprising, the observation of a third loop at low frequency was unexpected. The presence of the low frequency loop in the impedance spectra has been attributed mainly to the diffusion of oxygen in nitrogen [93]. The possibility of coolant contaminant effects and performance loss inside the cell during the long time operation of the fuel cell has been reported by Ahn et al. [91]. Their XRF (x-ray fluorescence) analysis of the coolant confirmed the existence of various inorganic materials causing performance loss of the fuel cell.

5.4 Conclusions

The durability of a PEM fuel cell is mainly affected by high temperature and low relative humidity, but the formation and/or presence of contamination can also have a significant impact. The development of innovative diagnostic methods, electrochemical methods, and analytical instruments will enable the local conditions inside the cell to be monitored during continuous operation. Attention must be continually focused on developing strategies to mitigate the degradation of cell components and performance. Despite great advances in chemically modified membranes to improve electrochemical stability, it is important to compare the durability and degradation of the newly developed membranes at different conditions. The prevention of crossover of hydrogen, ammonium ion, and other reformate gases is important for successful commercialization of fuel cells. The loss of ionomer at different relative humidities must be determined before steps to minimize it can be taken. The corrosion due to metal contaminants in the membrane and the effect on fluoride release depends on the location and transport of water inside the fuel cell. This can contribute to or indicate membrane failure. Therefore, measures and methods for contaminant reduction must be developed in order to minimize or eliminate its effects. The removal of external and internal contaminants, such as NO_x, SO_x, and CO_x, is necessary to increase the life of the fuel cell. Potentially useful strategies to mitigate these problems include the (1) development of CO- and sulfur-tolerant catalysts, (2) optimization of fuel cell operating conditions and catalyst loading, (3) use of thin membranes, (4) control of catalyst particle-size distribution and chemical composition to alleviate mass transfer problems, and (5) the use of efficient air filters or quartz fibers at the air inlet of the fuel cell to remove gases, notably CO, CO_2, NO, NO_2, SO_2, and dust.

The effects of contaminants on fuel cell lifetime performance under conditions closer to those of practical operations, such as start/stop cycles, cold startup, urban air quality, and dynamic loads, must also be investigated. A definite need exists to develop theoretical or empirical models that predict the extent and effects of fuel cell contamination. The development of these diagnostic tools or sensors for contamination would also help to increase the life of membranes and the stack as a whole. The contaminant levels that cause a critical loss in performance of the various membranes should be well-defined. Research on the development of contaminant filters, which can measure and remove the contaminants down to 0.001 ppm levels, which is potentially very useful, has recently been initiated [39].

Acknowledgments

The funding from the Natural Sciences and Engineering Research Council of Canada (NSERC) and University of Waterloo Institute of Nanotechnology (WIN) is greatly appreciated during the course of this work.

References

1. Carrette, L., Friedrich, K.A., Stimming, U. 2000. Fuel cells: principles, types, fuels and applications. *Chem Phys Chem*, 1, 162–193.
2. Dhanushkodi, S.R., Mahinpey, N., Srinivasan, A., Wilson, M. 2008a. Life cycle analysis of fuel cell technology. *Journal of Environmental Informatics*, 11, 36–44.
3. Kerres, J.A. 2001. Development of ionomer membranes for fuel cells. *Journal of Membrane Science*, 185, 3–27
4. Sambandam, S., Ramani, V. 2007. SPEEK/functionalized silica composite membranes for polymer electrolyte fuel cells. *Journal of Power Sources*, 170, 259–267.
5. Gebel, G., Aldebert, P., Pineri, M. 1993. Swelling study of perfluorosulphonated ionomer membranes. *Polymer*, 34, 333–339.
6. Gebel, G. 2000. Structural evolution of water swollen perfluorosulfonated ionomers from dry membrane to solution. *Polymer*, 41, 5829–5838.
7. Mauritz, K.A., Moore, R.B. 2004. State of understanding of Nafion. *Chemical Reviews*, 104, 4535–4586.
8. Vengatesan, S., Kim, H.-J., Kim, S.-K., Oh, I.-H., Lee, S.-Y., Cho, E., Ha, H.Y., Lim, T.-H. 2008. High dispersion platinum catalyst using mesoporous carbon support for fuel cells. *Electrochimica Acta*, 54, 856–861.
9. Vengatesan, S., Kim, H.J., Lee, S.Y., Cho, E., Ha, H.Y., Oh, I.H., Lim, T.H. 2007 Operation of a proton exchange membrane fuel cell under non-humidified conditions using a membrane-electrode assemblies with composite membrane and electrode. *Journal of Power Sources*, 167, 325–329.

10. Büchi, F.N., Gupta, B., Haas, O., Scherer, G.G. 1995. Study of radiation-grafted FEP-G-polystyrene membranes as polymer electrolytes in fuel cells. *Electrochimica Acta*, 40, 345–353.
11. Curtin, D.E., Lousenberg, R.D., Henry, T.J., Tangeman, P.C., Tisack, M.E. 2004. Advanced materials for improved PEMFC performance and life. *Journal of Power Sources*, 131, 41–48.
12. Fowler, M.W., Mann, R.F., Amphlett, J.C., Peppley, B.A., Roberge, P.R. 2003. Conceptual reliability analysis of PEM fuel cells, in *Handbook of Fuel Cells— Fundamentals, Technology and Applications*, chap. 56, ed. H.G. Wolf Vielstich, A. Lamm. Vol. 3. Chichester, U.K.: John Wiley & Sons.
13. Neburchilov, V., Martin, J., Wang, H., Zhang, J. 2007. A review of polymer electrolyte membranes for direct methanol fuel cells. *Journal of Power Sources*, 169, 221–238.
14. Wilkinson, D.P. and Pierre, J.S. 2003. *Handbook of Fuel Cells*, ed. A.L. Eds, W. Vielstich, H. Gasteiger, Vol. 3. Chichester, U.K.: John Wiley & Sons.
15. Baldwin, R., Pham, M., Leonida, A., McElroy, J., Nalette, T. 1990. Hydrogen-oxygen proton-exchange membrane fuel cells and electrolyzers. *Journal of Power Sources*, 29, 399–412.
16. Collier, A., Wang, H., Zi Yuan, X., Zhang, J., Wilkinson, D.P. 2006. Degradation of polymer electrolyte membranes. *International Journal of Hydrogen Energy*, 31, 1838–1854.
17. Contractor, A.Q., Lal, H. 1979. Two forms of chemisorbed sulfur on platinum and related studies. *Journal of Electroanalytical Chemistry*, 96, 175–181.
18. Mittal, V.O., Kunz, H.R., Fenton, J.M. 2007. Membrane degradation mechanisms in PEMFCs. *Journal of the Electrochemical Society*, 154, B652–B656.
19. Cleghorn, S.J.C., Kolde, J., Liu, W. 2003. Catalyst coated composites membranes, in *Handbook of Fuel Cells—Fundamentals, Technology and Applications*, eds. W. Vielstich, H.A. Gasteiger, and A. Lamm. Chichester, U.K.: John Wiley & Sons.
20. Huang, H., Fierro, C., Scherson, D., Yeager, E.B. 1991. *In situ* Fourier transform infrared spectroscopic study of carbon dioxide reduction on polycrystalline platinum in acid solutions. *Langmuir*, 7, 1154–1157.
21. Cleghorn, S.J.C., Mayfield, D.K., Moore, D.A., Moore, J.C., Rusch, G., Sherman, T.W., Sisofo, N.T., Beuscher, U. 2006. A polymer electrolyte fuel cell life test: 3 years of continuous operation. *Journal of Power Sources*, 158, 446–454.
22. Beuscher, U., Cleghorn, S.J.C., Johnson, W.B. 2005. Challenges for PEM fuel cell membranes. *International Journal of Energy Research*, 29, 1103–1112.
23. Liu, W., Ruth, K., Rusch, G. 2001. Membrane durability in PEM fuel cells. *Journal of New Materials for Electrochemical Systems*, 4, 227–232.
24. Lai, Y., Mittelsteadt, C.K., Gittleman, C.S., Dillard, D.A. 2005. Viscoelastic stress model and mechanical characterization of perfluorosulfonic acid (PFSA) polymer electrolyte membranes, *Proceedings of the Third International Conference on Fuel Cell Science, Engineering and Technology*, May 23–25, Ypsilanti, MI, p. 161.
25. Xie, G., Okada, T. 1995. Water transport behavior in Nafion 117 membranes. *Journal of the Electrochemical Society*, 142, 3057–3062.
26. LaConti, A.B., Hamdan, A., McDonald, R.C. 2003. Mechanisms of chemical degradation, in *Handbook of Fuel Cells: Fundamentals, Technology, and Applications*, W. Vielstich, A. Lamm, H. A. Gasteiger, Eds. Chichester, U.K.: John Wiley & Sons, Chap. 55.

27. Gittleman, G., Lai, Y.-H., Miller, D. 2005. Durability of perfluorosulfonic acid membranes for PEM fuel cells, Paper presented at the AIChE Meeting, November 4, Cincinnati, OH.
28. Han, J., Kim, I.-S., Choi, K.-S. 2002. High purity hydrogen generator for on-site hydrogen production. *International Journal of Hydrogen Energy*, 27, 1043–1047.
29. Lin, Y.-M., Rei, M.-H., 2000. Process development for generating high purity hydrogen by using supported palladium membrane reactor as steam reformer. *International Journal of Hydrogen Energy*, 25, 211–219.
30. He, C., Venkataraman, R., Kunz, H.R., Fenton, J.M. 1999. CO tolerant ternary anode catalyst development for fuel cell application. *Hazardous and Industrial Wastes—Proceedings of the Mid-Atlantic Industrial Waste Conference*, June 17, Boca Raton, FL: CRC press, pp. 663–668.
31. Hou, Z., Yi, B., Yu, H., Lin, Z., Zhang, H. 2003. CO tolerance electrocatalyst of PtRu-HxMeO$_3$/C (Me = W, Mo) made by composite support method. *Journal of Power Sources*, 123, 116–125.
32. Bellows, R.J., Marucchi-Soos, E., Reynolds, R.P. 1998. The mechanism of CO mitigation in proton exchange membrane fuel cells using dilute H$_2$O$_2$ in the anode humidifier. *Electrochemical and Solid-State Letters*, 1, 69–70.
33. Moore, J.M., Adcock, P.L., Lakeman, J.B., Mepsted, G.O. 2000. The effects of battlefield contaminants on PEMFC performance. *Journal of Power Sources*, 85, 254–260.
34. Dhanushkodi, S.R., Mahinpey, N., Wilson, M. 2008b. Kinetic and 2D reactor modeling for simulation of the catalytic reduction of NO$_x$ in the monolith honeycomb reactor. *Process Safety and Environmental Protection*, 86, 303–309.
35. Uribe, F.A., Gottesfeld, S., Zawodzinski, J.T.A. 2002. Effect of ammonia as potential fuel impurity on proton exchange membrane fuel cell performance. *Journal of the Electrochemical Society*, 149, A293–A296.
36. Rajalakshmi, N., Jayanth, T.T., Dhathathreyan, K.S. 2003. Effect of carbon dioxide and ammonia on polymer electrolyte membrane fuel cell stack performance. *Fuel Cells*, 3, 177–180.
37. Mohtadi, R., Lee, W.K., Van Zee, J.W. 2004. Assessing durability of cathodes exposed to common air impurities. *Journal of Power Sources*, 138, 216–225.
38. Yang, D., Ma, J., Xu, L., Wu, M., Wang, H. 2006. The effect of nitrogen oxides in air on the performance of proton exchange membrane fuel cell. *Electrochimica Acta*, 51, 4039–4044.
39. Cheng, X., Shi, Z., Glass, N., Zhang, L., Zhang, J., Song, D., Liu, Z.-S., Wang, H., Shen, J. 2007. A review of PEM hydrogen fuel cell contamination: Impacts, mechanisms, and mitigation. *Journal of Power Sources*, 165, 739–756.
40. Dicks, A.L. 1996. Hydrogen generation from natural gas for the fuel cell systems of tomorrow. *Journal of Power Sources*, 61, 113–124.
41. Castellanos, R.H., Ocampo, A.L., Moreira-Acosta, J., Sebastian, P.J. 2001. Synthesis and characterization of osmium carbonyl cluster compounds with molecular oxygen electroreduction capacity. *International Journal of Hydrogen Energy*, 26, 1301–1306.
42. Halseid, R., Vie, P.J.S., Tunold, R. 2006b. Effect of ammonia on the performance of polymer electrolyte membrane fuel cells. *Journal of Power Sources*, 154, 343–350.
43. Mohtadi, R., Lee, W.K., Cowan, S., Van Zee, J.W., Murthy, M. 2003. Effects of hydrogen sulfide on the performance of a PEMFC. *Electrochemical and Solid-State Letters*, 6, A272–A274.

44. Wilkinson, D.P. and St-Pierre, J. 2003. Durability, in *Handbook of Fuel Cells: Fundamentals, Technology, and Applications,* W. Vielstich, A. Lamm, H.A. Gasteiger, Eds. Chichester, U.K.: John Wiley & Sons.
45. Papageorgopoulos, D.C., de Bruijn, F.A. 2002. Examining a potential fuel cell poison. *Journal of the Electrochemical Society,* 149, A140–A145.
46. Michaelides, A., Hu, P. 2000. A first principles study of CH_3 dehydrogenation on Ni(111). *The Journal of Chemical Physics,* 112, 8120–8125.
47. Michaelides, A., Hu, P. 2001. Hydrogenation of S to H[sub 2]S on Pt(111): A first-principles study. *The Journal of Chemical Physics,* 115 (18), 8570–8574.
48. Zhang, J., Wang, H., Wilkinson, D.P., Song, D., Shen, J., Liu, Z.-S. 2005. Model for the contamination of fuel cell anode catalyst in the presence of fuel stream impurities. *Journal of Power Sources,* 147, 58–71.
49. Bonzel, H.P., Ku, R. 1973. Adsorbate interactions on a Pt(110) surface. I. Sulfur and carbon monoxide. *The Journal of Chemical Physics,* 58, 4617–4624.
50. Mathieu, M.-V., Primet, M. 1984. Sulfurization and regeneration of platinum. *Applied Catalysis,* 9, 361–370.
51. Kinoshita, S.P. 1977. Preparation and characterization of highly dispersed electrocatalytic materials, in *Modern Aspects of Electrochemistry,* J.O.M. Bockris and B.E. Conway, Eds., New York: Plenum Press.
52. Shi, W., Yi, B., Hou, M., Shao, Z. 2007. The effect of H_2S and CO mixtures on PEMFC performance. *International Journal of Hydrogen Energy,* 32, 4412–4417.
53. Granovskii, M., Dincer, I., Rosen, M.A. 2007. Greenhouse gas emissions reduction by use of wind and solar energies for hydrogen and electricity production: Economic factors. *International Journal of Hydrogen Energy,* 32, 927–931.
54. García, G., Silva-Chong, J.A., Guillén-Villafuerte, O., Rodríguez, J.L., González, E.R., Pastor, E. 2006. CO tolerant catalysts for PEM fuel cells: Spectroelectrochemical studies. *Catalysis Today,* 116, 415–421.
55. Arévalo, M.C., Gomis-Bas, C., Hahn, F., Beden, B., Arévalo, A., Arvia, A.J. 1994. A contribution to the mechanism of "reduced" CO_2 adsorbates electro-oxidation from combined spectroelectrochemical and voltammetric data. *Electrochimica Acta,* 39, 793–799.
56. Taguchi, S., Ohmori, T., Aramata, A., Enyo, M. 1994. Adsorption of CO and CO_2 on polycrystalline Pt at a potential in the hydrogen adsorption region in phosphate buffer solution: An *in situ* Fourier transform IR study. *Journal of Electroanalytical Chemistry,* 369, 199–205.
57. Janssen, G.J.M. 2004. Modelling study of CO_2 poisoning on PEMFC anodes. *Journal of Power Sources,* 136, 45–54.
58. Gilman, S. 1964. The mechanism of electrochemical oxidation of carbon monoxide and methanol on platinum. II. The reactant-pair mechanism for electrochemical oxidation of carbon monoxide and methanol. *The Journal of Physical Chemistry,* 68, 70–80.
59. Iwasita, T., Nart, F.C., Lopez, B., Vielstich, W. 1992. On the study of adsorbed species at platinum from methanol, formic acid and reduced carbon dioxide via *in situ* FT-IR spectroscopy. *Electrochimica Acta,* 37, 2361–2367.
60. Gu, T., Lee, W.K., Van Zee, J.W., Murthy, M. 2004. Effect of reformate components on PEMFC performance. *Journal of the Electrochemical Society,* 151, A2100–A2105.

61. Divisek, J., Oetjen, H.F., Peinecke, V., Schmidt, V.M., Stimming, U. 1998. Components for PEM fuel cell systems using hydrogen and CO containing fuels. *Electrochimica Acta*, 43, 3811–3815.
62. Baschuk, J.J., Li, X. 2003. Mathematical model of a PEM fuel cell incorporating CO poisoning and O_2 (air) bleeding. *International Journal of Global Energy Issues*, 20, 245–276.
63. Igarashi, H., Fujino, T., Watanabe, M. 1995. Hydrogen electro-oxidation on platinum catalysts in the presence of trace carbon monoxide. *Journal of Electroanalytical Chemistry*, 391, 119–123.
64. Cheng, X., Chen, L., Peng, C., Chen, Z., Zhang, Y., Fan, Q. 2004. Catalyst microstructure examination of PEMFC membrane electrode assemblies vs. time. *Journal of the Electrochemical Society*, 151, A48–A52.
65. Li, Q., He, R., Gao, J.-A., Jensen, J.O., Bjerrum, N.J. 2003. The CO poisoning effect in PEMFCs operational at temperatures up to 200°C. *Journal of the Electrochemical Society*, 150, A1599–A1605.
66. Borup, R., Inbody, M., Tafoya, J., Semelsberger, T., Perry, L. 2002. Durability studies: Gasoline/reformate durability. N.L.R.D. Meeting (Ed.).
67. Soto, H.J. 2005. The effect of ammonia as an anode impurity on the performance of PEM fuel cells. PhD disser. University of South Carolina, Columbia.
68. Halseid, R. 2004. Ammonia as hydrogen carrier—Effects of ammonia on polymer electrolyte membrane fuel cells. Ph.D. thesis, NTNU Trondheim, Norway.
69. Halseid, R., Vie, P.J.S., Tunold, R. 2004. Influence of ammonium on conductivity and water content of Nafion 117 membranes. *Journal of the Electrochemical Society*, 151, A381–A388.
70. Halseid, R., Vie, P.J.S., Tunold, R. 2006b. Effect of ammonia on the performance of polymer electrolyte membrane fuel cells. *Journal of Power Sources*, 154, 343–350.
71. Goodwin, J.G.J., Zhang, J., Hongsirikarn, K., Liu, Z., Rhodes, W., Colon-Mercado, H., Greenway, S. 2007. Effects of impurities on fuel cell performance and durability. Oral presentation, Washington, D.C.
72. Soto, H.J., Lee, W.-K., Van Zee, J.W., Murthy, M. 2003. Effect of transient ammonia concentrations on PEMFC performance. *Electrochemical and Solid-State Letters*, 6, A133–A135.
73. Halseid, R., Bystron, T., Tunold, R. 2006a. Oxygen reduction on platinum in aqueous sulphuric acid in the presence of ammonium. *Electrochimica Acta*, 51, 2737–2742.
74. Halseid, R., Wainright, J.S., Savinell, R.F., Tunold, R. 2007. Oxidation of ammonium on platinum in acidic solutions. *Journal of the Electrochemical Society* 154, B263–B270.
75. Pozio, A., Silva, R.F., De Francesco, M., Giorgi, L. 2003. Nafion degradation in PEFCs from end plate iron contamination. *Electrochimica Acta*, 48, 1543–1549.
76. Mazza, A.G., Frank, D.G., and Joos, N.I. Corrosion resistant end plate and method for producing same, Patent Application US 2004/0131917 A1.
77. St-Pierre, J., Wilkinson, D.P., Knights, S., Bos, M.L. 2000. Relationships between water management, contamination and lifetime degradation in PEFC. *Journal of New Materials for Electrochemical Systems*, 3, 99–106.
78. Okada, T., Xie, G., Tanabe, Y. 1996. Theory of water management at the anode side of polymer electrolyte fuel cell membranes. *Journal of Electroanalytical Chemistry*, 413, 49–65.

79. Okada, T., et al. 1998. Transport and equilibrium properties of Nafion® membranes with H^+ and Na^+ ions. *Journal of Electroanalytical Chemistry*, 442 (1–2), 137–145.
80. Shi, M., Anson, F.C. 1997. Dehydration of protonated Nafion® coatings induced by cation exchange and monitored by quartz crystal microgravimetry. *Journal of Electroanalytical Chemistry*, 425, 117–123.
81. Kundu, S., Fowler, M.W., Simon, L.C., Grot, S. 2006. Morphological features (defects) in fuel cell membrane electrode assemblies. *Journal of Power Sources*, 157, 650–656.
82. Yeager, H.L., Steck, A. 1981. Cation and water diffusion in Nafion ion exchange membranes: influence of polymer structure. *Journal of the Electrochemical Society*, 128, 1880–1884.
83. Samec, Z., Trojanek, A., Langmaier, J., Samcova, E. 1997. Diffusion coefficients of alkali metal cations in Nafion® from ion-exchange measurements. *Journal of the Electrochemical Society*, 144, 4236–4242.
84. Pourcelly, G., Oikonomou, A., Gavach, C., Hurwitz, H.D. 1990. Influence of the water content on the kinetics of counter-ion transport in perfluorosulphonic membranes. *Journal of Electroanalytical Chemistry*, 287, 43–59.
85. Okada, T., Ayato, Y., Yuasa, M., Sekine, I. 1999. The effect of impurity cations on the transport characteristics of perfluorosulfonated ionomer membranes. *Journal of Physical Chemistry B*, 103, 3315–3322.
86. Balanosky, E., Fernandez, J., Kiwi, J., Lopez, A., 1999. Degradation of membrane concentrates of the textile industry by Fenton like reactions in iron-free solutions at biocompatible pH values (pH [approximate] 7–8). *Water Science and Technology*, 40, 417–424.
87. Maletzky, P., Bauer, R. Lahnsteiner, J., Pouresmael, B. 1999. Immobilisation of iron ions on Nafion® and its applicability to the photo-fenton method. *Chemosphere* 38(10), 2315–2325.
88. Kundu, S., Simon, L.C. Fowler, M.W. 2008. Comparison of two accelerated Nafion(TM) degradation experiments. *Polymer Degradation and Stability*, 93(1), 214–224.
89. Inaba, M., Kinumoto, T., Kiriake, M., Umebayashi, R., Tasaka, A., Ogumi, Z. 2006. Gas crossover and membrane degradation in polymer electrolyte fuel cells. *Electrochimica Acta*, 51, 5746–5753.
90. Kelly, M.J., Fafilek, G., Besenhard, J.O., Kronberger, H., Nauer, G.E. 2005. Contaminant absorption and conductivity in polymer electrolyte membranes. *Journal of Power Sources*, 145, 249–252.
91. Ahn, S.Y., Shin, S.J., Ha, H.Y., Hong, S.A., Lee, Y.C., Lim, T.W., Oh, I.H. 2002. Performance and lifetime analysis of the kW-class PEMFC stack. *Journal of Power Sources*, 106, 295–303.
92. Gulzow, H.S., Wagner, N., Lorenz, M., Schneider, A., Schulze, M. 2000. Degradation of PEFC components. *Fuel Cell Seminar Abstracts*, Portland, p. 156.
93. Boillot, M., Bonnet, C., Jatroudakis, N., Carre, P., Didierjean, S., Lapicque, F. 2006. Effect of gas dilution on PEM fuel cell performance and impedance response. *Fuel Cells*, 6, 31–37.
94. Reeve, R.W., Christensen, P.A., Dickinson, A.J., Hamnett, A., Scott, K. 2000. Methanol-tolerant oxygen reduction catalysts based on transition metal sulfides and their application to the study of methanol permeation. *Electrochimica Acta*, 45, 4237–4250.

6

Cathode Contamination Modeling

Zheng Shi, Datong Song, Hui Li, Jianlu Zhang,
Zhong-Sheng Liu, and Jiujun Zhang

CONTENTS

6.1 Introduction

Fuel cell contamination is a serious issue, particularly for proton electrolyte membrane (PEM) fuel cell operation and applications. At the anode, the hydrogen fuel contains impurities, such as CO, H_2S, NH_3, organic sulfur carbon, and carbon–hydrogen compounds. These impurities originate primarily from hydrogen production processes, namely the reformation of hydrocarbons. For the cathode, at present the most practical and economical supply of oxygen is air. As a result, air quality directly impacts PEM fuel cell performance. The most common air pollutants are nitrogen oxides (NO_x, including NO and NO_2), sulfur oxides (SO_x, including SO_2 and SO_3), carbon oxides (CO_x, including CO and CO_2), ammonia, ozone, and volatile organic chemical (VOC) species (such as benzoic compounds) [1–3]. The major sources of these contaminants are automotive vehicle exhaust, industrial emissions, and agricultural activities. In an urban environment (e.g., Vancouver, Canada), the NO_x level can be as high as 100 ppb, which could cause a voltage drop of approximately 25 mV in single fuel cell performance, as shown in Figure 6.1 [4]. In a battlefield, the SO_x may be as high as 0.5 ppm, which could induce power failure in a fuel cell stack power supply [5].

For fuel cell technology advancement and commercialization, fuel and air quality specifications must be appropriately developed. This requires the ability to account for the effects of fuel and air quality on fuel cell performance, and to establish limiting conditions. As an integral part of contamination research, contamination modeling is a tool to predict the impact of fuel and air quality on fuel cell performance, and to project experimental tests and effects. It also contributes to the fundamental understanding of contamination mechanisms. With respect to fuel-side contamination modeling, numerous studies have been conducted, especially for CO contamination [6–17]. Models have also been developed for anode H_2S contamination [18,19]. However, the work on air-side contamination modeling is limited [20].

Contamination in a PEM fuel cell directly affects the kinetics, conductivity, and mass transport properties of the cell. In particular, the blocking of electrocatalysts by impurity adsorption can drastically reduce the effective surface area of the catalysts and, thus, slow down the kinetics and hinder cell performance. This chapter is devoted to cathode contamination modeling,

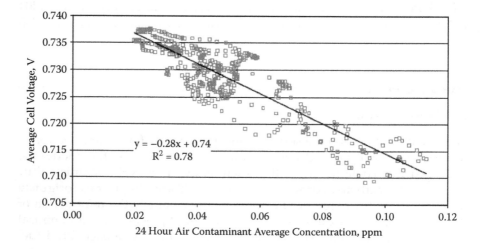

FIGURE 6.1

Average cell voltage as a function of 24-h average NO_2 concentration (Vancouver, Canada). (From S. Knights, N. Jia, C. Chuy, J. Zhang, *2005 Fuel Cell Seminar: Fuel Cell Progress, Challenges and Markets* abstract. P121, Palm Springs, CA, 2005. With permission.)

particularly kinetic modeling. Fuel cell contamination modeling generally involves the following steps: (1) understanding the contamination mechanism, which includes the electrochemical reactions and surface reactions; (2) developing a contamination kinetic model; and (3) validating the kinetic model. In this chapter, we will review cathode oxygen reduction mechanisms and oxygen reduction kinetic models, present our general cathode contamination model and its application to feed stream toluene contamination, discuss the effects of toluene contamination on fuel cell performance, and further review other published cathode contamination models, then conclude with a summary of cathode contamination research.

6.2 Oxygen Reduction Mechanism

The oxygen reduction reaction (ORR) at the PEM fuel cell cathode is a multielectron, multistep reaction with a sluggish kinetics; thus, a catalyst is generally required to accelerate the reaction. At present, platinum (Pt)-based catalysts are the most practical catalysts for the ORR in PEM fuel cells. The mechanism of the Pt-catalyzed ORR has been an active research area for about the past 40 years [21–24]. Yet, despite numerous studies, the detailed mechanism remains elusive. Figure 6.2 illustrates the simplified mechanism [24]. On Pt, the oxygen reduction reaction can proceed along several pathways; for example, a "direct" four-electron reduction to water, a two-electron pathway to hydrogen peroxide, and a "series" pathway with two- and four-electron reduction to water.

For the ORR on a Pt surface, the electrochemical polarization curves (current–potential curves), recorded in both acidic and alkaline solutions, normally give two Tafel regions. At low current densities (high potential), a Tafel slope of 60 mV/dec can be observed. At high current densities (low potential), a Tafel slope of 120 mV/dec is obtained [25]. The

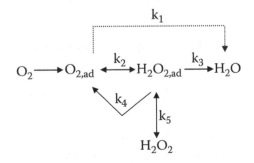

FIGURE 6.2
Simplified oxygen reduction reaction mechanism on Pt.

difference in Tafel slope indicates that the mechanism differs at different potential ranges. In the high-potential range, Pt is oxidized to PtO and the rate-determining step (rds) is a pseudo two-electron (2-e) procedure. In the low potential range (a pure Pt surface), the first electron transfer is the rate-determining step [26]. Damjanovic et al. [25,27,28] proposed that the charge transfer to the adsorbed oxygen molecule, with or without simultaneous proton transfer, should be the rate-determining step (the associative model):

$$M + O_2 \rightarrow M - O_2 \tag{6.1}$$

$$M - O_2 + H^+ + e^- \rightarrow M - O_2H \text{ (rds)} \tag{6.2}$$

$$M - O_2H + 3H^+ + 3e^- \rightarrow 2H_2O + M \tag{6.3}$$

where M represents a metal, such as Pt. Yeager [29] proposed that the first step should involve dissociative chemisorption of the O_2 molecule, which could occur simultaneously with the charge transfer (the dissociative model):

$$Pt + \frac{1}{2}O_2 + e^- \rightarrow Pt - O^- \tag{6.4}$$

Recently, many theoretical explorations of the ORR mechanism using quantum mechanics methods have been reported, as can be found in review articles [30,31]. These studies have provided insights and information regarding each elementary step, such as activation energies, reaction energies, and reversible potentials. Anderson et al. [32,33] investigated the activation barrier for each of the following electron transfer steps:

$$Pt - O_2 + H^+(aq) + e^- \rightarrow Pt - OOH \tag{6.5}$$

$$Pt - OOH + H^+(aq) + e^- \rightarrow Pt - (OHOH) \tag{6.6}$$

$$Pt - OHOH + H^+(aq) + e^- \rightarrow Pt - OH + H_2O \tag{6.7}$$

$$Pt - OH + H^+(aq) + e^- \rightarrow Pt - OH_2 \tag{6.8}$$

Their result indicated that the first electron transfer had a smaller barrier than that of oxygen dissociation and, thus, the electron would transfer first and the OOH would easily dissociate once formed.

Nørskov et al. [34,35] studied the thermodynamic properties of the following dissociative and associative mechanisms:

Dissociative mechanism:

$$\frac{1}{2}O_2 + * \rightarrow O^* \qquad (6.9)$$

$$O^* + H^+ + e^- \rightarrow HO^* \qquad (6.10)$$

$$HO^* + H^+ + e^- \rightarrow H_2O + * \qquad (6.11)$$

where * denotes a site on the surface.

Associative mechanism:

$$O_2 + * \rightarrow O_2^* \qquad (6.12)$$

$$O_2 + (H^+ + e^-) \rightarrow HO_2^* \qquad (6.13)$$

$$HO_2^* + (H^+ + e^-) \rightarrow H_2O + O^* \qquad (6.14)$$

$$O^* + (H^+ + e^-) \rightarrow HO^* \qquad (6.15)$$

$$HO^* + (H^+ + e^-) \rightarrow H_2O + * \qquad (6.16)$$

The theoretical results indicated that both mechanisms were possible and were controlled by electrode potential. At high electrode potential, the barrier for O_2 dissociation would increase, thus, the dissociative mechanism was less favorable. In another study, Jacob [36] also demonstrated that both associative and dissociative mechanisms might occur simultaneously, with a higher ratio for the associative pathway.

Besides the oxygen reduction reaction, other chemical, electrochemical, and side reactions occur at the Pt cathode, as formulated by Walch et al. [37]. These reactions are

Chemical reaction:

$$O_{2(ads)} \leftrightarrow O_{(ads)} + O_{(ads)} \tag{6.17}$$

$$O_2H_{(ads)} \leftrightarrow O_{(ads)} + OH_{(ads)} \tag{6.18}$$

$$H_2O_{(ads)} \leftrightarrow H_{(ads)} + HO_{(ads)} \tag{6.19}$$

$$OH_{(ads)} + OH_{(ads)} \leftrightarrow O_{(ads)} + H_2O_{(ads)} \tag{6.20}$$

$$OH_{(ads)} + O_{(ads)} \leftrightarrow O_{(ads)} + OH_{(ads)} \tag{6.21}$$

Electrochemical reaction:

$$O_{2(ads)} + H_{(aq)}^+ + e^- \leftrightarrow O_2H_{(ads)} \tag{6.22}$$

$$OH_{(ads)} + H_{(aq)}^+ + e^- \leftrightarrow H_2O_{(ads)} \tag{6.23}$$

$$O_{(ads)} + H_{(aq)}^+ + e^- \leftrightarrow OH_{(ads)} \tag{6.24}$$

$$H_2O_{(ads)} \leftrightarrow O_{(ads)} + 2H_{(aq)}^+ + 2e^- \tag{6.25}$$

Side reaction:

$$Pt \leftrightarrow Pt_{(aq)}^{2+} + 2e^- \tag{6.26}$$

$$Pt_{(aq)}^{2+} + H_2O \leftrightarrow PtO + 2H_{(aq)}^+ \tag{6.27}$$

$$Pt_{(aq)}^{2+} + H_2 \leftrightarrow Pt + 2H_{(aq)}^+ \tag{6.28}$$

Clearly, PEM fuel cell cathode reactions are very complex.

6.3 Review of Oxygen Reduction Kinetic Models

In attempts to simulate the experimentally observed oxygen reduction reaction, several kinetic models have been reported in the literature, based on various proposed ORR mechanisms [38–41]. These models employed either the associative or the dissociative model. The former was used by Antoine et al. [38] in their simulation of a polarization curve and impedance spectra, and by Du et al. [40] in a model-based electrochemical impedance spectroscopy study. Antoine et al. considered the following reaction steps:

$$O_2 + H^+ + e^- \rightarrow (O_2H)_{ads} \qquad (6.29)$$

$$(O_2H)_{ads} + H_2O \leftrightarrow 3(OH)_{ads} \qquad (6.30)$$

$$(OH)_{ads} + H^+ + e^- \rightarrow H_2O \qquad (6.31)$$

Kuhn and Scherer [39], on the other hand, applied the dissociative model to evaluate single electrode impedance measurements. For low current densities, they considered the following pathway:

$$O_2 \leftrightarrow 2O_{(ads)} \qquad (6.32)$$

$$O_{(ads)} + e^- \rightarrow O^-_{(ads)} \qquad (6.33)$$

$$O^-_{(ads)} + H^+ \rightarrow OH_{(ads)} \qquad (6.34)$$

$$OH_{(ads)} + H^+ + e^- \leftrightarrow H_2O \qquad (6.35)$$

Zhdanov and Kasemo [41] analyzed ORR kinetics corresponding to both associative and dissociative mechanisms. Their study evaluated the relationship between the reaction order and O_2 and O coverage, based on the two different reaction mechanisms and under different adsorption assumptions (Langmuir and lateral interaction). The study concluded that if the reaction steps were described by using Langmuir equations, the kinetic models based on both mechanisms could predict that the reaction order with respect to oxygen pressure was close to 1 only in situations when the O coverage was low. With O-O lateral interactions, the model based on the associative mechanism was able to predict the first-order kinetics for a wide range of pressures, even if the O coverage was appreciable. Thus, the associative mechanism provided better agreement between simulations, experiments, and theoretical calculations.

6.4 Cathode Contamination Model

We have developed a general cathode contamination model [20] for PEM fuel cells, which is presented in this section.

6.4.1 Cathode Contamination Reaction Mechanism

Our model is based on the associative mechanism, which has substantial experimental and theoretical support, and assumes that the first electron transfer is the rate-determining step. The following simplified reactions are involved:

$$Pt + O_2 \underset{k_{1b}}{\overset{k_{1f}}{\rightleftharpoons}} Pt - O_2 \tag{6.36}$$

$$Pt - O_2 + H^+ + e^- \underset{k_{2b}\exp((1-\alpha_2)n_2 F\eta_c/(RT))}{\overset{k_{2f}\exp(-\alpha_2 n_2 F\eta_c/(RT))}{\rightleftharpoons}} Pt - O_2 H \tag{6.37}$$

$$Pt - O_2 H + 3H^+ + 3e^- \underset{k_{3b}\exp((1-\alpha_3)n_3 F\eta_c/(RT))}{\overset{k_{3f}\exp(-\alpha_3 n_3 F\eta_c/(RT))}{\rightleftharpoons}} 2H_2O + Pt \tag{6.38}$$

where Pt is the active sites for oxygen adsorption, and may contain more than one Pt site; k_{jf} and k_{jb} are the forward and backward reaction rates of the $(35+j)$-th reaction ($j = 1, 2, 3$); α_j and n_j are the electron transfer coefficient and electron transfer number of the $(35+j)$-th electrochemical half-reaction ($j = 2, 3$); η_c is the cathode overpotential; and F, R, and T are Faraday's constant, the universal gas constant, and cell temperature, respectively.

The advantage of the above simplified ORR mechanism is that the coverage of intermediates is represented by $Pt - O_2H$ only. Therefore, the model variables and equations are significantly reduced.

Note that in this model, we only consider the scenario in which the first electron transfer (reaction (6.37)) is the rate-determining step, which is the general case. Under different potentials and operating conditions, it is possible that another step, either chemical or electrochemical, is the rate-determining step. Therefore, different cases can be developed to study other scenarios.

For a general contaminant P, the following possible electrode and surface reactions were considered:

$$nPt + P \underset{k_{p1b}}{\overset{k_{p1f}}{\rightleftharpoons}} Pt_n - P \tag{6.39}$$

$$P + nPt - O_2 \underset{k_{p2b}}{\overset{k_{p2f}}{\rightleftharpoons}} Pt_n - P + nO_2 \tag{6.40}$$

$$P + nPt - O_2H + 3ne^- + 3nH^+ \underset{k_{p3b}\exp((1-\alpha_{p3})n_{p3}F\eta_c/(RT))}{\overset{k_{p3f}\exp(-\alpha_{p3}n_{p3}F\eta_c/(RT))}{\rightleftharpoons}} 2nH_2O + Pt_n - P \tag{6.41}$$

$$Pt_n - P + mH_2O \underset{k_{p4b}\exp(-(1-\alpha_{p4})n_{p4}F\eta_c/(RT))}{\overset{k_{p4f}\exp(\alpha_{p4}n_{p4}F\eta_c/(RT))}{\rightleftharpoons}} Pt_n - P' + n_{p4}H^+ + n_{p4}e^- \tag{6.42}$$

$$Pt_n - P + lH_2O \underset{k_{p5b}\exp(-(1-\alpha_{p5})n_{p5}F\eta_c/(RT))}{\overset{k_{p5f}\exp(\alpha_{p5}n_{p5}F\eta_c/(RT))}{\rightleftharpoons}} nPt + P'' + n_{p5}H^+ + n_{p5}e^- \tag{6.43}$$

where n is the number of Pt sites occupied by contaminant P, k_{pjf} and k_{pjb} are the forward and backward reaction rates of the (38+j)-th reaction ($j = 1, 2, ..., 5$), α_{pj} and n_{pj} are the electron transfer coefficient and electron transfer number of the (38+j)-th reaction ($j = 3, 4, 5$), and P' and P" are the oxidized forms of contaminant P.

6.4.2 Cathode Contamination Model

Based on the proposed ORR and contamination mechanisms, the Pt surface consists of the following five species: Pt, $Pt-O_2$, $Pt-O_2H$, Pt_n-P, and Pt_n-P'. Thus, for the cathode catalyst surface coverage, one has

$$\theta_{Pt} + \theta_{Pt-O_2} + \theta_{Pt-O_2H} + \theta_{Pt_n-P} + \theta_{Pt_n-P'} = 1 \tag{6.44}$$

where θ_{Pt}, θ_{Pt-O_2}, θ_{Pt-O_2H}, θ_{Pt_n-P}, and $\theta_{Pt_n-P'}$, are the surface coverage of the corresponding five species, respectively.

Assuming that reaction (6.36) is fast and always at its equilibrium, one has

$$k_{1f}C_{O_2}\theta_{Pt}\Gamma = k_{1b}\theta_{Pt-O_2}\Gamma \tag{6.45}$$

where Γ is the total surface and C_{O_2} is the oxygen concentration in the cathode catalyst layer (CCL). Solving for θ_{Pt-O_2} leads to

$$\theta_{Pt-O_2} = \frac{k_{1f}C_{O_2}}{k_{1b}}\theta_{Pt} \tag{6.46}$$

Assuming reaction (6.38) is always at its equilibrium,

$$k_{3f}\theta_{Pt}C_{H^+}\Gamma C_{H^+}^3 \exp\left(-\frac{\alpha_3 n_3 F\eta_c}{RT}\right) = k_{3h}\theta_{Pt}\Gamma\exp\left(\frac{(1-\alpha_3)n_3 F\eta_c}{RT}\right) \tag{6.47}$$

where C_{H^+} is the proton concentration in the CCL. θ_{Pt-O_2H} can be obtained as

$$\theta_{Pt-O_2H} = \frac{k_{3b}}{k_{3f}C_{H^+}^3}\exp\left(\frac{n_3 F\eta_c}{RT}\right)\theta_{Pt} \tag{6.48}$$

Substituting equation (6.46) and equation (6.48) into equation (6.44) and solving the resultant equation for θ_{Pt} yields

$$\theta_{Pt} = A\left(1-\theta_{Pt_n-P}-\theta_{Pt_n-P'}\right) \tag{6.49}$$

where

$$A = \frac{1}{1+\dfrac{k_{1f}C_{O_2}}{k_{1b}}+\dfrac{k_{3b}}{k_{3f}C_{H^+}^3}\exp\left(\dfrac{n_3 F\eta_c}{RT}\right)} \tag{6.50}$$

Furthermore, the surface coverage of species $Pt-O_2$ and $Pt-O_2H$ can be expressed as functions of the surface coverage of the contaminant-related species,

$$\theta_{Pt-O_2} = B\left(1-\theta_{Pt_n-P}-\theta_{Pt_n-P'}\right) \tag{6.51}$$

and

$$\theta_{Pt-O_2H} = C\left(1-\theta_{Pt_n-P}-\theta_{Pt_n-P'}\right) \tag{6.52}$$

where

$$B = \frac{k_{1f}}{k_{1b}} C_{O_2} A \tag{6.53}$$

and

$$C = \frac{k_{3b}}{k_{3f}C_{H^+}^3} \exp\left(\frac{n_3 F \eta_c}{RT}\right) A \tag{6.54}$$

If there is no contaminant present in the CCL, the surface coverage of Pt, $Pt-O_2$, and $Pt-O_2H$ are only determined by the oxygen reduction reactions (equation (6.36) to equation (6.38)) and are given as

$$\theta_{Pt}^0 = \frac{1}{1+\dfrac{k_{1f}C_{O_2}}{k_{1b}}+\dfrac{k_{3b}}{k_{3f}C_{H^+}^3}\exp\left(\dfrac{n_3 F \eta_c}{RT}\right)} \tag{6.55}$$

$$\theta_{Pt-O_2}^0 = \frac{k_{1f}}{k_{1b}} C_{O_2} \theta_{Pt}^0 \tag{6.56}$$

$$\theta_{Pt-O_2H}^0 = \frac{k_{3b}}{k_{3f}C_{H^+}^3} \exp\left(\frac{n_3 F \eta_c}{RT}\right) \theta_{Pt}^0 \tag{6.57}$$

where θ_{Pt}^0, $\theta_{Pt-O_2}^0$, and $\theta_{Pt-O_2H}^0$ are the surface coverage of the corresponding species without contaminant.

Based on the proposed reaction mechanism, the surface coverage of species $Pt_n - P$ and $Pt_n - P'$ are governed by the following differential equations:

$$
\frac{d\theta_{Pt_n-P}}{dt} = k_{p1f}C_P\theta_{Pt}^n - k_{p1b}\theta_{Pt_n-P} + k_{p2f}C_P\theta_{Pt_n-O_2}^n - k_{p2b}C_{O_2}^n\theta_{Pt_n-P}
$$

$$
+k_{p3f}C_P\theta_{Pt-O_2H}^n C_{H^+}^3 \exp\left(-\frac{\alpha_{p3}n_{p3}F\eta_c}{RT}\right) - k_{p3b}\exp\left(\frac{(1-\alpha_{p3})n_{p3}F\eta_c}{RT}\right)\theta_{Pt_n-P}
$$

$$
-k_{p4f}\exp\left(\frac{\alpha_{p4}n_{p4}F\eta_c}{RT}\right)\theta_{Pt_n-P} + k_{p4b}C_{H^+}^{n_{p4}}\exp\left(-\frac{(1-\alpha_{p4})n_{p4}F\eta_c}{RT}\right)\theta_{Pt_n-P'}
$$

$$
-k_{p5f}\exp\left(\frac{\alpha_{p5}n_{p5}F\eta_c}{RT}\right)\theta_{Pt_n-P} + k_{p5b}C_{P'}C_{H^+}^{n_{p5}}\exp\left(-\frac{(1-\alpha_{p5})n_{p5}F\eta_c}{RT}\right)
$$

$$
= \left\{ k_{p1b} + k_{p2b}C_{O_2}^n + k_{p3b}\exp\left(\frac{(1-\alpha_{p3})n_{p3}F\eta_c}{RT}\right) \right. \tag{6.58}
$$

$$
\left. +k_{p4f}\exp\left(\frac{\alpha_{p4}n_{p4}F\eta_c}{RT}\right) + k_{p5f}\exp\left(\frac{\alpha_{p5}n_{p5}F\eta_c}{RT}\right) \right\}\theta_{Pt_n-P}
$$

$$
+\left\{ k_{p1f}C_P A^n + k_{p2f}C_P B^n + k_{p3f}C_P C_{H^+}^3 C^n \exp\left(-\frac{\alpha_{p3}n_{p3}F\eta_c}{RT}\right) \right.
$$

$$
\left. +k_{p5b}C_{P'}C_{H^+}^{n_{p5}} A^n \exp\left(-\frac{(1-\alpha_{p5})n_{p5}F\eta_c}{RT}\right) \right\}(1-\theta_{Pt_n-P} - \theta_{Pt_n-P'})^n
$$

$$
+k_{p4b}C_{H^+}^{n_{p4}}\exp\left(-\frac{(1-\alpha_{p4})n_{p4}F\eta_c}{RT}\right)\theta_{Pt_n-P'}
$$

and

$$
\frac{d\theta_{Pt_n-P'}}{dt} = k_{p4f}\exp\left(\frac{\alpha_{p4}n_{p4}F\eta_c}{RT}\right)\theta_{Pt_n-P} - k_{p4b}C_{H^+}^{n_{p4}}\exp\left(-\frac{(1-\alpha_{p4})n_{p4}F\eta_c}{RT}\right)\theta_{Pt_n-P'}
$$

$$
\tag{6.59}
$$

with the following initial conditions:

$$\theta_{Pt_n-P}\big|_{t=0} = 0 \tag{6.60}$$

and

$$\theta_{Pt_n-P'}\big|_{t=0} = 0 \tag{6.61}$$

where t denotes time, and C_P, and $C_{P'}$, are the concentrations of species P and P', respectively.

The cathode current density is determined by both forward and backward rates of reaction (6.37) [42], and calculated by

$$I_c = n_{O_2} F \gamma_c \left\{ k_{2f} \theta_{Pt\text{-}O_2} C_{H^+} \exp\left(-\frac{\alpha_2 n_2 F \eta_c}{RT}\right) - k_{2b} \theta_{Pt\text{-}O_2H} \exp\left(\frac{(1-\alpha_2)n_2 F \eta_c}{RT}\right) \right\} \tag{6.62}$$

where n_{O_2} is the electron transfer number of the ORR and γ_c is the ratio of the active surface to the geometric surface of the CCL.

Applying the condition $I_c = \eta_c = 0$ in equation (6.62) leads to the following relation between the forward and backward rates of reaction (6.37) when there is no contaminant:

$$k_{2b} = k_{2f} \frac{\theta^0_{Pt\text{-}O_2}}{\theta^0_{Pt-O_2H}} C_{H^+} \tag{6.63}$$

Substituting equation (6.51) into equation (6.57), and equation (6.63) into equation (6.62) results in

$$I_c = n_{O_2} F \gamma_c k_{2f} \frac{k_{1f}}{k_{1b}} C_{O_2} C_{H^+} A$$

$$\exp\left(-\frac{\alpha_2 n_2 F \eta_c}{RT}\right) \left\{1 - \exp\left(\frac{n_2 F \eta_c}{RT}\right)\right\}(1 - \theta_{Pt_n-P} - \theta_{Pt_n-P'}) \tag{6.64}$$

The cell voltage of a PEM fuel cell can be estimated as

$$V_{cell} = V^0 - \eta_a - \eta_c - R_0 I_c \tag{6.65}$$

where V^0 is the open circuit voltage (OCV) of the cell, η_a is the anode overpotential, and R_0 is the membrane resistance. The OCV is given by [43,44]

$$V^0 = 4.1868\left(\frac{70650 + 8T\log(T) - 92.84T}{2F}\right) \tag{6.66}$$

The anode overpotential is generally quite small compared to the cathode overpotential and can be neglected:

$$\eta_a \approx 0 \tag{6.67}$$

Therefore, equation (6.65) can be rewritten as

$$V_{cell} = V^0 - \eta_c - R_0 I_c \tag{6.68}$$

6.5 Model Validation and Results Discussion

For the purpose of validation, the general model was applied to a PEM single cell in which toluene contamination was present [20].

6.5.1 Toluene Contamination Mechanism

The studies of toluene adsorption on Pt [45–50] have revealed that: (1) in the gas phase toluene and benzene adsorb first as a π-complex (benzene ring parallel to the surface); (2) the toluene orientation depends also on its concentration; and (3) the substituent groups play an important role in the ordering of the overlayers, but have less effect on the adsorption strength [45]. While adsorbed aromatic compounds are in an electrochemical environment, their molecular orientation and packing density depend on factors such as chemical composition, concentration, temperature, electrode potential, and the interaction between electrode and anions. Rodrøguez and Pastor [51] suggested that all the aromatic compounds could interact via their aromatic ring parallel to the Pt surface, the benzene and toluene mainly adsorbed without dissociation. Zhu et al. [46] also indicated that the adsorbed toluene could be electrochemically oxidized, partially to higher oxidation state intermediates and partially to CO_2.

Based on the experimental evidence, we proposed that toluene adsorption (reaction (6.39)) should be the dominant mechanism for air-side toluene contamination in a PEM fuel cell. At high electrode potentials, toluene electrochemical oxidation takes place, yielding intermediates and carbon dioxides in higher oxidized states (as expressed in reaction (6.42) and reaction (6.43)) [20].

6.5.2 Experiment

The details of the testing platform setup for a single-cell toluene contamination experiment were described in [52]. The single-cell hardware was purchased from Teledyne (50 cm^2 CH-50). The membrane electrode assembly (MEA) had an active area of 50 cm^2 and consisted of a catalyst-coated membrane (CCM) from Ion Power and two gas diffusion layers (GDL) from SGL Carbon Group. The Ion Power CCM was made of Nafion® 211 membrane with 0.4 mg/cm^2 Pt loading on both sides. The flow field plates were designed and fabricated in-house with a single serpentine flow channel of 1.2 mm width, 1.0 mm channel depth, and 1.0 mm landing.

All tests were conducted on a Fideris 100 W fuel cell test station, and a fresh MEA was employed for each contamination test. The steady-state polarization curves were recorded using a load bank controlled in a constant-current pattern. The relative humidity (RH) of the fuel cell was set at 80% for both anode and cathode sides. The cell temperature and backpressure were held at 80°C and 30 psig, respectively.

6.5.3 Oxygen Concentration

The oxygen concentration in the cathode catalyst layer is given by the gas law as

$$C_{O_2} = \frac{p_{O_2}}{RT} \qquad (6.69)$$

where p_{O_2} is the oxygen partial pressure in the cathode catalyst layer.

The oxygen pressure in the CCL, which is different from that in the channel due to the pressure drop across the cathode gas diffusion layer, is roughly estimated by Fick's law as

$$p_{O_2} = 0.21 \times \left(p_{CGC} - p_V^{sat} \right) - RT \frac{I_c L_{CGDL}}{4FD_{O_2}^{eff}} \qquad (6.70)$$

where p_{CGC} and p_V^{sat} are the total gas (air) pressure in the cathode gas channel and the saturated water vapor pressure, respectively, L_{CGDL} is the thickness of the cathode gas diffusion layer (CGDL), and $D_{O_2}^{eff}$ is the effective diffusivity of oxygen in air. The saturated water vapor pressure (bar) is given as [53]

$$p_V^{sat} = \exp\left(11.6832 - \frac{3816.44}{T - 46.13}\right) \qquad (6.71)$$

The effective oxygen diffusivity is calculated by the Bruggemann relation,

$$D_{O_2}^{eff} = D_{O_2}\left(\varepsilon_{CGDL}\right)^{1.5} \qquad (6.72)$$

where D_{O_2} is the bulk diffusivity of oxygen in air and ε_{CGDL} is the porosity of the CGDL.

6.5.4 ORR Parameters

One of the difficulties in fuel cell contamination modeling is estimating the unknown ORR parameters. In the case of no toluene being present, we needed to know the forward and backward reaction rates or their ratios for reaction (6.36) to reaction (6.38). To do this, we simulated experimental baseline data free of toluene contamination. The parameters used for the simulation are listed in Table 6.1, and the modeling results and experimental polarization curves are shown in Figure 6.3.

Table 6.2 shows estimations of the ORR parameters from the baseline simulation. The estimated value of the rate constant for the rate-determining step (reaction (6.37)) is 1.64×10^{-11}. This value is quite similar to the value of 2.54×10^{-11}, which was derived from a reported experimental exchange current density obtained at fuel cell operating conditions in a high current density region, with a temperature of 80°C [54]. The estimated transfer coefficient for the same reaction (α_2) is 0.815.

When contaminant presents in the fuel cell, its concentration at the catalyst layer varies with both the inlet contaminant concentration and the current density, as discussed in Shi et al. [18]. Furthermore, the contaminant adsorption (desorption) rate constant is also related to the electrode potential. This variation of the contaminant concentration can be obtained by introducing the CGDL and cathode flow field into the model, which definitely increases its complexity. For simplicity here, we considered the product of the contaminant adsorption (desorption) rate constant and the contaminant concentration at the CCL, as a function of current density and contaminant inlet concentration ($k\,C_P \sim f\,(C_P^0, I)$), where C_P^0 is the contaminant inlet concentration in the cathode channel. Based on the experimental data at current densities of 0.2, 0.5, 0.75, and 1 A/cm², and contaminant inlet concentrations of

TABLE 6.1

Parameter Values Used in the Baseline Polarization Fitting

F	96485, C mol^{-1}
R	8.315, J K^{-1} mol^{-1}
T	353, K
n_{O_2}	4
γc	5000, cm^2 cm^{-2}
η_a	0, V
n	1
n_2	1
n_3	1
n_{p3}	1
n_{p4}	1
n_{p5}	1
α_{p3}	0.5
α_{p4}	0.5
α_{p5}	0.5
L_{CGDL}	3.53×10^{-2}, cm
P_{CGC}	30, psig
D_{O_2}	1.366×10^{-1}, cm^2 s^{-1}
ε_{CGDL}	0.8
C_{H^+}	1.7273×10^{-3}, mol cm^{-3}

Source: Shi, Z. et al. *Power Sources*, 186 (2009) 435.

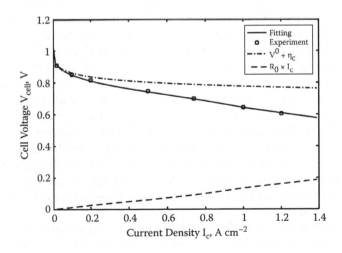

FIGURE 6.3

Comparison of calculated and experimental polarization curves in the absence of contaminant. (From Shi, Z. et al. *Power Sources*, 186 (2009) 435. With permission.)

TABLE 6.2

Estimated Parameter Values Based on the
Baseline Polarization Curve

k_{1f}/k_{1b}	5.8×10^4, cm^3 mol^{-1}
k_{3f}/k_{3b}	5.52×10^4, (cm^3)^3 (mol^{-1})^3
k_{2f}	1.64×10^{-11}, cm s^{-1} cm^3 mol^{-1}
α_2	0.815

Source: Shi, Z. et al. *Power Sources*, 186 (2009) 435.

1, 5, and 10 ppm, shown in Figure 6.4, we obtained the relation between contaminant adsorption (desorption) rate constants, toluene concentration at the flow channel inlet, and current density, which enabled us to predict the cell performance degradation and surface coverage change caused by toluene contamination.

6.5.5 Modeling Results and Discussion

By using the developed toluene contamination model and the ORR parameter relations obtained from experiments at ppm levels, we numerically studied the cell performance degradation when the toluene inlet concentration was at ppb levels, which closely resembled normal indoor and outdoor toluene levels [20]. One could also estimate the degree of cell performance degradation at a certain contaminant level and current density.

Figure 6.5 shows the transient cell performance behaviors at different current densities and with different toluene inlet concentrations. On the one hand, the effect of toluene contamination becomes more severe with a higher toluene concentration at the same cell current density; for example, Figure 6.5(d) indicates that at the same current density of 1.0 Acm^{-2}, the cell voltage drops due to toluene concentrations of 250, 500, and 750 ppb are 37, 42, and 48 mV, respectively. On the other hand, the toluene contamination increases steadily with increasing cell current density; for example, the voltage drops in response to 750 ppb toluene in the cathode flow channel are 9, 16, 27, and 48 mV, corresponding to cell current densities of 0.5, 0.75, and 1.0 Acm^{-2}, respectively, as shown in Figure 6.5. Furthermore, the time required for the cell voltage to reach steady state is also affected by both toluene concentration and current density, i.e., a larger toluene concentration and a lower current density result in a longer time before cell performance reaches steady state.

Figure 6.6 demonstrates the effects of toluene contamination on steady-state cell performance at different inlet concentration levels. Thus, the extent to which toluene contamination affects cell performance depends on both toluene concentration and current density. Based on this model, we can estimate the maximum allowable toluene concentration in order to limit the cell voltage drop to a specified range. For instance, to limit the contamination

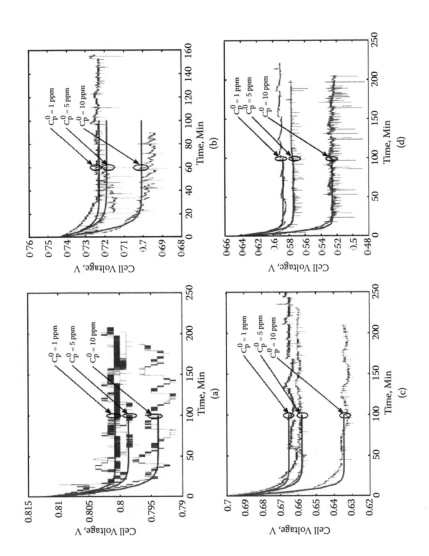

FIGURE 6.4
Experimental performance data and modeling results at different toluene concentrations for four different current densities: (a) $I_c = 0.2$ Acm^{-2}; (b) $I_c = 0.5$ Acm^{-2}; (c) $I_c = 0.75$ Acm^{-2}; (d) $I_c = 1.0$ Acm^{-2}. (From Shi, Z. et al. *Power Sources*, 186 (2009) 435. With permission.)

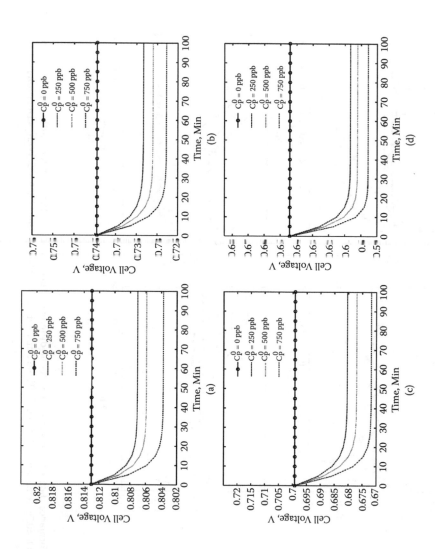

FIGURE 6.5
Effects of toluene contamination on transient fuel cell performance at different current densities: (a) $I_c = 0.2$ Acm^{-2}; (b) $I_c = 0.5$ Acm^{-2}; (c) $I_c = 0.75$ Acm^{-2}; (d) $I_c = 1.0$ Acm^{-2}. (From Shi, Z. et al. *Power Sources*, 186 (2009) 435. With permission.)

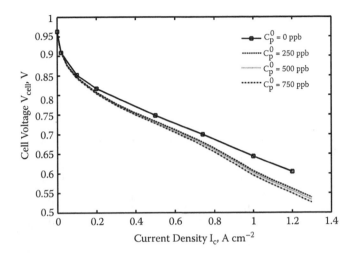

FIGURE 6.6
Effects of toluene contamination on steady-state fuel cell performance. (From Shi, Z. et al. *Power Sources*, 186 (2009) 435. With permission.)

potential drop to less than 10 mV, with a current density of 1.0 Acm², the toluene concentration should be less than 100 ppb.

This study also investigated the effect of toluene contamination on surface coverage [20]. Figure 6.7 demonstrates the effect of contaminant level on Pt surface coverage (θ_{Pt}) under different current densities. Three observations can be drawn from this figure: (1) the surface coverage of free Pt sites takes a shorter time to reach steady state in the case of high current density and low contamination level, compared to the case of low current density and high contamination level; (2) the coverage of free Pt sites at steady state decreases as the contamination level increases, indicating that toluene contamination blocks the free Pt sites; and (3) under the same toluene concentration, increasing current density will result in a low number of steady-state free Pt sites. For example, at a low current density of 0.2 Acm², the steady-state free Pt site coverage is reduced from 59 to 48% as the toluene concentration increases from 0 to 750 ppb, while at a high current density of 1.0 Acm², the coverage drops dramatically from 62 to 17%.

With respect to the contaminant's surface coverage, Figure 6.8 illustrates that increasing the contamination level and current density will in both cases increase the contaminant coverage, θ_{Pt-P}. Depending on the current density, with a toluene concentration of 750 ppb, the surface coverage θ_{Pt-P} can be as high as 73% at 1.0 Acm², indicating a significant catalyst poisoning effect.

The variations in the surface coverage of all the species with cathode overpotential are demonstrated in Figure 6.9 for different toluene concentrations. When toluene is absent, the catalyst surface is covered mainly (coverage

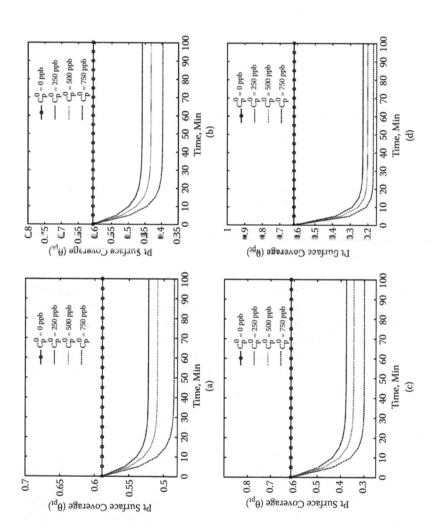

FIGURE 6.7
Free Pt surface coverage at different current densities: (a) $I_c = 0.2$ Acm^{-2}; (b) $I_c = 0.5$ Acm^{-2}; (c) $I_c = 0.75$ Acm^{-2}; (d) $I_c = 1.0$ Acm^{-2}. (From Shi, Z. et al. *Power Sources*, 186 (2009) 435. With permission.)

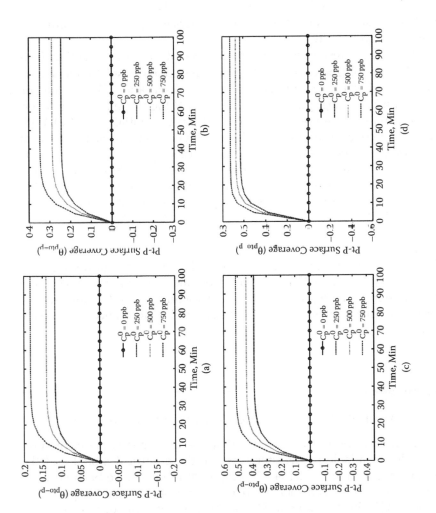

FIGURE 6.8

Toluene surface coverage at different current densities: (a) $I_c = 0.2$ Acm^{-2}; (b) $I_c = 0.5$ Acm^{-2}; (c) $I_c = 0.75$ Acm^{-2}; (d) $I_c = 1.0$ Acm^{-2}. (From Shi, Z. et al. *Power Sources*, 186 (2009) 435. With permission.)

>50%) by oxygen-containing intermediates in the overpotential range of 0 to 0.23 V, as shown in Figure 6.9(a). The simulation results also show that the coverage of the oxygen-containing intermediates is about 10% when the overpotential is around 0.29 V, which is the same coverage as reported in [27] at a cell potential of 0.84 V. Furthermore, at an overpotential greater than 0.26 V, the free Pt sites dominate (i.e., coverage >50%). By comparison, the free Pt potential reported by experiment is <0.8 V [27]. We also noticed that in free Pt-dominated potential the oxygen coverage (θ_{Pt-O_2}) is generally less than half of the free Pt site coverage.

When toluene is present, it is observable that at low overpotentials (less than 0.25 V), the toluene coverage (θ_{Pt-P}) is small regardless of toluene concentration. This is consistent with experimental observation [52] that toluene contamination had a negligible effect on the OCV. Significant toluene coverage (>10%) is mainly observed at large cathode overpotentials (>0.32 V). With higher toluene concentrations, significant toluene coverage will appear at relatively lower overpotentials; for instant, with 250 ppb toluene, 10% toluene coverage is reached at a cathode overpotential of 0.34 V, while with 750 ppb toluene, 10% toluene coverage is noticed at a cathode overpotential of 0.32 V. In addition, with the presence of contaminant, we see a narrowing potential window in which free Pt sites dominate the Pt surface. With 0 ppb toluene, this window is in the range of $\eta_c > 0.26$V; with 250 ppb toluene, the window's range is 0.12 V from 0.26 to 0.38 V; with 500 ppb, the window's range is 0.11 V (0.26V $< \eta_{<_c} < 0.37$V); and with 750 ppb toluene, the window's range is 0.10 V (0.26V $< \eta_c < 0.37$V). Furthermore, the toluene coverage increases significantly at high cathode overpotentials and, therefore, significantly reduces the coverage of free Pt sites. As a result, the fuel cell performance is adversely affected.

6.6 Discussion of Other Cathode Contamination Models

Besides the kinetic models we discussed above, the other cathode contamination models available in the literature are the empirical model and the competitive adsorption model [4,57]. Empirical models have been successfully used to describe PEM fuel cell performance at different temperatures and pressures [55,56]. Equation (6.73) was proposed by Kim et al. [56]:

$$E = E_0 - b_1 \times \log(i) - R_0 \times i - m \times \exp(n \times i) \tag{6.73}$$

$$E_0 = E_r + b \times \log(i_0) \tag{6.74}$$

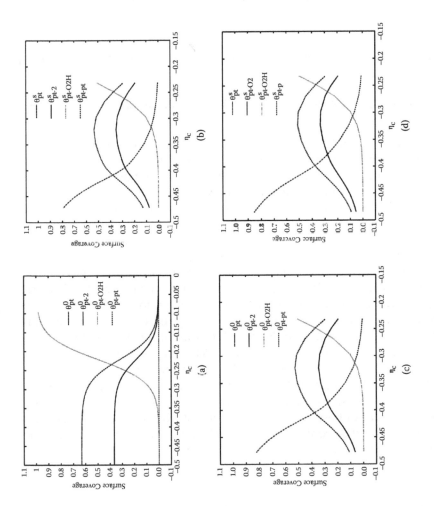

FIGURE 6.9
Variations of species surface coverage with overpotential in the CCL at different inlet toluene concentrations: (a) $C_P^0=0$ ppb; (b) $C_P^0=250$ ppb; (c) $C_P^0=500$ ppb; (d) $C_P^0=750$ ppb. (From Shi, Z. et al. *Power Sources*, 186 (2009) 435. With permission.)

where E_r is the reversible potential for the cell, i_0 and b are the Tafel parameters for oxygen reduction, R_0 is the ohmic resistance, and $m \times \exp(n \times i)$ pertains to the mass transport losses. By employing this equation in the study of feed stream contamination, researchers have proposed several empirical models for both anode and cathode sides [4]. Equation (6.75) is the model for cathode SO_2 and NO_2 contamination. Equation (6.76) is that for NH_3 contamination:

$$E_c = E_0 - \left(b_1 + K_{cK} \times C_p\right) \times \log(i) - R_0 \times i + K_{cK} \times C_p \times \log(i) \times \exp(-K_3 \times t) \quad (6.75)$$

$$E_c = E_0 - b_1 \times \log(i) - \left(R_0 + K_{cR} \times C_p\right) \times i + K_{cR} \times C_p \times i \times \exp(-K_4 \times t) \quad (6.76)$$

where K_{cK}, K_{cR}, K_3, and K_4 are constants that account for contamination, C_p is the concentration of the contaminant, and t is the contamination time. Based on the models, air-side SO_2 and NO_2 contamination mainly perturb the kinetic term of the empirical equation, while for NH_3 contamination, the major effect is on the ohmic term. These models reflect the contamination mechanisms of the particular impurities.

Empirical models are a valid approach for providing estimates of contamination effects on fuel cell performance. The major drawback of the empirical model approach is the lack of physical interpretation of various parameters.

For nonempirical models, NO_2 contamination has been reported in St-Pierre et al. [57]. This model is based on the competitive adsorption kinetics of O_2 and NO_2 on Pt catalysts. The species considered on the Pt surface are adsorbed O_2 and NO_2, and free Pt sites. With this simple model, the authors discussed NO_2 threshold concentrations. With the assumptions that the cell voltage was 0.8 V and the allowable performance loss for a fuel cell was 10%, the predicted allowable NO_2 concentration was 0.39 ppm. This is a preliminary model only. Considerable work is still anticipated to extend the model's capability of handling contamination at different current densities, and extend it to other contaminants, which involves electrochemical reactions as well.

6.7 Summary

Fuel cell contamination is a significant concern for PEM fuel cell operation and applications. Understanding the effects and mechanisms, and providing predictable tools to account for the influence of impurities on fuel cell performance, are the objectives of contamination modeling.

Currently, several air-side contamination models have been published in the literature, ranging from simple empirical and adsorption models to general kinetic models. These models have been applied to simulate and predict SO_2, NO_2, NH_3, and toluene contamination. The kinetic model is a very general one based on the associative oxygen reduction mechanism. It takes into account contaminant reactions, such as surface adsorption, competitive adsorption, and electrochemical oxidation, and has the capability of simulating and predicting both transient and steady state cell performance. The model can be applied to other cathode contaminants, e.g., SO_2 and NO_2.

In future work, more contamination modeling is needed, especially to account for the effect of different types of contaminants, such as anion and cation contaminants, on cell performance. In reality, of course, multiple contaminants are often the case, so a multicontaminant model needs to be developed that will take into account both anode and cathode contaminants, as well as membrane contamination.

References

1. P.R. Hayter, P. Mitchell, R.A.J. Dams, C. Dudfield, N. Gladding, The effect of contaminants in the fuel and air streams on the performance of a solid polymer fuel cell, Contract Report (ETSUF/02/00126/REP) Wellman CJB Limited, Portsmouth, U.K., 1997.
2. D. Brumbaugh, J. Guthrie, Literature Survey of Fuel Cell Contaminants, *US Fuel Cell Council* (ed.), 2004.
3. X. Cheng, Z. Shi, N. Glass, L. Zhang, J. Zhang, D. Song, Z. S. Liu, H. Wang, J. Shen, J. *Power Sources*, 165 (2007) 739.
4. S. Knights, N. Jia, C. Chuy, J. Zhang, *2005 Fuel Cell Seminar: Fuel Cell Progress, Challenges and Markets* abstract. P121, Palm Springs, CA, 2005.
5. J.M. Moore, P.L. Adcock, J.B. Lakeman, G.O. Mepsted, *J. Power Sources*, 85 (2000) 254.
6. T. Springer, T. Zawodzinski, S. Gottesfeld, in *Electrode Material and Processes for Energy Conversion and Storage IV*, J. McBreen, S. Mukherjee, S. Srinivasan (eds.), Pennington, NJ: Electrochemical Society, Inc., 1997.
7. T.E. Springer, T. Rockward, T.A. Zawodzinski, S. Gottesfeld, *J. Electrochem. Soc.*, 148 (2001) A11–A23.
8. X. Wang, I.M. Hsing, Y.J. Leng, P.L. Yue, *Electrochim. Acta*, 46 (2001) 4397.
9. P. Rama, R. Chen, R. Thring, *Proceedings of the Institution of Mechanical Engineers, Part A: J. Power and Energy*, 219 (2005) 255.
10. K.K. Bhatia, C-Y. Wang, *Electrochim. Acta*, 49 (2004) 2333.
11. J.J. Baschuk, A.M. Rowe, X. Li, *J. Energy Resour. Technol.*, 125 (2003) 94.
12. J.J. Baschuk, X. Li, *Int. J. Global Energy Issues*, 20 (2003) 245.
13. S.H. Chan, S.K. Goh, S.P. Jiang, *Electrochim. Acta*, 48 (2003) 1905.
14. J. Zhang, H. Wang, D.P. Wilkinson, D. Song, J. Shen, Z.-S. Liu, *J. Power Sources*, 147 (2005) 58.

15. A.A. Shah, P.C. Sui, G.S. Kim, S. Ye, *J. Power Sources*, 166 (2007) 1.
16. D.J.L. Brett, P. Aguiar, N.P. Brandon, A.R. Kucernak, *Int. J. Hydrogen Energy*, 32 (2007) 863.
17. C.G. Farrell, C.L. Gardner, M. Ternan, *J. Power Sources*, 171 (2007) 282.
18. Z. Shi, D. Song, J. Zhang, Z.S. Liu, S. Knights, R. Vohra, N. Jia, D. Harvey, *J. Electrochem. Soc.*, 154 (2007) B609–B615.
19. S.S. Shah, F.C. Walsh, *J. Power Sources*, 185 (2008) 287.
20. Z. Shi, D. Song, H. Li, K. Fatih, Y. Tang, J. Zhang, Z. Wang, S. Wu, Z. S. Liu, H. Wang, J. Zhang, *J. Power Sources*, 186 (2009) 435.
21. K. Kinoshita, in K. Kinoshita (ed.), *Electrochemical Oxygen Technology*, New York: John Wiley & Sons,1992, chap. 2.
22. R. Adzic, in J. Lipkowski, P.N. Ross (eds.), *Electrocatalysis*, New York: Wiley-VCH, 1998, chap. 5.
23. M. Gattrell, B. MacDougall, in W. Vielstich, H.A. Gasteiger, A. Lamm (eds.), *Handbook of Fuel Cells—Fundamentals, Technology and Applications*; Vol. 2: *Electrocatalysis*, Chichester, U.K.: John Wiley & Sons, 2003, chap. 30.
24. P.N. Ross Jr, in W. Vielstich, H.A. Gasteiger, A. Lamm (eds.), *Handbook of Fuel Cells—Fundamentals, Technology and Applications*, Vol. 2: *Electrocatalysis*, Chichester, U.K.: John Wiley & Sons, 2003, chap. 31.
25. A. Damjanovic, M.A. Genshaw, *Electrochim. Acta*, 15 (1970) 1281.
26. C. Song, J. Zhang, in J. Zhang (ed.), *PEM Fuel Cell Electrocatalysts and Catalyst Layers*, Heidelburg, Germany: Springer, 2008, chap. 2.
27. A. Damjanovic, V. Brusic, *Electrochimi. Acta*, 12 (1967) 615.
28. M.J. Kelly, G. Fafilek, J.O. Besenhard, H. Kronberger, G.E. Nauer, *J. Power Sources*, 145 (2005) 249.
29. E. Yeager, M. Razaq, D. Gervasio, A. Razaq, D. Tryk, in *Structure Effects in Electrocatalysis and Oxygen Electrochemistry*, Proc. Vol. 92–11, D. Scheerson, D. Tryk, M. Daroux, X. Xing (eds.), Pennington, NJ: The Electrochemical Society, 1992.
30. Z. Shi, J. Zhang, Z.S. Liu, H. Wang, D.P. Wilkinson, *Electrochimi. Acta*, 51 (2006) 1905.
31. Z. Shi, in J. Zhang (ed.), *PEM Fuel Cell Electrocatalysts and Catalyst Layers*, Heildelburg, Germany: Springer, 2008, chap. 5.
32. A.B. Anderson, T.V. Albu, *J. Electrochem. Soc.*, 147 (2000) 4229.
33. R.A. Sidik, A.B. Anderson, *J. Electroanal. Chem.*, 528 (2002) 69.
34. J.K. Norskov, J. Rossmeisl, A. Logadottir, L. Lindqvist, J.R. Kitchin, T. Bligaard, H. Jonsson, *J. Phys. Chem. B*, 108 (2004) 17886.
35. G.S. Karlberg, J. Rossmeisl, J.K. Norskov, *Phys. Chem. Chem. Phys.*, 9 (2007) 5158.
36. T. Jacob, *Fuel Cells*, 6 (2006) 159.
37. S. Walch, A. Dhanda, M. Aryanpour, H. Pitsch, *J. Phys. Chem. C*, 112 (2008) 8464.
38. O. Antoine, Y. Bultel, R. Durand, *J. Electroanal. Chem.*, 499 (2001) 85.
39. H. Kuhn, A. Wokaun, G.G. Scherer, *Electrochimi. Acta*, 52 (2007) 2322.
40. C.Y. Du, T.S. Zhao, C. Xu, *J. Power Sources*, 167 (2007) 265.
41. V.P. Zhdanov, B. Kasemo, *Electrochem. Commun.*, 8 (2006) 1132.
42. J. Newman, K.E. Thomas-Alyea, *Electrochemical Systems*, 3rd ed., New York: John Wiley & Sons, 2004, p. 210.
43. A. Parthasarathy, S. Srinivasan, A.J. Appleby, C.R. Martin, *J. Electrochem. Soc.*, 139 (1992) 2530.

44. A.Z. Weber, R.M. Darling, J. Newman, *J. Electrochem. Soc.*, 151 (2004) A1715–A1727.
45. J.L. Gland, G.A. Somorjai, *Surf. Sci.*, 41 (1974) 387.
46. J. Zhu, T. Hartung, D. Tegtmeyer, H. Baltruschat, J. Heitbaum, *J. Electroanal. Chem.*, 244 (1988) 273.
47. Z. Hlavathy, P. Tetenyi, *Applied Surf. Sci.*, 252 (2005) 412.
48. M.C. Tsai, E.L. Muetterties, *J. Am. Chem. Soc.*, 104 (1982) 2534.
49. S. Chiang, *Chem. Rev.*, 97 (1997) 1083.
50. A.T. Hubbard, *Chem. Rev.*, 88 (1988) 633.
51. J.L. Rodroguez, E. Pastor, *Electrochimi. Acta*, 45 (2000).
52. H. Li, J. Zhang, K. Fatih, Z. Wang, Y. Tang, Z. Shi, S. Wu, D. Song, J. Zhang, N. Jia, S. Wessel, R. Abouatallah, N. Joos, *J. Power Sources*, 185 (2008) 272.
53. F.P. Incropera, D.P. Dewitt, *Fundamentals of Heat and Mass Transfer*, 3rd ed., New York: John Wiley & Sons, 1990.
54. C. Song, Y. Tang, J.L. Zhang, J. Zhang, H. Wang, J. Shen, S. McDermid, J. Li, P. Kozak, *Electrochimi. Acta*, 52 (2007) 2552.
55. J.C. Amphlett, R.M. Baumert, R.F. Mann, B.A. Peppley, P.R. Roberge, T.J. Harris, *J. Electrochem. Soc.*, 142 (1995) 1.
56. J. Kim, S.M. Lee, S. Srinivasan, C.E. Chamberlin, *J. Electrochem. Soc.*, 142 (1995) 2670.
57. J. St-Pierre, N. Jia, R. Rahmani, *J. Electrochem. Soc.*, 155 (2008) B315–B320.

7

Anode Contamination Modeling

Nada Zamel and Xianguo Li

CONTENTS

7.1 Introduction

Increasing demand for energy, the need for energy security, and the need to minimize the impact on the environment related to energy are the major drivers for the research and development of alternative technologies. Due to their high energy efficiency and minimal-to-zero greenhouse gas emissions, polymer electrolyte membrane (PEM) fuel cells are considered to be one of the most promising candidates for the future clean power generation. They are electrochemical devices, which convert the chemical energy of hydrogen and oxygen directly and efficiently into electrical energy with only waste heat and liquid water as the byproducts. PEM fuel cells operate at a low temperature, use a solid electrolyte, and can obtain a power density competitive with the internal combustion engine [1], thus, PEM fuel cells are very promising for use in the transportation sector. However, many technical hurdles face the commercialization of PEM fuel cells. One such hurdle is the presence of impurities in the anode fuel stream, which inhibit hydrogen oxidation reaction and result in a severe decrease in energy conversion efficiency. Some of the most studied impurities, which attack the anode of PEM fuel cells, are carbon monoxide, carbon dioxide, hydrogen sulfide, and ammonia. According to the U.S. Department of energy (DOE), the fuel composition should not contain more than 2 μmol/mol CO_2, 0.2 μmol/mol CO, 0.004 μmol/mol sulfur species, and 0.1 μmol/mol NH_4 [2].

Due to the low operating temperature of PEM fuel cells and the technical and practical issues associated with the use of pure hydrogen as the fuel, the presence of these impurities becomes a significant issue in PEM fuel cells applications, specifically in the transportation applications. The use of pure hydrogen as the fuel is associated with many limitations from infrastructure to refueling processes to onboard and offboard storage issues. The infrastructure to deliver hydrogen is almost non-existent, the refueling process is very slow, and the onboard and offboard storage of hydrogen is very heavy. With these limitations, the practical range of vehicles powered by stored hydrogen will be less than that for typical gasoline powered vehicles [3]. Consequently, there exists the need to find other fuels. Some of the most promising alternatives is the use of liquid hydrocarbons and alcohol fuels [4]. The use of liquid hydrocarbons and alcohol fuels requires the use of a reformer to extract the hydrogen gas. There are two options to reforming these fuels, either onboard the vehicle or at the refueling station. With either option, impurities will be introduced into the fuel. For example, generating hydrogen by reforming methanol results in a fuel mixture consisting of approximately 74% hydrogen, 25% carbon dioxide, and 1 to 2% carbon monoxide by volume [5]. These impurities will cause large polarization losses during the electrooxidation of hydrogen in the cell [6], thus, decreasing the overall cell efficiency and power output. Hence, the need for managing these impurities arises. In order to find the best controls for poisoning by these impurities, experimental, analytical, and mathematical studies are used. The objective of this chapter is to give a comprehensive overview of the mathematical modeling of the many poisoning phenomena that occur in the anode of PEM fuel cells. As a guideline for the reader, the following general steps can be followed to model any physical problem in an engineering system.

1. Definition of the modeling domain
2. Mathematical formulation of the physical problem
   - List governing (conservation) equations, which apply to the physical problem—mass, momentum, energy, charge, species, etc.
   - List assumptions
   - Define the operating and physical parameters
   - Define boundary conditions
3. Numerical solution of the equation sets
   - Choose numerical method
   - Define convergence criteria
   - Define meshing procedure
   - Apply numerical method to the governing equations
   - Check for grid independency
4. Numerical and experimental results comparison
   - Compare numerical results to experimental results for the same physical problem

7.1.1 PEM Fuel Cell Components

To better understand the concepts introduced in this chapter, the fundamental components of a PEM fuel cell are shown in Figure 7.1. As illustrated, a PEM fuel cell consists of three major components: anode and cathode

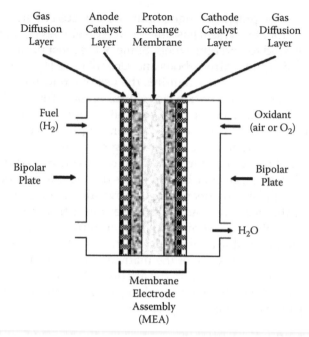

FIGURE 7.1

Basic components of polymer electrolyte membrane fuel cell. (From Cheng, X. et al., 2007. *Journal of Power Sources*, **165**:739–756. With permission.)

bipolar plates and an MEA (membrane electrode assembly). Gas channels are grooved into the bipolar plate for fuel and oxidant distribution. The MEA consists of two gas diffusion layers, two catalyst layers, and a proton conducting membrane. Hence, the anode consists of three parts: the bipolar plate, gas diffusion layer, and a catalyst layer.

The two main functions of the bipolar plate are to distribute the fuel through the grooved channels and to collect the electrons produced from the hydrogen oxidation reaction. The geometry of the channels has a significant effect on the method of fuel distribution. From the channels, the fuel travels through the gas diffusion layer. If a pressure drop exists between channels, then convective mass transfer becomes very important. However, if a pressure drop is almost negligible, then the mass diffusion through the electrode becomes the only means of mass transport of the fuel. The channel configuration governs the magnitude of convective mass transfer. The transport of the gaseous species in the electrode occurs in the void region, while the solid region of the electrode is used for electron transport. Once the hydrogen reaches the catalyst layer, the hydrogen oxidation reaction occurs as follows:

$$H_2 \Rightarrow 2H^+ + 2e^- \tag{7.1}$$

In order for the hydrogen oxidation reaction to occur, the hydrogen molecules must be adsorbed onto the platinum catalyst. However, with the presence of impurities this reaction will be altered. If reformate fuel is used, then carbon monoxide, carbon dioxide, hydrogen sulfide, and ammonia will be present in the fuel stream. Also, oxygen will be present in the fuel stream if mitigation of CO poisoning is accomplished with oxygen bleeding; nitrogen will be present if air bleeding is used. The presence of impurities in the fuel stream will affect the anode catalyst layer overpotential. The overpotential originates from three sources: ohmic, activation, and mass transfer. The resistance to electron migration in the solid phase of the catalyst layer and proton migration in the polymer electrolyte of the catalyst layer results in the ohmic component of the overpotential. The activation component is due to resistance to electrochemical reactions. The resistance to mass transfer in the anode gas flow channels, gas diffusion, and the catalyst layers all contribute to the mass transfer component of the overpotential.

The objective of this chapter is to give a full overview of modeling contamination in the anode of PEM fuel cells. Thus, the chapter is divided into six sections. In the introduction section (7.1), the origin of contaminants, a list of contaminants and their tolerances, as well as the basic components of a PEM fuel cell are discussed. The mathematical formulation of CO poisoning and its mitigation methods is developed in section 7.2. In the third section (7.3), numerical simulation of carbon dioxide is discussed. The reaction kinetics of the phenomenon are identified and some results from literature are given. The mathematical formulation of hydrogen sulfide poisoning and some results from literature are given in section 7.4. Section 7.5 is dedicated to the discussion of other contaminants in the cell. In this section, contamination mixtures and the progress in this area are analyzed. Finally, a summary of the chapter, as well as the direction of future work, are discussed in section 7.6.

7.2 Carbon Monoxide Poisoning

7.2.1 Review of CO Modeling in Literature

The poisoning of the anode of PEM fuel cells due to CO presence in the fuel stream can be said to be one of the most understood poisoning phenomena due to the abundant experimental, analytical, and mathematical studies in literature. The models found in literature are classified as either empirical or mathematical. Using curve-fitting schemes of experimental data, empirical models are usually developed. In PEM fuel cells, empirical models are used to understand the performance of the cell and are normally very specific. An empirical equation that correlates the cell voltage to the current density was developed by Kim et al. [8]. The cell voltage was found by subtracting the voltage losses due to ohmic, activation, and concentration polarization from the reversible voltage. Each of these losses is dependent on the current density. Another empirical model that represents the voltage losses as parametric equations based on current density, temperature, pressure, and reactant concentration was developed for the Ballard Mark IV cell by Amphlett et al. [9,10]. This model was later extended to take into account the effects of carbon monoxide poisoning in a PEM fuel cell stack by Rodrigues et al. [11]. In order to account for the effect of CO poisoning on the stack performance, the activation overpotential was modified by the addition of a term:

$$\eta_{act,a}^{CO} = \beta_{21} + \beta_{22}T + \beta_{23}T \ln\left[\frac{(1-\theta_{CO})^2 C_{H_2}}{I}\right] \quad (7.2)$$

where T is the temperature of the stack, C_{H_2} is the concentration of hydrogen at the catalyst layer, θ_{CO} is the coverage of the catalyst layer reaction sites by carbon monoxide, and β_i are constants. The coverage of CO is found using:

$$\theta_{CO} = \gamma_1 + \gamma_2 I + \gamma_3 T + \gamma_4 \ln I + \gamma_5 \ln T + \gamma_6 \ln(IT)$$
$$+\gamma_7 \ln\left(\frac{P_{CO}}{P_{H_2}}\right) + \gamma_8 T \ln\left(\frac{P_{CO}}{P_{H_2}}\right) + \gamma_9 I \ln\left(\frac{P_{CO}}{P_{H_2}}\right) \quad (7.3)$$

Their model successfully predicts the experimental data for CO concentrations up to 25 ppm. However, at 100 ppm, the model significantly over-predicts the cell potential due to the significance of chemisorption processes, which are not accounted for in the model. Bellows et al. [12] addressed CO poisoning with an inventory model that related CO_x impurity levels to performance losses in PEM fuel cells based on fundamental electrochemical half-cell experiments at ambient pressure and a temperature range from 23 to 66°C. The output current density of the cell was found by:

$$I_\delta^{H_2/CO} = I_\delta^{H_2}\left(1-\theta_{CO}\right)^2 \quad (7.4)$$

where $I_\delta^{H_2}$ is the current density of the cell with a pure hydrogen reactant stream and θ_{CO} is the coverage of the electrode by carbon monoxide, which can be found using:

$$\frac{d\theta_{CO}}{dt} = CO\,adsorption + CO_2\,reduction - CO\,oxidation \quad (7.5)$$

which states that the coverage of CO over time is a function of its rate of adsorption, rate of its production from carbon dioxide, and its rate of oxidation. In the half-cell experiments, the amount of CO production from carbon dioxide reduction is negligible for PEM fuel cell operating conditions. In order to study CO poisoning effects on the performance of the cell stack, Chu et al. applied a similar approach as shown in reference [13].

As much as empirical models can be helpful in understanding the performance of PEM fuel cells, they are still limited in that many parameters that affect the performance of the cell, such as the catalyst layer structure, are not included in the model. Thus, parametric studies cannot be done using these empirical models and the need for mathematical models arises. Mathematical models take into account the fundamental physical and chemical properties of the problem. The structure of the components of the fuel cell and the operating and physical conditions are also included. The first mathematical models that were developed for PEM fuel cells were one-dimensional and were developed by Bernardi and Verbrugge [14]. The model included the losses due to the anode and cathode reactions. The catalyst layer structure was modeled using a macrohomogeneous method and the polymer electrolyte membrane was assumed to be fully hydrated. Springer et al. [15,16] showed that the overpotential due to proton migration is dependent on the water content of the membrane.

Modeling carbon monoxide poisoning is normally categorized under mass transport studies. Wang and Savinell [17] created a model of a carbon monoxide-poisoned anode that used variable membrane conductivity data of Springer et al. [15, 16]. Only the mass transfer, proton migration, and electrochemical reaction of the anode catalyst layer were modeled. The local current production was found using:

$$\dot{w}_I = i_0^{H2} \left[\frac{1-\theta_{CO}}{\theta_0 + (1-\theta_0)\exp(F\eta/RT)} \right]^2 \left[\frac{C_{H2}}{C_{H2}^0} \exp\left(\frac{2F\eta}{RT} \right) - 1 \right] \qquad (7.6)$$

where i_0^{H2} is the exchange current density of hydrogen oxidation, θ_0 is the equilibrium coverage of hydrogen on the anode catalyst layer, C_{H2} is the concentration of hydrogen, C_{H2}^0 is the reference concentration of hydrogen, F is the Faraday constant, R is the universal gas constant, T is the temperature, and η is the overpotential of the catalyst layer. The current production is limited by the carbon monoxide coverage, θ_{CO}, which was found using the empirical relation derived by Dhar et al. [18]. The empirical relation for carbon monoxide coverage was derived from studies of a phosphoric acid fuel cell, which operates at a greater temperature (200°C) than a PEM fuel cell.

A more fundamental approach to modeling carbon monoxide poisoning of the anode is using the anode kinetics. One of the first models to be created with such an approach is that by Springer et al. [19,20]. In their model, only the anode kinetics were modeled. The hydrogen and carbon monoxide reactions were assumed to exist in the anode catalyst layer. Thus, the amount of carbon monoxide coverage was governed by the rates of reaction, adsorption, desorption, and oxidation. This modeling technique has the advantage of not relying on empirical data for the coverage of carbon monoxide. Baschuk and Li [21] followed a similar analysis. They modeled the adsorption, desorption, and oxidation of carbon monoxide and hydrogen and added the heterogeneous oxidation of carbon monoxide and hydrogen with oxygen. Thus, both carbon monoxide poisoning and oxygen bleeding were modeled. Their model was steady, one-dimensional, and half-cell. They included mass transport from the channels to the electrode backing/catalyst layer interface and within the catalyst layer. Proton and electron transport was also taken into consideration. The anode overpotential was compared to published, experimental data. In a study by Zhou and Liu [22], the effects of operating a PEM fuel cell on reformate were investigated using a three-dimensional mathematical model. They incorporated the adsorption and oxidation kinetics of CO on a platinum surface, which was proposed by Springer et al. [19,20]. A parametric study was done using the model. They studied the effects of flow rate, gas diffuser porosity, gas diffuser thickness, and the width of the collector plate shoulder. Chu et al. [23] presented a one-dimensional, transient model based on the model developed by Springer et al. [15,16] to simulate the carbon monoxide poisoning effect on the performance of a PEM fuel cell anode. They found that as time elapses, the hydrogen coverage on the catalyst sites decreases and, therefore, the performance of the cell is reduced. The same group [24] extended their early study [23] to investigate the effects of CO poisoning on liquid water transport in the PEM fuel cell. Their results indicated that the saturation of liquid water is reduced in the catalytic layers. The distribution of liquid water depends more strongly on the CO concentration than on the dilution of hydrogen in the MEA of the cell. Zamel and Li [25] created a one-dimensional transient model of the anode catalyst layer in order to study the effects of CO poisoning and oxygen bleeding. They studied the effect of carbon monoxide and oxygen concentration, temperature, and pressure on the overall transient behavior of the cell. A model developed by Barbir et al. [26] concentrated on modeling various components of a stack system to determine power output and system efficiency. In their model, selective oxidation of CO was modeled in order to mitigate CO poisoning.

From the above survey of literature, it is clear that the phenomenon of CO poisoning is well investigated. In the next sections, a full understanding of the CO poisoning phenomenon will be established. The phenomenon is well defined, the reaction kinetics are put forward, and the mathematical model is introduced. Further, the mitigation methods of CO poisoning and their incorporation into numerical modeling are discussed. After reviewing this part of the chapter, the reader should be able to simulate CO poisoning as well as its mitigation methods numerically.

7.2.2 CO Contamination and Mitigation Methods

At low temperatures, the adsorption of carbon monoxide onto platinum is very significant. Carbon monoxide poisoning is said to occur when CO molecules cover the platinum sites in the catalyst layer. Their coverage of platinum, thus, decreases the available area necessary for the hydrogen reduction reaction to occur. This phenomenon is a transient process and the understanding of its effects is crucial for optimizing the performance of PEM fuel cells. In the next section, the electrochemical kinetics of CO poisoning are given in detail.

7.2.2.1 Reaction Pair Mechanism of CO Poisoning

Numerical modeling relies directly on the physical problem at hand. Before simulating the physical problem, it is necessary to establish the mathematical model with all the governing equations. In the case of carbon monoxide poisoning, the electrochemical kinetics of this phenomenon should first be established. In this section, the adsorption, desorption, and the electrooxidation of hydrogen and carbon monoxide are analyzed.

The electrooxidation reaction of hydrogen consists of two steps [27]. The rate-limiting step is the dissociation of the hydrogen molecule, which requires two adjacent bare platinum (Pt) sites. This reaction is also sometimes referred to as the Tafel reaction and is given as follows [27]:

$$H_2 + 2Pt \Leftrightarrow 2(H-Pt) \tag{7.7}$$

The second step is the discharge step and is relatively rapid. This reaction is also sometimes referred to as the Volmer reaction and is expressed as [27]:

$$2(H-Pt) \Leftrightarrow 2Pt + 2H^+ + 2e^- \tag{7.8}$$

Carbon monoxide poisoning occurs when the CO molecules are adsorbed onto the platinum sites. The adsorption of CO on the bare Pt sites happens according to [27]:

$$CO + Pt \Leftrightarrow CO - Pt \tag{7.9}$$

The adsorbed CO can then be electrooxidized at higher electrode potentials via the "reactant pair" mechanism [28]:

$$CO - Pt + H_2O \Leftrightarrow Pt + CO_2 + 2H^+ + 2e^- \tag{7.10}$$

Reaction (7.7) through reaction (7.10) illustrate the mechanism of CO poisoning [29]. The CO molecule chemisorbs on the platinum reaction sites creating an obstacle for the hydrogen molecules to reach the reaction sites. This is especially true since the CO–Pt bond is much stronger than that of the H–Pt bond [30]. The removal of carbon monoxide via electrooxidation once the CO is adsorbed is very minimal since electrooxidation occurs within a potential range of 0.6 to 0.9 V. This potential range is somewhat difficult to achieve since, with fewer available reaction sites for hydrogen, adsorption results in a reduction in the overall electrooxidation. Higher anode overpotential is required to achieve the same rate of reaction as a CO-free environment.

7.2.2.2 Mitigation Methods

Due to the CO poisoning severity, many mitigation methods have been investigated. There are five methods that have proved successful and are commonly used to lessen the effects of CO poisoning. They include:

1. Oxygen or air bleeding
2. CO-tolerant catalysts
3. Advanced reformer designs
4. High temperature membranes
5. Multilayered gas diffusion layers

The mechanism of oxygen or air bleeding refers to the introduction of oxygen into the fuel stream to lessen the effects of CO poisoning in low temperature-operating PEM fuel cells. This method was first suggested by Gottesfeld and Pafford [31]. Their suggested approach came from the knowledge that the level of CO in a gas mixture from reformed methanol at high temperatures (150°C) can be reduced by introducing low levels of O_2 concentration (2% by weight) or by passing the gas mixture over a 1% Pt/Al_2O_3 catalyst maintained at 150°C. Effective CO oxidation by molecular oxygen at Pt catalysts at the solid/gas interface is known to require higher temperatures than that of the operating fuel cell (more than 100°C) [32]. Nonetheless, Gottesfeld and Pafford were able to show that, even at a low temperature of 80°C, a CO-free cell performance can be completely restored by injecting oxygen at a level of 2 to 5% (O_2/H_2) and with 100 ppm CO. With these oxygen levels, an almost complete recovery was demonstrated in a similar cell for CO levels as high as 500 ppm.

With the addition of oxygen in the fuel stream, either as pure oxygen, air, or via a H_2O_2 mixture, the reaction kinetics discussed earlier in equation (7.7) through equation (7.10) change and the heterogeneous oxidation of carbon monoxide and hydrogen by oxygen is taken into consideration. At low operating temperatures, the gas phase oxidation of carbon monoxide and hydrogen by oxygen is assumed to be negligible and the oxidation is mainly through the heterogeneous catalysis [33]. Chemically, the heterogeneous oxidation of CO on platinum is described by a Langmuir–Hinshelwood mechanism [33]:

$$O_2 + 2Pt \Leftrightarrow O_2 - Pt + Pt \Rightarrow 2(O - Pt) \tag{7.11}$$

$$CO + Pt \Leftrightarrow CO - Pt \tag{7.12}$$

$$CO - Pt + O - Pt \Rightarrow CO_2 + 2Pt \tag{7.13}$$

The heterogeneous oxidation of carbon monoxide by oxygen on the catalyst surface is important to lessen the effects of the CO poisoning. However, oxygen will also react with the adsorbed hydrogen, which results in the reduction of the overall hydrogen available for reaction. The heterogeneous oxidation of hydrogen by oxygen is approximated by a Langmuir–Hinshelwood mechanism [34]:

$$H_2 + 2Pt \Leftrightarrow 2(H - Pt) \tag{7.14}$$

$$O_2 + 2Pt \Leftrightarrow O_2 - Pt + Pt \Rightarrow 2(O - Pt) \tag{7.15}$$

$$O - Pt + 2(H - Pt) \Rightarrow H_2O + 3Pt \tag{7.16}$$

The use of CO-tolerant catalysts is considered as promising as the use of oxygen bleeding in mitigating the effect of CO poisoning due to their efficiency and minimal associated problems. The CO-tolerant catalysts consist of binary systems with platinum as one of the main components. The other element is present in this binary system in the form of an alloy or a co-deposit. Studies found in literature have shown that Ru, Sn, Co, Cr, Fe, Ni, Pd, Os, Mo, and Mn are all elements that help in lessening the CO poisoning effect. The most commonly used catalyst is the PtRu on carbon support. There are many studies in literature that focus on investigating the use of CO-tolerant catalysts to mitigate the effect of CO poisoning. Table 7.1 is given as a summary of the catalyst used and some references.

The studies listed in Table 7.1 are all experimental studies. They are focused on investigating the CO tolerance of PEM fuel cells running with different catalysts. The catalyst type referred to as "other systems" consists of binary and higher systems. Metals, such as Mo, Nb, and Ta, are used along with platinum on carbon support as the catalyst. In some other cases, these same metals are used in combination with PtRu on carbon support. Other high order catalysts are PtRu-H_xMO_3/C (with M = Mo and W).

TABLE 7.1

Summary of CO-Tolerant Catalysts

Catalyst Type	Reference
PtRu/C	35–38
PtSn/C	39–47
PdAu/C	48–50
Other systems	51–62

In order to be able to numerically model the effect of CO-tolerant catalysts, the reaction kinetics should be established. For example, the use of PtRu on carbon support requires the activation of the Ru surface with water. Hence, the reaction kinetics would be as follows [63]:

$$CO + Pt \underset{k_{-1}}{\overset{k_1}{\Longleftrightarrow}} Pt - CO \tag{7.17}$$

$$Ru \mid H_2O \underset{k_{-2}}{\overset{k_2}{\Longleftrightarrow}} Ru \ OH \mid H^+ \mid e^- \tag{7.18}$$

$$Pt - CO + Ru - OH \underset{k_{-3}}{\overset{k_3}{\Longleftrightarrow}} Pt + Ru + CO_2 + H^+ + e^- \tag{7.19}$$

Equation (7.17) to equation (7.19) suggest that for a PtRu alloy catalyst there are two catalyst sites that can be occupied. Carbon monoxide adsorbs onto platinum, while the hydroxide ions adsorb onto ruthenium. Therefore, to understand the rates of reaction mathematically, the coverage of platinum and ruthenium by molecules should be considered. Enbäck and Lindbergh [64] developed a steady-state model to simulate the reaction kinetics of CO poisoning in the presence of a PtRu/C catalyst. The reaction kinetic parameters were obtained from fitting the model predictions to the experimental measurements and are listed in Table 7.2. The following equation set was used to describe their mathematical model:

$$r_1 = k_1 p_{CO}\left(1 - \theta_{CO}^{Pt}\right)\exp\left(-\frac{\beta_1^F \gamma_1}{RT}\theta_{CO}^{Pt}\right) - k_{-1}\theta_{CO}^{Pt}\exp\left(\frac{(1-\beta_1^F)\gamma_1}{RT}\theta_{CO}^{Pt}\right) \tag{7.20}$$

where r is the reaction rate in $mol.s^{-1}.m^{-2}$, p_{CO} is the partial pressure of carbon monoxide, θ_{CO}^{Pt} is the coverage of CO on platinum, β is a symmetry factor, and γ is an interaction coefficient in $J.mol^{-1}$.

TABLE 7.2

Kinetic Parameters Used in the Model in Enbäck and Lindbergh [64]

Parameter	Average of Two Sets	Units
$K_1 = k_1/k_{-1}$	2.75	
$K_2 = k_2/k_{-2}$	7.65	
k_1	2.7×10^{-8}	$mol.m^{-2}PtRu.s^{-1}.atm^{-1}$
$k_2 a_{H_2O}$	5.8×10^{-8}	$mol.m^{-2}PtRu.s^{-1}$ at actual activity of water
k_3	9.1×10^{-12}	$mol.m^{-2}PtRu.s^{-1}$
γ_1	17995	$J.mol^{-1}$
γ_2	670	$J.mol^{-1}$
β_2	0.255	
β_3	0.345	
C_{dl}	0.18	$F.m^{-1} PtRu/C$
β_F	0.5	
S_{PtRu}	100	$m^2 PtRu.g^{-1}$
$S_{PtRu/C}$	700	$m^2 PtRu/C.g^{-1} PtRu$
σ_{PtRu}	2.1	$C.m^2 PtRu$
m_{spec}	0.5	$g.PtRu.m^{-2}$ electrode

Note: The surface of the electrode refers to the geometric surface. Langmuir isotherm was assumed for water adsorption on Ru.

Source: Data from Enbäck, S., G. Lindbergh, 2005. *Journal of the Electrochemical Society*, 152:A23–A31.

$$r_2 = k_2 a_{H_2O}\left(1 - \theta_{OH}^{Ru}\right)\exp\left(\frac{\beta_2 F}{RT}E\right)\exp\left(-\frac{\beta_2^F \gamma_2}{RT}\theta_{OH}^{Ru}\right)$$

$$-k_{-2}a_{H^+}\theta_{OH}^{Ru}\exp\left(\frac{(1-\beta_2)F}{RT}E\right)\exp\left(\frac{(1-\beta_2^F)\gamma_2}{RT}\theta_{OH}^{Ru}\right) \tag{7.21}$$

where a_{H_2O} is the activity of water, θ_{OH}^{Ru} is the coverage of OH on Ru, E is the voltage versus reversible hydrogen electrode in V, and F is Faraday's constant.

$$r_3 = k_3 \theta_{OH}^{Ru}\theta_{CO}^{Pt}\exp\left(\frac{\beta_3 F}{RT}E\right)\exp\left(\frac{\beta_1^F \gamma_1}{RT}\theta_{CO}^{Pt}\right)\exp\left(\frac{\beta_2^F \gamma_2}{RT}\theta_{OH}^{Ru}\right)$$

$$-k_{-3}p_{CO_2}a_{H^+}\left(1-\theta_{CO}^{Pt}\right)\left(1-\theta_{OH}^{Ru}\right) \tag{7.22}$$

$$\times\exp\left(-\frac{(1-\beta_3)F}{RT}E\right)\exp\left(-\frac{(1-\beta_1^F)\gamma_1}{RT}\theta_{CO}^{Pt}\right)\exp\left(-\frac{(1-\beta_2^F)\gamma_2}{RT}\theta_{OH}^{Ru}\right)$$

The current density is calculated as follows:

$$i = m_{spec} S_{PtRu} F(r_2 + r_3)$$ (7.23)

where m_{spec} is the specific weight of catalyst loading in $gPtRu.m^{-2}$ electrode and S_{PtRu} is the specific surface area in $m^2.g^{-1}$.

Once the current density is obtained, the voltage can be found by:

$$\frac{dE}{dt} = -\frac{i}{m_{spec} S_{PtRu/c} C_{dl}}$$ (7.24)

where C_{dl} is the double layer capacitance in $F.m^{-2} PtRu/C$.

The coverages of CO and OH are:

$$\frac{d\theta_{CO}^{Pt}}{dt} = \frac{F}{0.5\sigma_{PtRu}}(r_1 - r_3)$$ (7.25)

where σ_{PtRu} is the charge density in $C.m^{-2}$

$$\frac{d\theta_{OH}^{Ru}}{dt} = \frac{F}{0.5\sigma_{PtRu}}(r_2 - r_3)$$ (7.26)

In the case of PtMo on carbon support catalysts, Molybdenum is also dispersed in the gas diffusion layer. The use of such a technique promotes the water–gas shift reaction (WGSR) in the gas diffusion layer. Thus, the concentration of CO in the gas channel is lowered [65,66]. To study the use of PtMo/C as the catalyst numerically, the WGSR (equation (7.27)) should be incorporated in the model.

$$CO + H_2O \Leftrightarrow CO_2 + H_2$$ (7.27)

The reader is encouraged to examine the work of Christoffersen et al. [67], which introduces the density functional theory. This theory is used to evaluate the effects of the interaction between platinum and other metals. Understanding this interaction then can be used to choose the best materials for the design of the anode.

Finally, it should be pointed out that it is often the case that the effect of using CO-tolerant catalysts is modeled numerically by increasing the electrooxidation rate of CO. It is argued that this assumption is valid because the use of CO-tolerant catalysts allows for a more rapid reaction of CO. In section 7.2.4.3, results of both methods from literature are discussed.

Even though the two mitigation methods discussed earlier are very promising, there is much effort being exerted to develop new methods to decrease the effects of CO poisoning in PEM fuel cells. One of the most obvious processes is the use of advanced reformer design. Currently available reformers are capable of producing a CO content of 50 ppm or less after a warm-up period of up to 2 h [68]. Studies given in Choudhary and Goodman [69] and Lee et al. [70,71] have shown that fitting auxiliary processors, such as shift converters and selective oxidizers, will further clean up the fuel stream. However, these additional stages will increase the complexity and cost of the fuel cell system.

An alternative method to increasing the tolerance of PEM fuel cells to the presence of CO in the fuel stream is increasing the operating temperature. As mentioned earlier, CO adsorption on platinum sites is increased at low temperatures. However, at high temperatures, there is a risk of degradation and dehydration of the membrane. Hence, to achieve high performance at higher temperatures, high temperature (~140°C) membranes are developed [72]. Much work has been focused on developing phosphoric acid-doped polybenzimidazole [73,74] in order to avoid life-time problems and dehydration obstacles at high temperatures. Nevertheless, there are still some problems to be overcome with membrane cycle life and hydration.

In order to numerically model the effect of using high temperature membranes, high operating temperature is used. However, the physical properties of the membrane also should be taken into account. In the case of using high temperature membranes to mitigate the effects of CO poisoning, other transport mechanisms, such as liquid water, should also be modeled. It is crucial to understand the overall effects on the cell rather than on just one phenomenon.

The final mitigation technique that is discussed in this chapter relates the diffusion process and reaction dynamics to the electrode structure. Given that the diffusion coefficients of hydrogen and carbon monoxide are different, a special composite electrode structure is designed. The structure given in Figure 7.2 was investigated by Yu et al. [75]. As it can be seen, there are two catalyst layers in this electrode: an outer layer and an inner layer. Poisonous CO is forced to react with CO active electrocatalysts, PtRu, in advance at the outer layer. The hydrogen can then react at the inner layer with a traditional platinum electrocatalyst, Pt. The choice of electrocatalyst assignment is made due to the fact that hydrogen diffuses faster than carbon monoxide in the gas diffusion layer. Using this electrode structure, Yu et al. [75] found that the negative effect of CO poisoning was decreased dramatically. Other

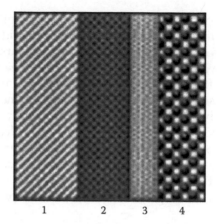

FIGURE 7.2
Schematic structure of composite anode: (1) gas diffusion layer, (2) outer catalyst layer, (3) inner catalyst layer, (4) Nafion membrane. (From Yu, H. et al., 2002. *Journal of Power Sources*, **105**:52–57. With permission.)

studies done on similar structures are found in Shi et al. [76] and Wan and Zhuang [77]

The modeling domain should be adjusted to numerically model the effects of using a multilayer electrode. In this case, the electrochemical reaction must occur at two different layers: the inner and outer catalyst layers.

7.2.3 Modeling of CO Poisoning

Carbon monoxide poisoning occurs due to the presence of CO in the fuel stream. To numerically model this phenomenon, mass transport in the gas flow channel, bipolar plates, gas diffusion layer, and catalyst layer should be taken into consideration. Because carbon monoxide poisoning affects the performance of the cell, the transport of electrons and protons can be hindered and, thus, should be investigated. The following sections (7.2.3.1 to 7.2.3.4) give details about the mathematical equations, which should be solved to understand carbon monoxide poisoning. The mathematical formulation of oxygen bleeding is also put forward.

7.2.3.1 Catalyst Layer Model

The reactant gases enter the catalyst layer after being transported through the void region of the gas diffusion layer. In the catalyst layer, the reactants are consumed to produce protons and electrons. The electrons are transported through the solid region of the catalyst layer and the protons diffuse through the electrolyte membrane region of the catalyst layer. In order to be able to formulate the governing equations of the transport phenomena, the reaction kinetics of the anode catalyst layer should be utilized. In other words, the rates of the electrochemical reactions are considered. For the electrochemical reaction to proceed, three chemical reactions occur: adsorption, desorption, and electrooxidation of the species. Each part of this reaction proceeds at a different rate and contributes to a specific transport phenomenon. The reaction rates are dependent on the concentration and surface coverage of the species as well as the overpotential of the anode catalyst layer. The Butler–Volmer equation with the anodic and cathodic transfer coefficients being equal is used to describe the electrooxidation of carbon monoxide and hydrogen. The rate of hydrogen adsorption on the platinum site is dependent on the hydrogen concentration and coverage and given as follows:

$$q_{H\_ads} = K_{H\_ads} C_{H_2} \left(\theta_{pt}\right)^2 - B_{H\_ads} K_{H\_ads} \left(\theta_H\right)^2 \tag{7.28}$$

where K_{H_ads} is the rate constant of hydrogen adsorption, B_{H_ads} is the rate constant of hydrogen desorption, C_{H_2} is the concentration of hydrogen, and $\theta_{pt} = 1 - \theta_H - \theta_{CO} - \theta_O$ is the platinum coverage with θ_H, θ_{CO}, and θ_O as the coverage on the platinum sites of hydrogen, CO, and oxygen molecules, respectively. In the case when oxygen bleeding is not considered in the numerical model, θ_O is set to zero.

The rate of electrooxidation of the adsorbed hydrogen is dependent on the hydrogen coverage and the anode overpotential and is estimated by:

$$q_{H\_ox} = 2K_{H\_ox} \theta_H \sinh\left(\frac{F\eta_a}{2RT}\right) \tag{7.29}$$

where K_{H_ox} is the rate of hydrogen oxidation and η_a is the anode overpotential. The carbon monoxide adsorption rate follows:

$$\begin{aligned} q_{CO\_ads} = K_{CO\_ads} C_{CO} \left(\theta_{pt}\right) \exp\left(\frac{-\beta r \theta_{CO}}{RT}\right) \\ -B_{CO\_ads} K_{CO\_ads} \theta_{CO} \exp\left(\frac{-[1-\beta] r \theta_{CO}}{RT}\right) \end{aligned} \tag{7.30}$$

where B_{CO_ads} and K_{CO_ads} are the rate of desorption and adsorption of carbon monoxide, respectively, β is a symmetry factor and has a value between 0 and 1, and r is an interaction parameter that represents the effect of lateral-interaction, which depends on the temperature as follows [78]:

$$r = \begin{cases} 39.7 \ kJ/mole & T < 373K \\ 41.4 \ kJ/mole & 373K \leq T < 388K \\ 56.5 \ kJ/mole & T \geq 388K \end{cases} \tag{7.31}$$

The electrooxidation of carbon monoxide is dependent on the coverage of CO and the anode overpotential and is given by:

$$q_{CO\_ox} = 2K_{CO\_ox}\theta_{CO}\sinh\left(\frac{F\eta_a}{2RT}\right) \tag{7.32}$$

where K_{CO_ox} is the rate of the electrooxidation of carbon monoxide.

The rate of oxygen adsorption is given in equation (7.33) and the heterogeneous oxidation rates of adsorbed hydrogen and carbon monoxide are given by equation (7.34) and equation (7.35), respectively.

$$q_{O\_ads} = K_{O\_ads}C_{O_2}\left(\theta_{pt}\right)^2 - B_{O\_ads}K_{O\_ads}\left(\theta_O\right)^2 \tag{7.33}$$

$$q_{H-O\_ox} = K_{H-O\_ox}\theta_O\left(\theta_H\right)^2 \tag{7.34}$$

$$q_{CO-O\_ox} = K_{CO-O\_ox}\theta_O\theta_{CO} \tag{7.35}$$

where K_{O_ads} and B_{O_ads} are the rates of adsorption and desportion of oxygen, respectively, and K_{H-O_ox} and K_{CO-O_ox} are the rates of the heterogeneous oxidation of hydrogen and carbon monoxide, respectively.

The reaction rate constants are determined using Arrhenius' law as follows:

$$K_i = K_i^o\exp\left(\frac{-E_i^k}{RT}\right)$$

$$B_i = B_i^o\exp\left(\frac{-E_i^b}{RT}\right) \tag{7.36}$$

where the preexponential constants, K_i^0 and B_i^0, and the activation energy, E_i^k and E_i^b, are given in Table 7.3.

The reader is encouraged to visit the study by Vogel et al. [79]. In their study, they examined the kinetics of CO poisoning on the anode electrode. They were able to find the rates of reactions of the different steps of this electrochemical reaction.

After establishing the rates of reaction, the governing equations of the gas, electron, and proton transport can be formulated. Due to the porous structure of the catalyst layer, Fick's second law is usually used to describe the transfer of gaseous species as follows:

$$\varepsilon_c \frac{\partial C_i}{\partial t} = D_{i-c}^{eff} \nabla^2 \cdot C_i - A_v q_i \tag{7.37}$$

TABLE 7.3

Kinetic Parameters Used in the Models in Baschuk and Li [21] and Zamel and Li [25]

Parameter	Value	Unit
$K_{H,ads}^{\nu 0}$	298	$m \cdot s^{-1}$
$E_{H,ads}^k$	10.4	$kJ \cdot mole^{-1}$
$K_{H,ox}^0$	23.1	$mole \cdot m^{-2}$
$E_{H,ox}^k$	16.7	$kJ \cdot mole^{-1}$
$B_{H,ads}^0$	4.18×10^{11}	$mole \cdot m^{-3}$
$E_{H,ads}^b$	87.9	$kJ \cdot mole^{-1}$
$K_{CO,ads}^0$	2×10^5	$m \cdot s^{-1}$
$K_{CO,ads}^k$	47.3	$kJ \cdot mole^{-1}$
$B_{CO,ads}^0$	6.87×10^2	$mole \cdot m^{-3}$
$E_{CO,ads}^b$	100	$kJ \cdot mole^{-1}$
$K_{CO,ox}^0$	3.40×10^{10}	$mole \cdot m^{-3} \cdot s^{-1}$
$E_{CO,ox}^k$	127	$kJ \cdot mole^{-1}$
$K_{O,ads}^0$	2	$m \cdot s^{-1}$
$E_{O,ads}^k$	14.2	$kJ \cdot mole^{-1}$
$B_{O,ads}^0$	1.36×10^{36}	$mole \cdot m^{-3}$
$E_{O,ads}^b$	250	$kJ \cdot mole^{-1}$
$K_{H-O,ads}^0$	3.28×10^4	$mole \cdot m^{-2} \cdot s^{-1}$
$E_{H-O,ads}^k$	65.9	$kJ \cdot mole^{-1}$
$K_{CO-O,ox}^0$	5.0×10^8	$mole \cdot m^{-2} \cdot s^{-1}$
$E_{CO-O,ox}^k$	90	$kJ \cdot mole^{-1}$

Source: Data from Baschuk, J.J., X. Li, 2003. *International Journal of Energy Research*, **27**:1095–1116; Zamel, N., X. Li, 2007. *International Journal of Hydrogen Energy*, **33**: 1335–1344.

where i is the gaseous species (hydrogen, carbon monoxide, or oxygen), C is the concentration, q_i is the rate of species adsorption and is dependent on the kinetics formulation, D_{i-c}^{eff} is the effective diffusion coefficient in the catalyst layer, A_v is the active area, and ε_c is the catalyst layer porosity.

The rate of species adsorption, q_i, is:

$$q_i = \begin{cases} q_{H\_ads} & i = H_2 \\ q_{CO\_ads} & i = CO \\ q_{O\_ads} & i = O_2 \end{cases} \tag{7.38}$$

The corresponding rates for hydrogen, carbon monoxide, and oxygen were given earlier in equation (7.28), equation (7.30), and equation (7.33), respectively.

The electrooxidation reaction occurs for hydrogen and carbon monoxide only. Thus, the transport of electrons and protons is dependent on the electrooxidation rates of hydrogen and CO and is governed by Ohm's second law, as given below:

$$\nabla \cdot i_s = -\sigma_{s-c}^{eff} \nabla^2 \cdot \phi_s = -A_v F\left(q_{H\_ox} + 2q_{CO\_ox}\right) \tag{7.39}$$

$$\nabla \cdot i_m = -\sigma_m^{eff} \nabla^2 \cdot \phi_m = A_v F\left(q_{H\_ox} + 2q_{CO\_ox}\right) \tag{7.40}$$

where i_s and i_m are the electronic and protonic current density, respectively, and ϕ_s and ϕ_m are the electronic and protonic potential.

In order for the electrochemical reactions to proceed, the species should be first adsorbed onto the platinum (catalyst) sites. In other words, the platinum sites should be covered with the species. When considering the formulation of CO poisoning in the anode catalyst layer, the coverage of hydrogen and carbon monoxide is needed. They are found using the following equations:

$$\xi \frac{\partial \theta_{H_2}}{\partial t} = 2q_{H\_ads} - q_{H\_ox} - 2q_{H-O\_ox} \tag{7.41}$$

$$\xi \frac{\partial \theta_{CO}}{\partial t} = q_{CO\_ads} - q_{CO\_ox} - q_{CO-O\_ox} \tag{7.42}$$

$$\xi \frac{\partial \theta_{O_2}}{\partial t} = 2q_{O\_ads} - q_{H-O\_ox} - 2q_{CO-O\_ox} \qquad (7.43)$$

where ξ is the molar area density of platinum catalyst sites. In the case that oxygen bleeding is not considered, then equation (7.43) is neglected and the heterogeneous oxidation of hydrogen and carbon monoxide by oxygen should also be neglected. In other words, q_{H-O_ox} and q_{CO-O_ox} are set to zero.

Equation (7.37) through equation (7.43) form the set of governing equations for the transport phenomenon taking place in the catalyst layer. Before proceeding on to discuss the parameters needed to close this set of equations, it should be pointed out that the transport of electrons and protons respond almost instantaneously to the change in electrical potential compared to the slow process in the transport of gas species. Thus, in equation (7.39) and equation (7.40), the transient terms are neglected. This behavior has been studied by Wu et al. [80].

Now, to be able to solve equation (7.37) through equation (7.43), many physical conditions should be obtained. The porosity of the catalyst layer, ε_c, is directly dependent on the amount of platinum and carbon fractions as illustrated by [81]:

$$\varepsilon_c = 1 - \left(\frac{1}{\rho_{pt}} + \frac{1 - f_{pt}}{f_{pt}\rho_c} \right) \frac{m_{pt}}{l_c} \qquad (7.44)$$

where ρ_{pt} is the density of platinum, ρ_c is the density of carbon, f_{pt} is the mass ratio of platinum on carbon support, m_{pt} is the mass loading of platinum per unit area, and l_c is the catalyst layer thickness.

D^{eff}_{i-c} is the effective diffusion coefficient of species i in the catalyst layer. The diffusion coefficient in the catalyst layer should take into account the diffusion of the reactant gases in the electrolyte membrane, gas pores, and liquid water, which are present in the catalyst layer. Thus, the effective diffusion coefficient of species i can be described as follows [21]:

$$\frac{1}{D^c_{i,eff}} = \left(\frac{RT}{H_i} \right) \left(\frac{1 - \delta_m - \delta_{H_2O}}{D^{eff}_{i-g}} \right) + \frac{\delta_{H_2O}}{D^{eff}_{i-H_2O}} + \frac{\delta_m}{D^{eff}_{i-m}} \qquad (7.45)$$

where R is the universal gas constant, T is the temperature, H_i is Henry's constant of species i, δ_m is the fraction of the electrolyte membrane in the catalyst layer, and δ_{H_2O} is the fraction of liquid water in the catalyst layer. In order to evaluate the effective diffusion coefficient of species i in the void region (D^{eff}_{i-g}), electrolyte (D^{eff}_{i-m}), and liquid water ($D^{eff}_{i-H_2O}$), Bruggemann's correction factor, which relates the effective diffusion coefficient to the bulk diffusion coefficient, is used:

$$D_{i-j}^{eff} = \left(\varepsilon_c\right)^{1.5} D_{i-j} \tag{7.46}$$

The diffusion coefficients of hydrogen and oxygen in the polymer electrolyte are evaluated as follows [14]:

$$D_{H_2-m} = 4.1 \times 10^{-7} \exp\left(\frac{-2602}{T}\right) m^2 s^{-1} \tag{7.47}$$

$$D_{O_2-m} = 3.1 \times 10^{-7} \exp\left(\frac{-2768}{T}\right) m^2 s^{-1} \tag{7.48}$$

The diffusion of the carbon monoxide in the electrolyte is taken to be equal to that for oxygen. To evaluate Henry's constant in the catalyst layer, its relation to the amount of liquid water and Nafion® in the catalyst layer should be taken into account as shown below [21]:

$$\frac{1}{H_i^c} = \frac{(1-\lambda)H_i^{H_2O} + \lambda H_i^{Nafion}}{H_i^{H_2O} H_i^{Nafion}} \tag{7.49}$$

where $H_i^{H_2O}$ and H_i^{Nafion} are Henry's constants of species i in water and Nafion, respectively, and λ is evaluated as follows:

$$\lambda = \frac{\delta_{H_2O}}{1-\delta_m} \tag{7.50}$$

Henry's constant for hydrogen and oxygen in Nafion are found from [20]:

$$H_{H_2}^{Nafion} = 4.5 \times 10^4 \, atm.cm^3 / mole \tag{7.51}$$

$$H_{O_2}^{Nafion} = \exp\left(\frac{-666}{T} + 14.1\right) atm.cm^3 / mole \tag{7.52}$$

The values of Henry's constant for hydrogen, carbon monoxide, and oxygen in liquid water can be found in Dean [82].

A_v is the active area, which can be calculated as follows:

$$A_v = \frac{m_{pt}A_s}{l_c}$$ (7.53)

where m_{pt} is the mass loading of platinum per unit area, A_s is the surface per unit mass, which is related to the mass ratio of platinum on carbon support, f_{pt}, as tabulated in Table 7.4 [83], and l_c is the catalyst layer thickness.

The effective conductivities of the conducting solid and electrolyte are related to their bulk values through Bruggemann's correction factor [84]:

$$\sigma_{s-c}^{eff} = (1-\varepsilon_c)^{1.5}\sigma_s$$ (7.54)

$$\sigma_m^{eff} = (\delta_m\varepsilon_c)^{1.5}\sigma_m$$ (7.55)

7.2.3.2 Transport in the Gas Diffusion Layer

The gas diffusion layer (GDL) consists of two regions: (1) a void region for gas transport and (2) a solid region for the transport of electrons. Due to the porous nature of the gas diffusion layer, the transport of gaseous species is also governed by Fick's second law of diffusion as follows:

$$\varepsilon_{GDL}\frac{\partial C_i}{\partial t} = D_{i-GDL}^{eff}\nabla^2 \cdot C_i$$ (7.56)

TABLE 7.4

Catalyst Surface Area Per Unit Mass of Catalyst versus Mass Ratio of Platinum to Carbon Support

f_{pt}	A_s
0.1	140
0.2	112
0.3	88
0.4	32
0.6	72
0.8	11
1.0	28

Source: E-TEK. *Gas Diffusion Electrodes and Catalyst Materials* catalog, 1995; Marr, C., X. Li, 1999. *Journal of Power Sources*, 77:17–27. With permission.

where ε_{GDL} is the porosity of the gas diffusion layer; C_i is the concentration of species i with i as hydrogen, carbon monoxide, and oxygen; and D^{eff}_{i-GDL} is the effective diffusion coefficient of species i in the gas diffusion layer, which can be obtained as follows [84]:

$$D^{eff}_{i-GDL} = \varepsilon^{1.5}_{GDL} D_{i-bulk}$$

(7.57)

The transport of electrons follows Ohm's law as seen below:

$$\nabla \cdot i_s = -\sigma^{eff}_{s-GDL} \nabla^2 \cdot \phi_s = 0$$

(7.58)

where

$$\sigma^{eff}_{s-GDL} = \left(1 - \varepsilon_{GDL}\right)^{1.5} \sigma_s$$

7.2.3.3 Transport in the Flow Channel and Bipolar Plate

It is important to consider the mass transport in the flow channels as well as the bipolar plate. The flow channels are used to transport the gaseous species to the gas diffusion layer, while the electrons travel through the bipolar plate toward the gas diffusion layer. Ohm's law governs the transport of electrons in the bipolar plate as follows:

$$\nabla \cdot i_s = -\sigma_{s-BP} \nabla^2 \cdot \phi_s = 0$$

(7.59)

where σ_{s-BP} is the electrical conductivity of the bipolar plate.

The transport of the gas species is not as straightforward and is highly dependent on the computational domain of interest. The flow channels are grooved into the bipolar plate and their design can vary depending on the performance desired. When solving for the mass transport of gaseous species in the flow channels, the momentum and mass should be conserved as follows:

$$\frac{\partial(\rho \vec{u})}{\partial t} + \nabla \cdot (\rho \vec{u} \vec{u}) = -\nabla p + \nabla \cdot (\mu \nabla \vec{u})$$

(7.60)

$$\frac{\partial C_i}{\partial t} + \vec{u} \nabla \cdot C_i = D_{bulk} \nabla^2 \cdot C_i$$

(7.61)

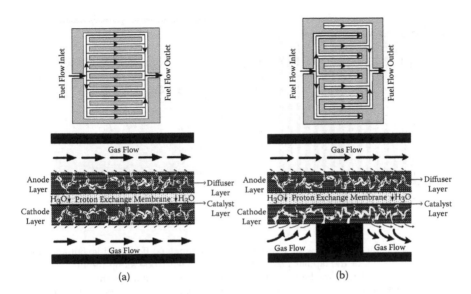

FIGURE 7.3
The physical schematic diagrams of the fuel flow within the PEM fuel cell with different flow fields: (a) conventional flow field and (b) interdigitated flow field. (From Yan, W.-M. et al., 2006. *Journal of Power Sources*, **160**:284 292. With permission.)

The conservation equations are applied to the gas channel despite its design. For PEM fuel cell applications, there are three common flow fields: parallel (also known as conventional) (Figure 7.3a), interdigitated (Figure 7.3b), and serpentine (Figure 7.4). There are numerous studies in literature that are dedicated to understanding these designs and how they influence the overall performance of the cell. Some of these studies can be found in the following references [85–90].

Simulating mass transport in the full geometry of the flow channels along with the other components of the PEM fuel cells can be very computationally expensive. Hence, depending on the objectives of the investigation at hand, some assumptions are usually made. It is very common to use one of the following options for obtaining the concentration distribution of gases in the gas channel:

1. Simulate the flow in one channel along with the other components of the fuel cell. For this option, the computational domain is represented by Figure 7.5.

2. Simulate the flow in the channel flow field and use the results of the simulation for boundary conditions.

3. Estimate the concentration of species in the flow channels and use this estimate for boundary conditions. More details about this option are given in the next section.

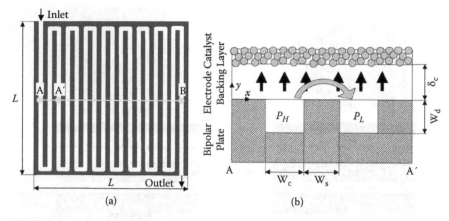

FIGURE 7.4
Schematic of the bipolar plate with serpentine flow channel: (a) flow field layout and (b) the cross-sectional view of the bipolar plate and the porous electrode along the line A-A', illustrating the cross-leakage flow between the two adjacent flow channels through the porous electrode structure as presented by the thick arrow. (From Kanezaki, T. et al., 2006. *Journal of Power Sources*, **162**:415–425. With permission.)

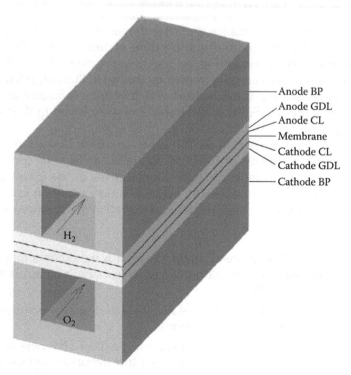

FIGURE 7.5
Computational domain for option 1.

The objective of the next section is to give the reader the mathematical formulation for estimating the concentration of gases at the flow channel/gas diffusion layer boundary. The use of this estimation for understanding the poisoning phenomena is justified since no chemical reactions occur in the flow channels. The effects of the gas diffusion layer and the catalyst layer on the mass transport are of greater interest and impact.

7.2.3.4 Estimation of Flow Channel Concentration

In this section, the gas flow in the anode channel with a serpentine gas channel configuration is considered. The species present in the gas feed are hydrogen, carbon monoxide, and oxygen. The formulation presented in this section is based on that adopted by Baschuk and Li [21].

In order to be able to understand the mass transfer in the gas channel, several assumptions are needed and listed as follows:

1. The gaseous species are ideal gases, the flow is steady, isothermal, and one-dimensional.
2. The species have the same velocity as the bulk velocity.
3. The inlet conditions, such as molar flow rate, species concentration, and pressure, are known.

Consider Figure 7.6 for the formulation of the concentration of gaseous species at the channel/gas diffusion layer interface.

The molar flow rate $N_{i,1}$ is the total molar flux provided to the cell and is a known quantity. In the gas channel, some of species i is consumed at a rate of $N_{i,c}$ as the convective mass transfer to the gas diffusion layer from the channel. The unconsumed species i exits the channel at a flow rate of $N_{i,3}$. The

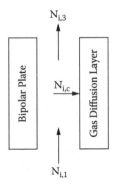

FIGURE 7.6
Schematic of mass transfer for gas channel with $N_{i,1}$ as the molar flux at the inlet of the channel, $N_{i,c}$ is the convective molar flux toward the gas diffusion layer, and $N_{i,3}$ is the molar flux at the outlet of the channel.

amount of species i consumed is a function of the current density of the cell and is described by Faraday's law [84]:

$$\frac{I_\delta A_{cell}}{2F} = N_{H_2,c} + N_{CO,c} - 2N_{O_2,c} \tag{7.62}$$

where I_δ is the cell current density, A_{cell} is the active area of the cell, and F is Faraday's constant.

The molar flow rate and the concentration are related by:

$$N_i = C_i V A_{fc} \tag{7.63}$$

where V is the velocity and $A_{fc} = w_c h_c$ is the cross-sectional area of the gas channel with w_c and h_c as the width and height of the channel, respectively. The velocity in the gas channel is:

$$V = \frac{\dot{m}}{\rho A_{fc}} \tag{7.64}$$

where $\dot{m}$ is the mass flow rate and ρ is the density of the fluid in the channel. The mass flow rate is related to the summation of the molar flow rates of all the species involved through:

$$\dot{m} = \sum \hat{M}_i N_i \tag{7.65}$$

where $\hat{M}_i$ is the molecular weight of species i. The density of the fluid in the gas channel is a combination of the density of the gas species [91] and can be evaluated using:

$$\rho = \sum \frac{x_i P \hat{M}_i}{RT} \tag{7.66}$$

where x_i is the mole fraction of species i.

Thus, the concentration at the channel outlet ($C_{i,3}$) can be found using:

$$C_{i,3} = \frac{V_1}{V_3} C_{i,1} - \frac{N_{i,c}}{V_3 A_{fc}} \tag{7.67}$$

The inlet conditions are known and, thus, the only unknown is the gas channel outlet velocity, V_3. The gas channel outlet is a function of the outlet pressure, which is found using:

$$P_3 = P_1 - \Delta P_f - \Delta P_m \tag{7.68}$$

$$\Delta P_f = \frac{1}{2} \xi \rho V_{ave}^2 \tag{7.69}$$

$$\Delta P_m = \frac{1}{A_{fc}} (\dot{m}_3 V_3 - \dot{m}_1 V_1) \tag{7.70}$$

where ΔP_f is the pressure drop due to the wall shear as well as the bend in the channel [92] and ΔP_m is the pressure drop due to the change in mass flow rate (momentum) between the inlet and outlet of the gas channel. The loss coefficient, ξ, is evaluated as:

$$\xi = \xi_f + 2\xi_b \tag{7.71}$$

$$\xi_f = \frac{56.9}{Re_{d_h}} \cdot \frac{L_{ch}}{d_h} \tag{7.72}$$

$$\xi_b = \frac{0.21}{(r_0 / d_h)^{0.25}} + 50.4 \frac{r_0}{d_h} \left(Re_{d_h} \sqrt{\frac{2r_0}{d_h}} \right)^{-\frac{2}{3}} \tag{7.73}$$

where ξ_f and ξ_b are the friction and bend loss coefficients, L_{ch} is the channel length, r_0 is the bend radius, and d_h is the hydraulic diameter, which is a function of the bipolar plate geometry:

$$d_h = \frac{4A_c}{\wp} \tag{7.74}$$

where $\wp$ is the wetted perimeter of the gas channel. Reynolds number, Re_h, of the flow is based on the hydraulic diameter and the average properties of the fluid as:

$$Re_{d_h} = \frac{\rho_{ave} V_{ave} d_h}{\mu_{ave}} \tag{7.75}$$

The average values of the velocity, density, and viscosity are denoted by V_{ave}, ρ_{ave}, and μ_{ave}, respectively, and are estimated by:

$$V_{ave} = \frac{1}{2A}\left(\frac{\dot{m}_1}{\rho_1} + \frac{\dot{m}_3}{\rho_3}\right) \tag{7.76}$$

$$\rho_{ave} = \frac{1}{2}(\rho_1 + \rho_3) \tag{7.77}$$

$$\mu_{ave} = \frac{1}{2}(\mu_1 + \mu_3) \tag{7.78}$$

The viscosity of the gas mixture can be found using the Wilke correlation [93]:

$$\mu_g = \sum_{i=1}^{N} \frac{x_i \mu_i}{\sum_{j=1}^{N} x_j \phi_{ij}} \tag{7.79}$$

$$\phi_{ij} = \frac{1}{\sqrt{8}}\left(1 + \frac{\hat{M}_i}{\hat{M}_j}\right)^{-1/2}\left[1 + \left(\frac{\mu_i}{\mu_j}\right)^{1/2}\left(\frac{\hat{M}_j}{\hat{M}_i}\right)^{1/4}\right]^2 \tag{7.80}$$

Under the assumption that the concentration at the gas diffusion layer/channel interface, $C_{i,c}$ is constant, it can be found via:

$$N_{i,c} = \frac{Sh D_{i-fc} A_x}{d_h} \Delta C_{i,\ln} \tag{7.81}$$

$$\Delta C_{i,\ln} = \frac{C_{i,1} - C_{i,3}}{\ln\left(\dfrac{C_{i,1} - C_{i,c}}{C_{i,3} - C_{i,c}}\right)} \tag{7.82}$$

where $\Delta C_{i,\ln}$ is the log-mean concentration difference and A_x is the area of the gas diffusion layer, which is exposed to the gas channel and is equal to:

$$A_x = w_c L_{ch} \tag{7.83}$$

Invoking the heat and mass transfer analogy, whereby the heat and mass transfer relations for a particular geometry are interchangeable, the Sherwood number, Sh, is defined as [94]:

$$Sh = \begin{cases} 2.98 & Re_{d_h} \leq 2000 \\[2ex] \dfrac{(C_f/2)(Re_{d_h} - 1000)Sc}{1 + 12.7(C_f/2)^{1/2}(Sc^{2/3} - 1)} & Re_{d_h} > 2000 \end{cases} \tag{7.84}$$

where Sc is the Schmidt number and is equal to:

$$Sc = \frac{\mu_i}{\rho_i D_{i-fc}} \tag{7.85}$$

The diffusion mass transfer coefficient of species i in the gas channel is denoted by D_{i-fc} and is found via [95]:

$$D_{i-fc} = \frac{1 - x_i}{\displaystyle\sum_{j \neq i} \frac{x_j}{D_{i-j}}} \tag{7.86}$$

where D_{i-j} is the binary diffusion coefficient between species i and j and can be calculated in cm^2/s by [95]:

$$D_{i-j} = \frac{BT^{3/2}\sqrt{\left(\dfrac{1}{\hat{M}_i}\right) + \left(\dfrac{1}{\hat{M}_j}\right)}}{Pr_{ij}^2 I_D} \tag{7.87}$$

$$B = \left[10.85 - 2.50\sqrt{\left(\dfrac{1}{\hat{M}_i}\right) + \left(\dfrac{1}{\hat{M}_j}\right)} \right] \tag{7.88}$$

where r_{ij} is the collision diameter in angstroms and I_D is the collision integral for diffusion.

The friction coefficient, C_f, is given by [95]:

$$C_f = \begin{cases} 16\left(Re_{d_h}\right)^{-1/2} & Re_{d_h} \le 2000 \\ 0.079\left(Re_{d_h}\right)^{-1/4} & Re_{d_h} > 2000 \end{cases} \tag{7.89}$$

7.2.3.5 Validation Procedure

Once the mathematical problem has been formulated and solved, the numerical results should be compared to experimental results. In PEM fuel cell modeling, experimental data of the polarization curve are normally used for comparison reasons. The polarization curve is a measure of the performance of the PEM fuel cell. There are two types of performance measures, a full cell polarization curve and a half-cell polarization curve. In many cases, which is also the case in CO poisoning, mass transport is simulated in only half of the cell. One of the most widely used experimental datasets for CO poisoning was collected by Lee et al. [96]. In their study, they investigated the performance of the cell exposed to CO and studied the behavior of the performance depending on the electrode used. Baschuk and Li [21] compared their numerical results to the results from Lee et al. [96]. The comparison is shown in Figure 7.7 and Figure 7.8.

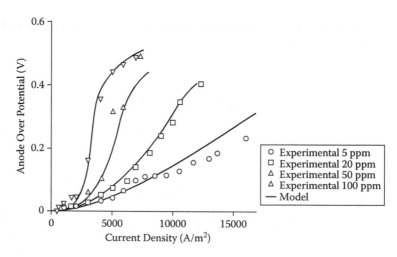

FIGURE 7.7
Comparison between the model prediction of Baschuk and Li [21] and experimental data of Lee et al. [96] at various CO concentrations and a temperature of 358 K. (From Baschuk, J., X. Li. 2003. *International Journal of Energy Research*, 27:1095–1116S; Lee, S. et al., 1999. *Electrochimica Acta*, 44:3283–3293.)

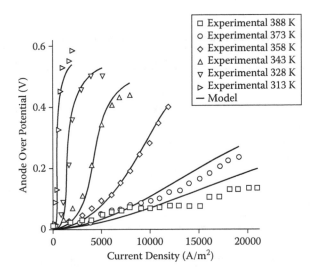

FIGURE 7.8

Comparison between the model prediction of Baschuk and Li [21] and experimental data of Lee et al. [96] at various temperatures and a CO concentration of 20 ppm. (From Baschuk, J., X. Li., 2003. *International Journal of Energy Research*, **27**:1095–1116S; Lee, S. et al., 1999. *Electrochimica Acta*, **44**:3283–3293.)

It is seen that the numerical and experimental results are comparable in most cases. However, the model over-predicts the anode overpotential at low concentrations. Baschuk and Li [21] explained that this over-prediction could be due to oxygen crossover from the cathode while the experimental data was collected. Oxygen crossover and its effects on CO poisoning has been investigated experimentally by Zawodzinski et al. [97]. The slight discrepancy between the model predictions and the experimental results at high temperatures is mostly attributed to error in reproducing the experimental data points from the published article.

As usually is the case, a complete set of information in regards to the cell used to obtain experimental results is not given. Hence, some parameters in the model should be carefully chosen to obtain an agreement between the model prediction and the experimental results. Baschuk and Li [21] gave a detailed description of the design parameters and operating conditions required for this agreement. Their operating and physical parameters are given in Table 7.5.

7.2.4 Results and Discussion

7.2.4.1 CO Poisoning

Numerical simulations of CO poisoning in literature are abundant as shown by the literature review given earlier in the chapter. Many of these investigations are concerned with performing parametric studies on CO poisoning.

TABLE 7.5

Physical and Operating Parameters Used in the Model by Baschuk and Li [21]

Parameter	Value	Units
Platinum mass loading, m_{pt}	0.004	kg/m^2
Mass ratio, f_{pt}	0.2	
Electronic conductivity, σ_s	2700	S/m
Protonic conductivity, σ_m	17	S/m
Catalyst layer thickness, δ_c	2.0465×10^{-5}	m
Membrane fraction, l_m	0.9	
Temperature range	313–388	K
Operating pressure	1	atm
Relative humidity	100%	

Source: Baschuk, J.J., X. Li, 2003. *International Journal of Energy Research,* **27**:1095–1116.

The effect of varying different operating and physical parameters on the CO poisoning phenomenon is investigated. Some of these parameters include operating temperature and pressure, flow rate of fuel feed, porosity and thickness of layers, and design of gas channels. In the case of transient models, the supply of transient CO concentration is usually investigated. These parametric studies can be done to understand the effect of these parameters not only on the performance of the cell, but also on the coverage of the platinum sites. Thus, an overall understanding of the electrochemical reactions is gained. Further, the rate constants of reaction are also investigated in order to understand the electrochemical phenomenon that occurs.

Springer et al. [20] created a one-dimensional, steady-state model of a PEM fuel cell, which is operated on a reformate feed. In their study, they investigated the effects of CO poisoning, dilution of feed, and utilization. They looked into the effects of fuel utilization on the overall poisoning effect of carbon monoxide. In order to do so, they used four levels of utilization (0, 0.5, 0.8, and 0.9) and two levels of CO concentration (10 ppm and 50 ppm). They also extended the study to investigate the effect of hydrogen dilution. They used 75% and 40% hydrogen as the fuel feed. Their results are shown in Figure 7.9, Figure 7.10, and Figure 7.23. From Figure 7.9 and Figure 7.10, it can be seen that the effects of CO poisoning are amplified when the inlet feed is diluted and the fuel utilization is high, even at low CO levels. It is found that with 10 ppm CO, the critical current density at a platinum anode catalyst is 0.8 and 0.4 A/cm² with a fuel feed of 75% H₂ and 0% and 90% fuel utilization, respectively. At the same CO level and a more dilute feed of 40% H₂, the critical current density is 0.5 and 0.2 A/cm² at 0% and 90% fuel utilization, respectively. Mitigation of CO poisoning methods are required since the operation with "critical" anode current densities as low as 0.5 A/cm², or below, is quite impractical [20]. Some of these methods consist of either improving the platinum catalyst or using the method of oxygen bleeding.

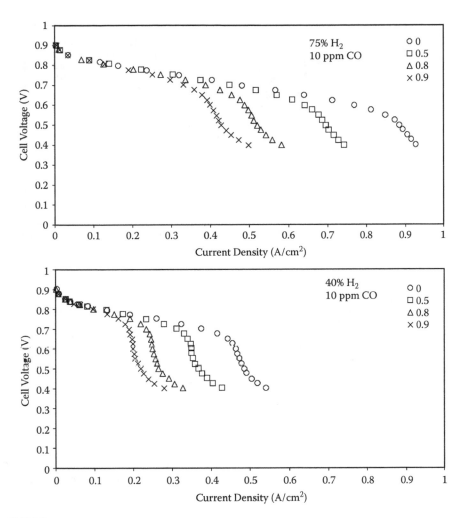

FIGURE 7.9
Fuel utilization effect on cell polarization in presence of CO, calculated for 1 atm, 75% (top) or 40% H_2 (bottom) inlet, with 10 ppm CO. In each case, the polarization curve is calculated for four fuel utilization levels of 0, 0.5, 0.8 and 0.9. (From Springer, T. et al., 2001. *Journal of the Electrochemical Society*, **148**:A11–A23. With permission.)

In addition, the numerical models can be used in order to understand the overall effect of CO poisoning on other transport phenomena, such as liquid water transport. In a study by Wang and Chu [24], they developed a transient, one-dimensional, two-phase numerical model of the electrolyte membrane and anode and cathode catalyst layers. Their model was used to look into the effect of CO poisoning on the water distribution in the catalyst layers and the electrolyte membrane. With 100% H_2 (i.e., the hydrogen feed was not dilute), 10 ppm CO level, and a cell voltage of 0.6 V, they investigated the liquid water saturation in the catalyst layers and the water content in

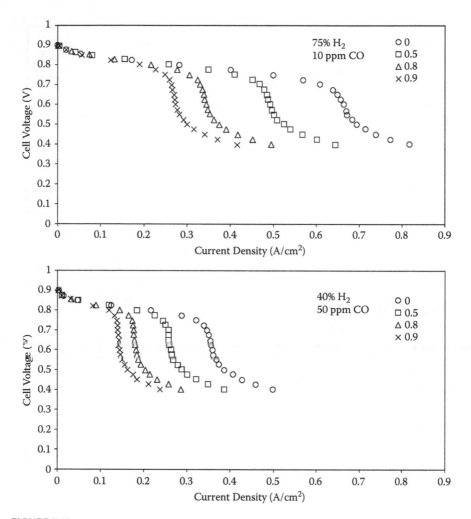

FIGURE 7.10
Fuel utilization effect on cell polarization in presence of CO, calculated for 1 atm, 75% (top) or 40% H_2 (bottom) inlet, with 50 ppm CO. In each case, the polarization curve is calculated for four fuel utilization levels of 0, 0.5, 0.8, and 0.9. (From Springer, T. et al., 2001. *Journal of the Electrochemical Society*, **148**:A11–A23. With permission.)

the membrane. Due to poisoning, the current density is reduced, weakening the effect of the electroosmotic drag. Further, the oxygen reduction reaction is also suppressed, reducing the diffusion of water from the cathode to the anode. Hence, with time the liquid water saturation decreases in the anode catalyst layer, as shown in Figure 7.11. The same trend is also seen in the cathode catalyst layer, as illustrated in Figure 7.12. The small electroosmotic drag and the generation of less liquid water cause the saturation level to drop with time. It is also important to notice here that the amount of liquid water in the cathode is higher than that in the anode.

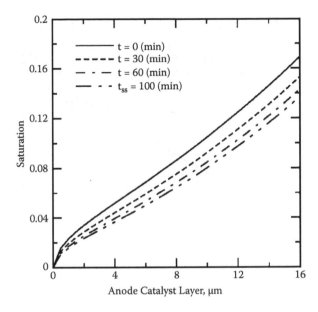

FIGURE 7.11
The transient evolution of the liquid water saturation profile across the anode catalyst layer with 100% H_2, 10 ppm CO, 0.6 V. (From Wang, C.P., H.S. Chu, 2006. *Journal of Power Sources*, 159:1025–1033. With permission.)

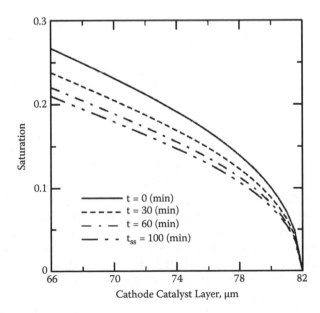

FIGURE 7.12
The transient evolution of the liquid water saturation profile across the cathode catalyst layer with 100% H_2, 10 ppm CO, 0.6 V. (From Wang, C.P., H.S. Chu, 2006. *Journal of Power Sources*, 159:1025–1033. With permission.)

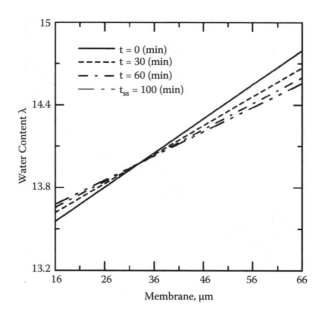

FIGURE 7.13
The transient evolution of the water content profile across the membrane with 100% H₂, 10 ppm CO, 0.6 V. (From Wang, C.P., H.S. Chu, 2006. *Journal of Power Sources*, **159**:1025–1033. With permission.)

Across the membrane, the gradient of the liquid water distribution also declines with time, as shown in Figure 7.13. This occurs since the rate of the reaction on both the anode and the cathode sides decreases as the poisoning increases. This is further amplified by the effect of the electroosmotic drag and the diffusion of liquid water from the cathode to the anode, which are also weakened.

Other studies were also extended to investigate the actual kinetics of the poisoning phenomenon. In the study done by Baschuk and Li [21], which assumes a one-dimensional steady flow, they considered the use of either Langmuir or Temkin kinetics in describing CO adsorption and desorption, as seen in Figure 7.14. They found that the use of Langmuir adsorption and desorption kinetics resulted in a poor fit with the experimental data they used for comparison, which were generated by Lee et al. [96]. They explained that this poor fit arises because the Langmuir model does not account for the interference effect of the previously adsorbed CO on the adsorption process.

The same model [21] was used in order to understand the difference between the use of a gas mixture of reformate and CO and the use of a gas mixture of pure hydrogen and CO. Their results are illustrated in Figure 7.15. They showed that CO poisoning is more severe when reformate fuel is utilized, indicating that the relative concentration of CO to H₂ influences this poisoning phenomenon. This is also in agreement with experimental measurements done by Dhar et al. [18].

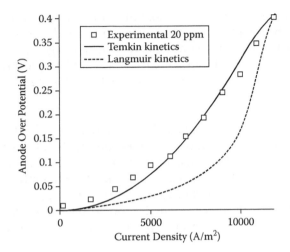

FIGURE 7.14
Comparison of Temkin and Langmuir adsorption and desorption kinetics for a CO concentration of 20 ppm. (Experimental data is from Baschuk, J., X. Li. 2003. *International Journal of Energy Research*, **27**:1095–1116S; Lee, S. et al., 1999. *Electrochimica Acta*, **44**:3283–3293.)

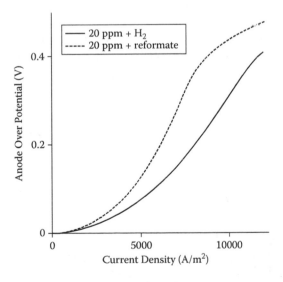

FIGURE 7.15
Effect of reformate fuel (fully humidified, 75% H_2 + 25% CO_2) on anode overpotential for a CO concentration of 20 ppm. (From Baschuk, J., X. Li., 2003. *International Journal of Energy Research*, **27**:1095–1116S.)

Further, the model in Baschuk and Li [21] was used to study the effect of the temperature on the performance of the cell because CO poisoning is an important issue in PEM fuel cells due to its low operating temperature.

Their results are given in Figure 7.16. Since an increase in the temperature results in an enhanced hydrogen electrooxidation and a decrease in CO coverage, the effect of CO can be mitigated. The coverage of CO decreases at higher temperatures due to an increased rate of CO desorption. The transient behavior of the temperature effect on CO poisoning is also similar, as shown in Figure 7.17.

Figure 7.17 was obtained using a model developed by Zamel and Li [25], which is based on the mathematical formulation used by Baschuk and Li [21]. The model is a transient, one-dimensional model of the anode catalyst layer. As can be seen, the transient response with higher temperatures is faster. Thus, the overall performance at a higher temperature is enhanced. Although high temperature PEM fuel cell operation appears advantageous, high temperatures can lead to dehydration of the membrane. The operation at high temperatures can only be possible with the use of improved membranes.

The operating pressure of the cell is another common factor to study. The effect of pressure on the anode overpotential is illustrated in Figure 7.18. This figure is taken from the study by Baschuk and Li [21] and has been obtained using a cell temperature of 358 K. As illustrated through the figure, increasing the pressure mitigates the effect of CO poisoning, yet these gains diminish as pressure increases. An increase in pressure increases the concentration of hydrogen more than that of carbon monoxide; however, the sticking probability of CO on platinum is 15 times higher than that of hydrogen. So, in other words, the high sticking probability of carbon monoxide offsets the

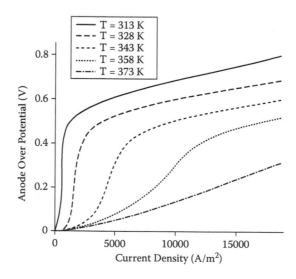

FIGURE 7.16

Effect of temperature on the anode overpotential of a PEM fuel cell poisoned by 20 ppm CO. Cell operating pressure is 159 kPa. (From Baschuk, J., X. Li., 2003. *International Journal of Energy Research*, **27**:1095–1116S.)

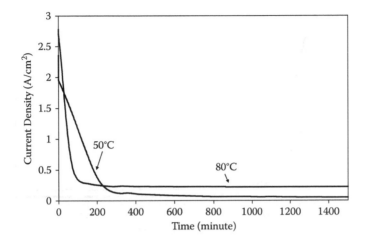

FIGURE 7.17
Effect of temperature on transient behavior of the current density in the anode catalyst layer
– P = 1 atm, η_a = 0.1 V, 100 ppm CO. (From Zamel, N., X. Li, 2007. *International Journal of Hydrogen Energy*, **33**: 1335–1344. With permission.)

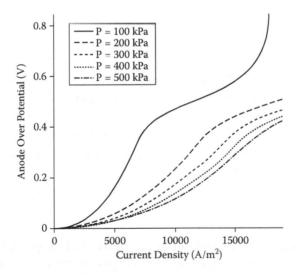

FIGURE 7.18
Effect of pressure on the anode overpotential of a PEM fuel cell poisoned by 20 ppm CO. Cell operating temperature is 358 K. (From Baschuk, J., X. Li, 2003. *International Journal of Energy Research*, **27**:1095–1116S.)

benefits of increasing the pressure. The transient behavior of the effect of pressure on the anode current density is given in Figure 7.19 where it is seen that increasing the pressure slightly increases the transient response.

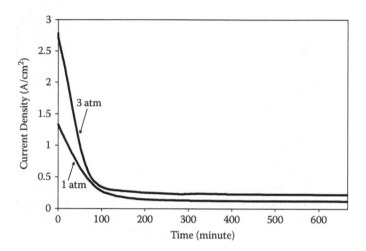

FIGURE 7.19
Effect of pressure on transient behavior of the current density in the anode catalyst layer – T = 80°C, η_a = 0.1 V, 100 ppm CO. (From Zamel, N., X. Li, 2007. *International Journal of Hydrogen Energy*, **33**: 1335–1344. With permission.)

Zamel and Li [25] also studied the effects of exposure of the cell to transient CO concentration. In order to do so, the transient concentration under investigation followed:

- For the time period 0 to 30 min, the anode feed is pure hydrogen.
- For the time period 30 to 50 min, the anode feed is hydrogen + 100 ppm carbon monoxide.
- For the time period 50 to 110 min, the anode feed is pure hydrogen.
- For the time period 110 to 130 min, the anode feed is hydrogen + 100 ppm carbon monoxide.
- For the time period 130 to 190 min, the anode feed is pure hydrogen.

The results of their investigation are illustrated through Figure 7.20. It is obvious that, even though the allowed poisoning time is a third of the recovery time, a full recovery is not achieved. The bonds created between the platinum and carbon monoxide are much stronger than those created by the platinum and the hydrogen molecules. Nevertheless, the poisoning effect is reduced with the use of transient concentrations of CO. After 190 min, the current density is 1.5 A/cm^2 with the use of transient concentrations, while it is 0.26 A/cm^2 after 190 min when the anode is exposed to 100 ppm CO + H$_2$.

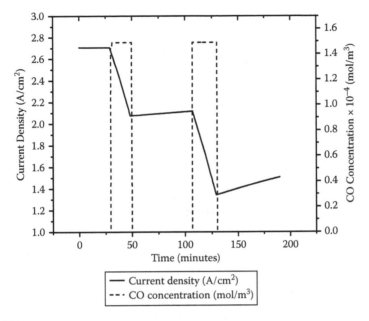

FIGURE 7.20
Current density profile over time of anode exposed to transient CO concentrations – P = 3 atm, T = 80°C, η_a = 0.1 V, 100 ppm CO. (From Zamel, N., X. Li, 2007. *International Journal of Hydrogen Energy*, **33**: 1335–1344. With permission.)

7.2.4.2 Oxygen Bleeding

The introduction of oxygen in the fuel stream results in lessening the effects of CO poisoning in PEM fuel cells. Using the reaction kinetics and the governing equations listed earlier, oxygen bleeding and its effects can be simulated numerically. The model developed by Baschuk and Li [21] incorporated the effects of oxygen bleeding on CO poisoning. Their results are shown in Figure 7.21. This figure shows the anode overpotential versus the current density for the case of pure hydrogen, addition of 100 ppm CO to the fuel feed, and the addition of 2% oxygen to the fuel feed. It is seen that the introduction of 2% of oxygen allows for a significant performance recovery. However, a full recovery is not obtained at high current densities. CO poisoning occurs because the platinum sites required for hydrogen adsorption and electrooxidation are blocked. With the introduction of oxygen, heterogeneous catalysis of CO occurs, resulting in available reaction sites for hydrogen adsorption and electrooxidation. However, the heterogeneous catalysis requires catalyst sites; hence, the amount of available sites for adsorption is reduced from the CO-free case. The difference between low current densities and high current densities occurs due to the rate at which the hydrogen electrooxidation occurs. At low current densities, the electrooxidation of hydrogen is rapid and any decrease in available sites has negligible effect

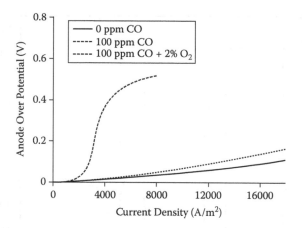

FIGURE 7.21
Simulation of 2% oxygen bleeding to mitigate 100 ppm CO poisoning. (From Baschuk, J., X. Li, 2003. *International Journal of Energy Research*, **27**:1095–1116S.)

on the anode overpotential. To reduce cost, the amount of platinum is usually decreased. However, this reduction of platinum in the anode catalyst layer may not be practical for operation with CO poisoned fuel if oxygen or air bleeding is used as the mitigation method. The transient behavior of the current density with the use of oxygen bleeding is show in Figure 7.22.

7.2.4.3 CO-Tolerant Catalysts

Numerical modeling of CO-tolerant catalysts can be done either from fully understanding the electrochemical kinetics associated with these catalysts or by modifying the electrooxidation rate of carbon monoxide. Springer et al. [20] utilized the assumption that the advantage of improved catalysts is a significantly higher rate of CO electrooxidation at low anodic potentials to simulate the performance of the cell with CO-tolerant catalysts. Their results are presented in Figure 7.23. In this figure, the rate of the electrooxidation of CO is increased from 1×10^{-8} (used in the production of Figure 7.16 and Figure 7.17) to 1×10^{-6}.

Milder drops of potential within each polarization curve are observed. The increased rate of CO electrooxidation results in a more rapid CO oxidation providing more catalyst sites for the hydrogen electrooxidation to occur. Polarization curves with similar behaviors have been reported for PEM fuel cells with PtRu/C and PtMo/C anode catalysts [98].

The model developed by Enbäck and Lindbergh [64] took into account the reaction kinetics of CO on Pt and OH on Ru. The main objective of their study was to obtain an understanding of adsorption of CO and OH on the catalyst sites. They found that the adsorption of CO onto the platinum can either be linear or bridge bonded to the surface (Figure 7.24). However, their

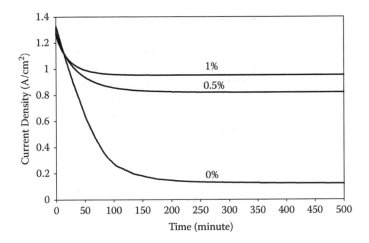

FIGURE 7.22
Effect of oxygen bleeding on the anode current density – P = 1 atm, T = 80°C, η_a = 0.1 V, 100 ppm.

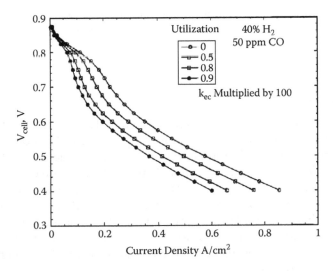

FIGURE 7.23
Fuel utilization effect on cell polarization in the case of 40% H_2 feed stream containing 50 ppm CO. (From Springer, T. et al., 2001 *Journal of the Electrochemical Society*, **148**:A11–A23. With permission.)

work could not make the distinction between the two bonding mechanisms. Further, they concluded that the oxidation of CO is rate determining with other reactions in pseudo-equilibrium up to about 0.5 V. The adsorption of CO, however, is not rate determining. Finally, they proposed that their model could be used to obtain the CO coverage on platinum and combined with another model for hydrogen oxidation.

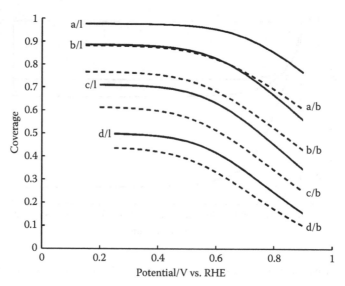

FIGURE 7.24
Simulated coverage of CO_{ads} on Pt with model with linear bounded CO_{ads}, continuous lines, and bridge bounded CO_{ads}, dashed lines. The letters a to d correspond to 10%, 1%, 1000 ppm, 100 ppm CO. (From Enbäck, S., G. Lindbergh, 2005. *Journal of the Electrochemical Society*, 152:A23–A31. With permission.)

7.2.5 Summary

In this part of the chapter, a comprehensive literature review of the modeling and experimental efforts of CO poisoning and its mitigation methods is given. The mathematical formulation of the reaction kinetics of CO poisoning, oxygen bleeding, and the use of a PtRu/C catalyst is given. The governing equations for mass transport in the catalyst layer, gas diffusion layer, and the bipolar plate and flow channels of the anode of PEM fuel cells are put forward. Further, a simplification for estimating the concentration of gases in a one-dimensional flow channel is also explained. The validation procedure for numerical modeling is also presented. Finally, results from literature are presented and explained.

After examining this part of the chapter, the reader should be able to formulate the mathematical model for CO poisoning and its mitigation methods.

7.3 CO_2 Contamination Kinetics

Carbon dioxide exists in high concentrations (>20%) in the fuel stream when reformate fuel is used. These high concentration values can affect the performance of the cells due to the reduction of CO_2 to CO and the formation of CO from CO_2 by the reverse WGSR. It has been reported by Giner [99] that carbon dioxide is electroreduced by platinum hydrides at low potentials following the reaction:

$$CO_2 + 2Pt - H \rightarrow Pt - CO + H_2O + Pt \qquad (7.90)$$

The *Pt–CO* bond created by the reduction of carbon dioxide affects hydrogen electrooxidation in the same manner as the *Pt–CO* formed via CO_2 reduction. Although, the reduction of carbon dioxide polarizes hydrogen electrooxidation, more studies are tailored to understanding the reverse water-gas shift (RWGS) reaction and its effects. The WGSR is:

$$CO + H_2O \Leftrightarrow CO_2 + H_2 \qquad \Delta H^0_{298} = -41.1\ kJ\ mol^{-1} \qquad (7.91)$$

The above reaction is often used for purification of hydrogen produced from reforming hydrocarbons and alcohols. (For further information in regards to this cleaning process, the reader is encouraged to visit the following references [100-104].) The reverse reaction occurs when carbon dioxide is converted to carbon monoxide inside the cell. The effects of CO_2 poisoning on the performance of the cell can be very significant even at a very low rate of RWGS reaction [105]. The effects of the RWGS reaction have been studied numerically by numerous groups. Simulating the poisoning effects of CO_2 is similar to that of CO. The reaction kinetics should be established before being able to solve the numerical model. In the next section, the catalyst layer kinetics with the presence of CO_2 are given.

7.3.1 CO_2 Reaction Kinetics

In this section, the reaction kinetics of the anode catalyst layer in the presence of carbon dioxide are established. These kinetics can be used in the simulation of mass transport in a similar manner to the CO kinetics. The reaction kinetics formulation given below is the same as that used in the models developed in Janssen [106] and Minutillo and Perna [107]. The Tafel–Volmer mechanism, as shown in equation (7.92) and equation (7.93), is used to describe the hydrogen oxidation.

$$H_2 + 2Pt \Leftrightarrow 2Pt - H \qquad (7.92)$$

$$2Pt - H \Rightarrow 2H^+ + 2e^- + 2Pt \qquad (7.93)$$

The CO adsorption and desorption are modeled by a Langmuir process based on the adsorption of one CO per catalyst site, as shown below:

$$CO + Pt \Leftrightarrow CO - Pt \tag{7.94}$$

The CO oxidation by water is represented by:

$$Pt - CO + H_2O \Rightarrow CO_2 + 2H^+ + 2e^- + Pt \tag{7.95}$$

The CO_2 reduction reaction is treated as suggested by Bellows et al. [12] and given below:

$$CO_2 + 2Pt - H \Rightarrow Pt - CO + H_2O + Pt \tag{7.96}$$

From the reaction kinetics, it is clear that hydrogen and carbon monoxide adsorb onto the active catalyst sites. In order to obtain the current density in the anode catalyst layer, the coverages of hydrogen and carbon monoxide are needed. The coverages of hydrogen and carbon monoxide are found using equation (7.97) and equation (7.98), respectively.

$$F\rho \frac{d\theta_H}{dt} = k_a p_{H_2} \left(1 - \theta_H - \theta_{CO}\right)^2 - k_d \theta_H^2 - k_{rs} p_{CO_2} \theta_H^2 - 2k_{eh}\theta_H \sinh\left(\frac{\eta}{b_h}\right) \tag{7.97}$$

where k_a is the H_2 adsorption rate constant times $2F$ in $A.cm^{-2}.bar^{-1}$, k_d is the H_2 desorption rate constant times $2F$ in $A.cm^{-2}$, k_{rs} is the rate constant of the reverse water gas shift reaction in $A.cm^{-2}$, k_{eh} is the H preexponential electrochemical oxidation rate in $A.cm^{-2}$, and b_h is the Tafel slope of the H electrooxidation reaction in V.

$$2F\rho \frac{d\theta_{CO}}{dt} = k_{ac} p_{CO}\left(1 - \theta_H - \theta_{CO}\right) - k_{dc}\theta_{CO} + k_{rs} p_{CO_2}\theta_H^2 - 2k_{ec}\theta_{CO} \sinh\left(\frac{\eta}{b_c}\right) \tag{7.98}$$

where k_{ac} is the CO adsorption rate constant times $2F$ in $A.cm^{-2}.bar^{-1}$, k_{dc} is the CO desorption rate constant times $2F$ in $A.cm^{-2}$, k_{ec} is the CO preexponential electrochemical oxidation rate in $A.cm^{-2}$, and b_c is the Tafel slope of the CO electrooxidation reaction in V.

The total current density in the anode catalyst layer is a sum of the current, I, generated during the electrooxidation reaction of hydrogen and carbon monoxide and is found using:

$$I = I_H + I_{CO} \tag{7.99}$$

where I_H and I_{CO} are the current densities due to hydrogen and carbon monoxide oxidation, respectively, and found by:

$$I_H = 2k_{eh}\theta_H \sinh\left(\frac{\eta}{b_h}\right) \tag{7.100}$$

$$I_{CO} = 2k_{ec}\theta_{CO} \sinh\left(\frac{\eta}{b_c}\right) \tag{7.101}$$

The Tafel slopes due to the hydrogen, b_h, and carbon monoxide, b_c, oxidation reactions are calculated as:

$$b_h = b_c = \frac{2RT}{F} \tag{7.102}$$

The reaction rates are given in Table 7.6 and they have been taken from Janssen [106]. The reaction rates can be found using the Arrhenius law as functions of temperature:

$$k_i = A_i \exp\left(-\frac{E_i}{RT}\right) \tag{7.103}$$

7.3.2 Discussion

Using numerical simulations, Janssen [106] investigated the phenomena of carbon dioxide poisoning and the RWGS reaction. He found that RWGS is the main effect of CO_2 poisoning resulting in a large part of the catalytic surface being inactive for hydrogen dissociation. It was also shown that the readsorption of desorbed CO molecules is very minor due to the blockage of the platinum surface by the RWGS reaction. Janssen also looked into the effects of reaction rates on the overall performance of the cell. It was found that a high rate constant of RWGS reaction increases the anode polarization losses, which is a similar behavior as when the rate constant of hydrogen dissociation reaction is reduced. He also extended the model to consider three different catalysts, Pt/C, PtRu/C, and PtMo/C, and plotted the results in Figure 7.25. This finding is also consistent with the experimental studies in Grgur et al. [55], Bruijn et al. [105], and Ball et al. [108].

TABLE 7.6

Preexponential Factors and Activation Energies for Rate Constants
Used in the Model in Minutillo and Perna [107]

H_2 Adsorption (k_{ah})	H_2 Desorption (k_{dh})
$A_i = 1.386 \times 10^3$ A.cm^{-2}.bar^{-1}	$A_i = 5.159 \times 10^3$ k_{ah}bar
$E_i = 10.4$ kJ mole^{-1}	$E_i = 87.9$ kJ mole^{-1}
H_2 Electrooxidation (k_{eh})	CO Adsorption (k_{ac})
$A_i = 1.187 \times 10^3$ A.cm^{-2}	$Ai = 2.01 \times 10^8$ A.cm^{-2}.bar^{-1}
$E_i = 16.7$ kJ mole^{-1}	$E_i = 47.3$ kJ mole^{-1}
CO Desorption (k_{dc})	CO-Electrooxidation (k_{ec})
$A_i = 1.91 \times 10^9$ k_{ac}bar	$A_i = 1.9 \times 10^{14}$ A.cm^{-2}
$E_i = 100$ kJ mole^{-1}	$E_i = 127$ kJ mole^{-1}

Reverse Water Gas Shift Reaction (k_{rs})

$Ai = 5.86 \times 10^{19}$ A.cm^{-2}.bar^{-1}

$E_i = 145$ kJ mole^{-1}

Source: Minutillo, M., A. Perna, 2008. *International Journal of Energy Research,*
32:1297–1308.

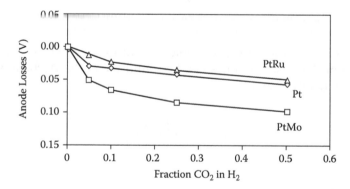

FIGURE 7.25
CO_2 poisoning effect for various anodes using a H_2/CO_2 feed. The figure shows the anode
losses at a fixed current density of 350 mA/cm^2 as a function of CO_2 content. Hydrogen stoichiometry: 1.5, 80°C, 1.5 bar gas pressure. (From Janssen, G.J.M., 2004. *Journal of Power Sources,*
136:45–54. With permission.)

The numerical model developed by Karimi and Li [109] is used to understand
the effects of CO_2 poisoning on the performance of a fuel cell stack. The result
of their model in the form of polarization curves is shown in Figure 7.26.

Karimi and Li [109] found that the presence of carbon dioxide in the fuel
could lead to a significant degradation of the PEM fuel cell stack performance.
This degradation is mainly due to the RWGS reaction. They investigated air
and oxygen bleeding as a mitigation method. The use of this mitigation method
is necessary since the presence of carbon dioxide in the fuel stream brings the

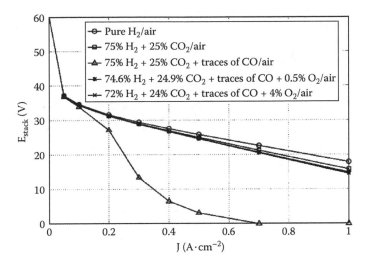

FIGURE 7.26
Polarization curves for a fuel cell stack operating with different fuel compositions. (From Karimi, G., X. Li, 2006. *Journal of Power Sources*, **159**:943–950. With permission.)

fuel cell stack to a halt at a current density of 0.7 A/cm^2. A maximum reduction in the stack voltage of about 12% at 1 A/cm^2 is reached with oxygen bleeding, which is in agreement with the findings of Chinchen et al. [110]. A negligible difference in recovery is found with increasing the oxygen concentration.

Minutillo and Perna [107] also modeled the effects of RWGS reaction on the performance of PEM fuel cells. The steady-state results of the model were compared against experimental data and showed good agreement. They studied the effects of hydrogen dilution, carbon dioxide concentration, and temperature on the cell performance. They showed that the effects of the reverse WGSR are very significant and negatively impact the cell performance. They finally suggested that their model could be extended and made more complex.

7.4 Poisoning of S Compounds

7.4.1 Review of H$_2$S Modeling in Literature

The presence of sulfur compounds in the fuel stream of PEM fuel cells is due to the use of reformate as the fuel. The most common compound is hydrogen sulfide. Hydrogen sulfide poisoning of the catalyst sites is sometimes more severe than carbon monoxide poisoning and occurs even with trace amounts of sulfur. At low temperatures, the adsorption of sulfur onto the catalyst sites becomes very significant, thus, blocking the access for the hydrogen molecules to these sites. Heinzel et al. [111] reported a decrease

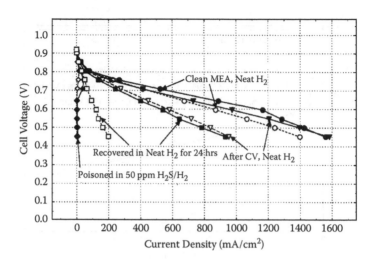

FIGURE 7.27
Effect of 50 ppm H_2S/H_2 on cell performance. Open symbols and dashed lines correspond to Pt-Ru alloy (i.e., PRIMEA MEA series 5621) and closed symbols and solid lines correspond to Pt only catalyst (i.e., PRIMEA MEA series 5510). (From Mohtadi, R. et al., 2003. *Electrochemistry and Solid State Letters*, 6:A272–A274. With permission.)

in cell performance due to the presence of hydrogen sulfide and reported the concentration decrease was as low as 0.05 ppm in the feed. In a study by Mohtadi et al. [112], they reported the severity of the effects of 50 ppm hydrogen sulfide in hydrogen on the performance of PEM fuel cells at an operating temperature of 70 to 80°C. They used this high concentration in order to observe a signal in a reasonable time. They also reported a somewhat reversible, yet not complete, recovery of the performance upon removal of the hydrogen sulfide impurity from hydrogen, as shown in Figure 7.27.

A study by Knights et al. [113] showed the effect of hydrogen sulfide poisoning at various current densities and H_2S concentrations over time. They measured the poisoning effect of hydrogen sulfide experimentally and compared it to an empirical relation, which was found by Mohtadi et al. [114], Kim et al. [115], and Amphlett et al. [116] and is given as:

$$E_c = E_0 - (b_1 + K_{cK}C)\log i - R_0 i + K_{cK}C\log i\left(\frac{1}{1 + K_1\exp(K_2 t)}\right) \quad (7.104)$$

where K_{cK}, K_1, K_2 are constants, which accounts for contamination; C is the concentration of the contaminant; t is the contamination time; i is the current density; and E_0 is cell potential and given by Srinivasan et al. [117]:

$$E_0 = E_r + b\log i_0 \quad (7.105)$$

where E_r is the reversible cell potential, and b and i_0 are the Tafel parameters for oxygen reduction reaction.

They found that at 0.1 A/cm^2 and 80°C a 1.2 ppm H$_2$S could cause a cell voltage drop greater than 300 mV within 25 h of operation. For an even longer period of time, 300 h of operation, and a higher current density, 0.5 A/cm^2, at 0.1 ppm level of H$_2$S in the fuel stream, a voltage drop of 250 mV was observed. Large performance losses were measured at H$_2$S concentrations as low as 50 ppm at 70°C when the fuel cell anode was exposed to the H$_2$S containing fuel for 3.8 h.

The use of mathematical modeling in literature is also seen to be as important in studying the fundamentals of hydrogen sulfide poisoning. The mathematical approach is very similar to that of carbon monoxide poisoning, which is based on the anode kinetics. These anode kinetics were used by Shi et al. [118] and Shah and Walsh [119] in order to understand the reaction mechanism in the presence of hydrogen sulfide in the fuel feed. The study by Shi et al. was based on a transient model of the anode catalyst layer of a PEM fuel cell. In their study, they used experimental data in order to estimate the reaction rates, which were considered as functions of cell current density and contamination level. They found that at a constant current density an increase in contamination (hydrogen sulfide) concentration led to a faster and more severe cell performance degradation. A similar trend was found when the concentration of hydrogen sulfide was kept constant and the current density was increased. A one-dimensional, full cell, transient, nonisothermal, two-phase model was created by Shah and Walsh [119] to study the overall effect of hydrogen sulfide poisoning on a PEM fuel cell. They found that the temperature and water activity in the cell have a significant effect on the poisoning phenomenon in the anode. They showed that due to the dependency of the oxidation reaction of sulfur on water, the water levels in the anode can decrease significantly due to this reaction. Hence, a decrease in the electrolyte membrane conductivity, as well as the cell current density, was noticed. The decrease in current density then affects the water production in the cathode and limits the back diffusion of water via proton migration. They also showed that an increase of the cell temperature lessens the degree of poisoning.

The above mathematical studies are based on experimental studies. Experimental studies of hydrogen sulfide poisoning are more abundant than mathematical studies in literature. Some of the first experimental work, which considers hydrogen sulfide poisoning of platinum, was done by Loučka [120] in 1971. In this study, the adsorption and oxidation of sulfur dioxide on a smooth platinum electrode using the potentiodynamic method were investigated. It was shown that the charges consumed during the oxidation of the adsorbed sulfur and sulfur dioxide are higher than those required for the oxidation of a monolayer of sulfur and sulfur dioxide. Further, the potential range at which the adsorbed sulfur or sulfur dioxide can be reduced was shown to be the same as that which is needed for the adsorption of hydrogen to take place. In a later study, Loučka [121] investigated the effect of adsorbed sulfur on the adsorption of methanol and formic acid. He also examined the kinetics of adsorption of hydrogen sulfide on platinum electrodes. The kinetics of sulfide oxidation on platinum electrodes have also been investigated in other studies. Using cyclic voltammetry and potentiostatic electrolysis, Najdeker and Bishop [122] were able to explain the formation and behavior of platinum sulfide on platinum electrodes. In a study by Jayaram et al. [123], they investigated the effect the preadsorbed sulfur layer on the oxidation of formic acid on a platinum electrode. They found that the adsorption of sulfur from a hydrogen sulfide solution involves either 2-site or 1-site adsorption depending on the coverage. With low sulfur coverage, 2-site adsorption occurs, while 1-site adsorption is noticed with high sulfur coverage. Further, their results show the possibility of multilayer adsorption under appropriate conditions. In a later study [124], the same group investigated the effect of temperature on the oxidation of adsorbed sulfur on a platinum electrode. They showed that there are two forms of sulfur bonds, a strong and a weak bond. Since hydrogen sulfide poisoning is a major degradation mechanism for PEM fuel cells, Mohtadi et al. [114] studied the effect of temperature on this poisoning mechanism in PEM fuel cells. They found that sulfur more strongly adsorbs onto platinum at lower temperatures. Further, assuming a rapid hydrogen sulfide dissociation reaction, the rate constant of formation for a Pt-S species resulting from the exposure to H_2S was determined. The rate of formation of the Pt-S species at 50°C was found to be 69% lower than that at 90°C. However, after long exposure time, the sulfur appeared to eventually saturate the Pt at all temperatures. In another study by Mohtadi et al. [112], they investigated the recovery of the cell after exposure to hydrogen sulfide. They found that unlike CO poisoning, full recovery is not applicable with the use of pure hydrogen after exposure to hydrogen sulfide. Further, the use of a Pt-Ru alloy electrode does not increase the tolerance of the MEA toward hydrogen sulfide. The recovery of the cell from hydrogen sulfide poisoning was also investigated by Imamura and Hashimasa [125]. In their study, they also reported that full recovery is not obtainable by supplying neat hydrogen to the cell after exposure to hydrogen sulfide. However, their study suggested that close to the initial performance after exposure is obtainable by holding the open circuit voltage while supplying pure hydrogen to the cell.

7.4.2 H₂S Contamination

At low temperatures and with the use of hydrogen reformate as the fuel in PEM fuel cells, a trace amount of sulfur can result in a severe loss in the PEM fuel cell performance. This loss occurs due to the adsorption of sulfur compounds onto the catalyst sites, thereby reducing the active available sites for the hydrogen reduction reaction. In the following sections, a full description of the kinetics of H_2S poisoning is discussed.

7.4.2.1 Mechanism of H_2S Poisoning

The kinetics of hydrogen sulfide on platinum in the anode catalyst layer of PEM fuel cells is very similar to that of CO on platinum. The reactions governing the dissociation and the electrooxidation of the hydrogen molecule follow the same steps as listed earlier in equation (7.7) and equation (7.8) and are given again here:

$$H_2 + 2Pt \Leftrightarrow 2(H - Pt) \qquad (7.106)$$

$$2(H - Pt) \Leftrightarrow 2Pt + 2H^+ + 2e^- \qquad (7.107)$$

The kinetics of hydrogen sulfide poisoning were investigated by Contractor and Lal [124] who suggested that there are two adsorbed forms of sulfur on platinum with one strongly bonded and the other weakly bonded. Mathieu and Primet [126] proposed that the surface reactions, which hydrogen sulfide follows are:

$$H_2S + Pt \Leftrightarrow Pt - S + H_2 \qquad (7.108)$$

$$Pt - H + H_2S \Leftrightarrow Pt - S + \frac{3}{2}H_2 \qquad (7.109)$$

The above two reactions result in the poisoning of the platinum sites. In other words, the adsorption of the sulfur on the platinum sites results in a smaller area for hydrogen adsorption. Thus, the above two reactions are in competition with the adsorption and electrooxidation reactions of hydrogen given in equation (7.106) and equation (7.107). The formation of the platinum sulfide (Pt-S) can either follow equation (7.108) and equation (7.109) or could be formed electrochemically as well, as suggested by Mohtadi et al. [112], according to the following equation:

$$H_2S + Pt \Leftrightarrow Pt - S + 2H^+ + 2e^- \qquad (7.110)$$

Further, Najdeker and Bishop [122] proposed that platinum disulfide (Pt-S_2) could also be formed according to the reaction:

$$H_2S + Pt - S \Leftrightarrow Pt - S_2 + 2H^+ + 2e^- \qquad (7.111)$$

According to cyclic voltammetry measurements, however, the main species to form is Pt-S, as shown in Mohtadi et al. [112] and Contractor and Lal
[124]. The presence of liquid water in the anode catalyst layer can also take
part in the electrochemical reaction. The adsorbed sulfur can oxidize via
reaction with water. This mechanism was proposed by Loučka [127] and
confirmed by Mohtadi et al. [112] and Wang et al. [128] and is given as:

$$Pt - S + 3H_2O \Leftrightarrow SO_3 + 6H^+ + 6e^- + Pt \tag{7.112}$$

$$Pt - S + 4H_2O \Leftrightarrow SO_4^{2-} + 8H^+ + 6e^- + Pt \tag{7.113}$$

7.4.3 Modeling of H₂S Poisoning

Numerical modeling of hydrogen sulfide poisoning is similar to that of CO
poisoning. The reaction kinetics, which have already been fully defined earlier, are used to understand the mass transport in the catalyst layer. Further,
the mass transport in the GDL, as well as the bipolar plate and the flow channels, should be modeled. In the next section, the mathematical formulation of
the hydrogen sulfide poisoning is established

7.4.3.1 Transport in the Catalyst Layer

In the catalyst layer, the transport of reactant gases, electrons, and protons
should be investigated. The transport of all these species in the presence
of H₂S will be hindered. Similar to the case of CO poisoning, to be able to
formulate the governing equations, the rate of the electrochemical reaction
should be considered. The rate in which the electrochemical reaction proceeds depends on the adsorption, desorption, and electrooxidation of the
reactant gases. The reaction rates are dependent on the concentration and
surface coverage of the species. The competitive adsorption of H₂S and desorption of hydrogen, which is given by the chemical equation (7.109), follows
the Langmuir isotherms:

$$q_1 = r_{1f}\theta_H C_{H_2S} - r_{1b}\theta_s \left(C_{H_2}\right)^{1.5} \tag{7.114}$$

where θ_s is the coverage of the platinum sites by sulfur molecules, and r_{1f}
and r_{1b} are the reaction rates for the forward and backward reactions,
respectively.

The Langmuir isotherms are also used to calculate the rate of the dissociative adsorption of hydrogen, equation (7.106), as shown below:

$$q_2 = r_{2f}\left(1 - \theta_H - \theta_s\right)^2 C_{H_2} - r_{2b}\theta_H^2 \tag{7.115}$$

where r_{2f} and r_{2b} are the reaction rates for the forward and backward reactions, respectively.

The electrooxidation of hydrogen is governed by Butler–Volmer kinetics:

$$q_{H_2-ox} = r_3\theta_H \sinh\left(\frac{\alpha_3 F\eta_a}{RT}\right) \tag{7.116}$$

where r_3 is the rate of the electrooxidation of hydrogen and α_3 is the charge transfer coefficient.

The rate of the sulfur adsorption and desorption given in equation (7.108) follows a Frumkin isotherm:

$$q_{S-ads} = r_{4f}C_{H_2S}(1-\theta_H-\theta_S)\exp\left(-\frac{\alpha_4 i\theta_S}{RT}\right) - r_{4b}C_{H_2S}\theta_S \exp\left(\frac{(1-\alpha_4)i\theta_S}{RT}\right) \tag{7.117}$$

where r_{4f} and r_{4b} are the reaction rates for the forward and backward reactions, respectively, and α_4 is the charge transfer coefficient.

The oxidation reactions in equation (7.110) and equation (7.112) follow a Butler–Volmer-type kinetics:

$$q_{S-ox}^{(1)} = r_{5f}C_{H_2S}(1-\theta_H-\theta_S)\exp\left(\frac{2\alpha_5 F\eta_a}{RT}\right) - r_{5b}C_{H^+}^2\theta_S \exp\left(-\frac{2(1-\alpha_5)F\eta_a}{RT}\right) \tag{7.118}$$

$$q_{S-ox}^{(2)} = r_{6f}\left(C_{H_2S}\right)^3\theta_S \exp\left(\frac{6\alpha_6 F\eta_a}{RT}\right)$$
$$- r_{6b}C_{H^+}^6 C_{SO-}^d (1-\theta_H-\theta_S)\exp\left(-\frac{6(1-\alpha_6)F\eta_a}{RT}\right) \tag{7.119}$$

where r_{5f} and r_{6f} are the reaction rates of the forward reactions in equation (7.110) and equation (7.112), respectively, r_{5b} and r_{6b} are the reaction rates for the backward reactions in equation (7.110) and equation (7.112), respectively, and α_5 and α_6 are the charge transfer coefficients.

Similarly to the carbon monoxide poisoning, in order to better understand the mechanism of hydrogen sulfide poisoning numerically, the rate of reactions should be represented mathematically. The kinetic parameters are given in Table 7.7 and were taken from the studies by Shi et al. [118] and Shah and Walsh [119].

TABLE 7.7

Kinetics Parameters for H_2S Poisoning Used in the Models in Shi et al. [118] and Shah and Walsh [119]

Parameter	Value	Unit
α_3	0.5	
α_4	0.5	
α_5	0.5	
α_6	0.5	
r_{1f}	1×10^{-2}	$m \cdot s^{-1}$
r_{1b}/r_{1f}	1×10^{-6}	$m^{3/2} \cdot mole^{-1/2}$
r_{2f}	$3e^{-10400/RT}$	$m \cdot s^{-1}$
r_{2b}/r_{2f}	$4.18 \times 10^{11} e^{-87900/RT}$	$mole \cdot m^{-3}$
r_3	$23.1 e^{-16700/RT}$	$mole \cdot m^{-2} \cdot s^{-1}$
r_{4f}	3×10^{-2}	$m \cdot s^{-1}$
r_{5f}	1×10^{-4}	$m \cdot s^{-1}$
r_{6f}	4×10^{-12}	$m^7 \cdot s^{-1} \cdot mole^{-2}$

Source: Shi, Z. et al., 2007. *Journal of the Electrochemical Society,* **154**:B609–B615; Shah, A.A., F.C. Walsh, 2008. *Journal of Power Sources,* 185:287–301.

After establishing the rates of reaction, the governing equations of the gas, electron, and proton transport can be formulated. Due to the porous structure of the catalyst layer, Fick's second law is usually used to describe the transfer of gaseous species as follows:

$$\varepsilon_c \frac{\partial C_i}{\partial t} = D_{i-c}^{eff} \nabla^2 \cdot C_i - A_v q_i \qquad (7.120)$$

where i is the gaseous species (hydrogen and hydrogen sulfide), C is the concentration, q_i is the rate of species adsorption and is dependent on the kinetics formulation, D_{i-c}^{eff} is the effective diffusion coefficient in the catalyst layer, A_v is the active area, and ε_c is the catalyst layer porosity. D_{i-c}^{eff}, ε_c, and A_v are found earlier in equation (7.44), equation (7.45), and equation (7.53), respectively.

The rate of species adsorption, q_i, is:

$$q_i = \begin{cases} q_2 - \dfrac{3}{2} q_1 & i = H_2 \\[2mm] q_{S-ads} + q_1 + q_{S-ox}^{(1)} & i = H_2S \end{cases} \qquad (7.121)$$

From equation (7.121), it is clear that the rate of formation of the Pt-H and Pt-S sites affects the overall concentration gradient in the catalyst layer. It is important to note here that concentration of hydrogen sulfide is affected by the desorption and adsorption of sulfur as well as the electrochemical oxidation of sulfur. This has been discussed earlier in section 7.4.2.1. Equation (7.110) suggests that Pt-S sites can be formed electrochemically.

Now, in case hydrogen sulfide poisoning of the catalyst sites is considered, the coverage of hydrogen and hydrogen sulfide should be found. This is achieved using this set of equations:

$$\xi \frac{\partial \theta_H}{\partial t} = -q_1 + 2q_2 - q_{H_2-ox} \tag{7.122}$$

$$\xi \frac{\partial \theta_S}{\partial t} = q_{S-ads} + q_1 + q_{S-ox}^{(1)} + q_{S-ox}^{(2)} \tag{7.123}$$

For this reaction, the transport of electrons and protons can then be solved for as follows:

$$\nabla \cdot i_s = -\sigma_{s-c}^{eff} \nabla^2 \cdot \phi_s = -A_v F \left(q_{H_2-ox} + 2q_{S-ox}^{(1)} + 6q_{S-ox}^{(2)} \right) \tag{7.124}$$

$$\nabla \cdot i_m = -\sigma_m^{eff} \nabla^2 \cdot \phi_m = A_v F \left(q_{H_2-ox} + 2q_{S-ox}^{(1)} + 6q_{S-ox}^{(2)} \right) \tag{7.125}$$

where the effective electric conductivities are evaluated using equation (7.54) and equation (7.55).

Finally, it should be pointed out that the diffusion of hydrogen sulfide in electrolyte is $4.38 \times 10^{-9} m^2 s^{-1}$. The value of Henry's constant for hydrogen sulfide can be found in Dean [82].

7.4.3.2 Transport in the Gas Diffusion Layer

The transport in the gas diffusion layer is similar to that described earlier for the CO poisoning case. The transport of gaseous species and electrons should be considered. In the void region, the gases are transported following Fick's second law of diffusion:

$$\varepsilon_{GDL} \frac{\partial C_i}{\partial t} = D_{i-GDL}^{eff} \nabla^2 \cdot C_i \tag{7.126}$$

where ε_{GDL} is the porosity of the gas diffusion layer, C_i is the concentration of species i with i as hydrogen and hydrogen sulfide, and D_{i-GDL}^{eff} is the effective diffusion coefficient of species i in the gas diffusion layer, which can be obtained as follows [84]:

$$D^{eff}_{i-GDL} = \varepsilon^{1.5}_{GDL} D_{i-bulk}$$

(7.127)

The transport of electrons occurs in the solid matrix of the GDL and following Ohm's law as stated below:

$$\nabla \cdot i_s = -\sigma^{eff}_{s-GDL} \nabla^2 \cdot \phi_s = 0$$

(7.128)

where $\sigma^{eff}_{s-GDL} = (1-\varepsilon_{GDL})^{1.5}\sigma_s$.

7.4.3.3 Transport in the Flow Channel and Bipolar Plate

The mass transport in the flow channels, as well as the bipolar plate, is considered. The flow channels are used to transport the gaseous species to the gas diffusion layer, while the electrons travel through the bipolar plate toward the gas diffusion layer. Ohm's law governs the transport of electrons in the bipolar plate as follows:

$$\nabla \cdot i_s = -\sigma_{s-BP} \nabla^2 \cdot \phi_s = 0$$

(7.129)

where σ_{s-BP} is the electrical conductivity of the bipolar plate.

As mentioned earlier in section 7.2.3.3, the transport in the flow channel is highly dependent on the channel configuration and the momentum and mass conservation equations should be considered as follows:

$$\frac{\partial(\rho\vec{u})}{\partial t} + \nabla \cdot (\rho\vec{u}\vec{u}) = -\nabla p + \nabla \cdot (\mu\nabla\vec{u})$$

(7.130)

$$\frac{\partial C_i}{\partial t} + \vec{u}\nabla \cdot C_i = D_{bulk} \nabla^2 \cdot C_i$$

(7.131)

The reader is encouraged to study section 7.2.3.3 for more information on the different flow channel configurations as well as the estimation procedure that could be followed.

7.4.3.4 Validation Procedure

Once the mathematical model is formulated, it can be easily solved numerically. There are various numerical methods that could be used. In order to understand the validity of the numerical results, they are usually compared against experimental results. As mentioned earlier, the polarization curve is very often used for the purpose of this comparison in PEM fuel cell modeling.

In certain cases, the validation procedure is used to obtain the reaction rates. This is normally done in cases that are not very well understood and researched. Therefore, there is a need to find these reaction rates. This is true in the case of hydrogen sulfide poisoning of platinum in PEM fuel cells. The poisoning phenomenon of H_2S is not as well researched as CO poisoning, hence, many of the reaction rates are not as well defined. In order to overcome this issue, Shi et al. [118] used experimentally measured transient polarization curves as shown in Figure 7.28 to Figure 7.30. It is seen that the cell voltage was measured at a constant current density over time. The numerical model was then used to model this transient behavior. The experimental data were used to fit the reaction rates. A very good agreement between the numerical and measured results is shown.

7.4.4 Results and Discussion

Numerical modeling of the effects of hydrogen sulfide poisoning on PEM fuel cells is not as abundant as that of CO poisoning in literature. In this

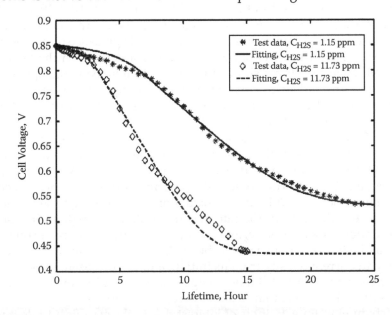

FIGURE 7.28
Experimental results for two different concentration levels of H_2S at a current density of 0.1 A/cm². (From Shi, Z. et al., 2007. *Journal of the Electrochemical Society*, **154**:B609–B615. With permission.)

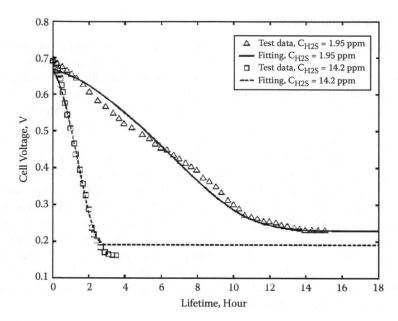

FIGURE 7.29
Experimental results for two different concentration levels of H₂S at a current density of 0.5 A/cm². (From Shi, Z. et al., 2007. *Journal of the Electrochemical Society*, 154:B609–B615. With permission.)

section, some of the results of the study by Shah and Walsh [119] are given. The model they created is a one-dimensional, two-phase, transient model of a full cell, which takes into account hydrogen sulfide presence in the fuel feed. They investigated the effect of the anode current density on the poisoning of the cell. Their results are shown in Figure 7.31. It is seen that, operating at high anode current densities, the extent of poisoning is increased. The extent of poisoning is also increased as the concentration of hydrogen sulfide is increased, as illustrated in Figure 7.32.

Figure 7.33 is a pictorial demonstration of the effect of the channel temperature on the extent of the poisoning phenomenon. As the temperature is increased, the initial current density increases. Shah and Walsh explained that this trend is due to a decrease in the concentration of reactants, a decrease of oxygen reduction and hydrogen oxidation reacts through the Arrhenius dependence, a decrease in the rate of condensation through an increase in the saturation vapor pressure, and an increase in the membrane conductivity. From the figure, it is also noticeable that, as the temperature is increased, the drop in current density from its initial value is decreased, thus reducing the degree of poisoning. At higher temperatures, the sulfur molecules have a lower terminal coverage than that of hydrogen molecules.

The channel water activity is of interest due to the significant role it plays in controlling the amount of water in the cell. Improper amounts of water in the cell can lead to either the flooding or the dehydration of the cell, thus

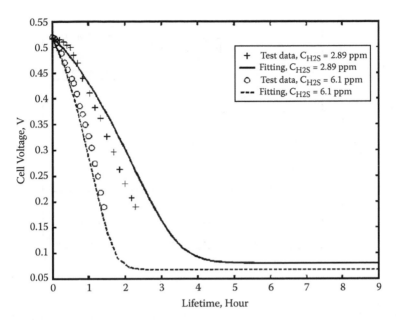

FIGURE 7.30
Experimental results for two different concentration levels of H₂S at a current density of
1.0 A/cm². (From Shi, Z. et al., 2007. *Journal of the Electrochemical Society*, **154**:B609–B615. With
permission.)

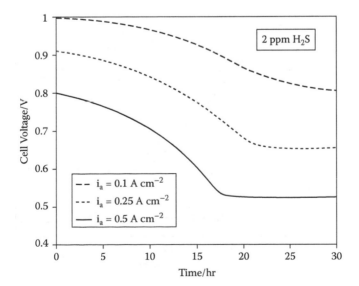

FIGURE 7.31
Transient behavior of the cell voltage at different current densities with a hydrogen sulfide con-
centration of 2 ppm. (From Shah, A.A., F.C. Walsh, 2008. *Journal of Power Sources*, **185**:287–301.
With permission.)

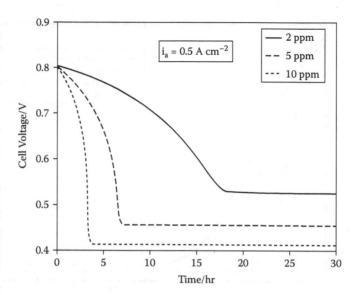

FIGURE 7.32

Transient behavior of the cell voltage for different concentration levels of hydrogen sulfide at an anode current density of 0.5 A/cm². From Shah, A.A., F.C. Walsh, 2008. *Journal of Power Sources,* 185:287–301. With permission.)

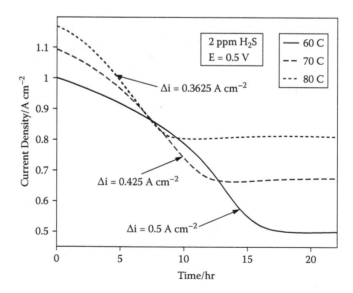

FIGURE 7.33

The effect of channel temperature on the current density for a cell voltage of 0.5 V and a hydrogen sulfide concentration of 2 ppm. From Shah, A.A., F.C. Walsh, 2008. *Journal of Power Sources,* 185:287–301. With permission.)

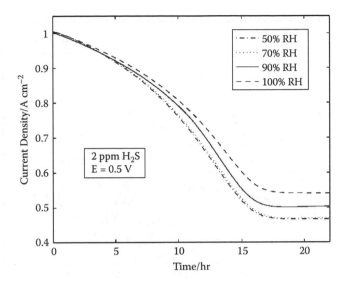

FIGURE 7.34
The effect of channel water activity on the current density for a cell voltage of 0.5 V and a hydrogen sulfide concentration of 2 ppm. From Shah, A.A., F.C. Walsh, 2008. *Journal of Power Sources*, **185**:287–301. With permission.)

hindering the performance. The effect of variations in the water activity on the current density at a cell voltage of 0.5 V and with 2 ppm H_2S in 40% H_2/ N_2 is illustrated in Figure 7.34. It is seen that the performance is improved as the water activity increases. However, for water activities below 0.7, very little change in performance is observed. Hence, to better understand this effect, the water saturation along the length of the cell is plotted for a water activity of 1 and 0.7, as shown in Figure 7.35.

It is immediately noticeable that the saturation levels, with a lower water activity, are lower mainly due to the reduced rates of condensation. The levels in the anode are approximately zero and, thus, are not displayed in the plots. Further, due to the poisoning process, the saturation levels decrease with time. This is due to reduced rate of water production and reduced back diffusion caused by the drag of water molecules from the anode to the cathode.

7.4.5 Summary

The mechanism of hydrogen sulfide contamination and its effects on the overall performance of PEM fuel cells are discussed in this part of the chapter. The reaction kinetics as well as the mathematical formulation of the poisoning phenomenon of hydrogen sulfide are put forward. Further, numerical results of the effects of hydrogen sulfide on the performance of PEM fuel cells are given. The importance of comparing the numerical results to experimental data is further emphasized in this section.

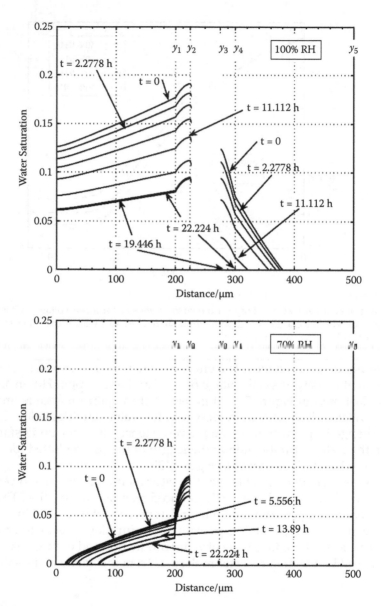

FIGURE 7.35
The evolution of the water saturation during calculations at water activities of 1 and 0.7. The left hand boundary, y = 0, corresponds to the cathode channel/gas diffusion layer interface and the right hand boundary, y = 500 μm, to the anode channel/gas diffusion layer interface. The region 225 μm <y <275 μm corresponds to the membrane. (From Shah, A.A., F.C. Walsh, 2008. *Journal of Power Sources*, **185**:287–301. With permission.)

After reading the sections provided in this part of the chapter, the reader should be able to formulate a numerical model that accounts for hydrogen sulfide poisoning of the anode of PEM fuel cells.

7.5 Modeling of Other Contaminants

This section is designed to give the reader an overview of the modeling efforts in understanding the effects of other contaminants. Numerical studies focused on understanding other contaminants are scarce or even nonexistent in literature. Other chapters of this book are tailored to understanding the experimental works on the contamination of the anode. However, the authors of this chapter feel that an overview of these experimental works should be given in this chapter. Experimental and numerical studies are closely related. The experimental work discussed in this section should give the reader an overview of the direction of future work in the area of contamination modeling.

7.5.1 Contamination of Ammonia

The effects of ammonia are severe and can be noticed even at very low concentrations. Understanding the effects of ammonia on the performance of PEM fuel cells is very crucial. The presence of ammonia in the fuel cell can be the result of various processes. The use of reformed hydrogen as the fuel is the most obvious source of NH_3. During the reforming processes, levels of up to 150 ppm of ammonia are reported with the chance of this formation increasing if the fuel itself contains nitrogen containing species [129]. Other studies have reported that ammonia can be introduced to a PEM fuel cell via metal hydride alloys used for hydrogen storage [130].

Uribe et al. [131] investigated the performance of PEM fuel cells with platinum catalysts on both electrodes exposed to 30 ppm NH_3 for 15 h. The recovery time for a 1 h exposure to 30 ppm ammonia was found to be around 18 h. Their study on required recovery time showed that if the exposure time to ammonia exceeded 3 h, the performance could no longer be recovered. The performance of the cell after long exposure times of around 17 h could not be fully recovered even after 96 h of operation at pure hydrogen. They suggested that the observed losses in performance are mainly due to loss of proton conductivity as well as an increase in ohmic cell resistance. The effect of transient ammonia concentrations on PEM fuel cells was also the focus of the study by Soto et al. [132]. In their study, they looked into the effect of using a platinum alloy catalyst on the anode side, PtRu/C. They found that, even with exposure to 200 ppm NH_3 for 10 h, the cell resistance was increased by about 35% compared to the results by Uribe et al. [131]. Soto et al. [132] measured the voltage versus current density and showed that the concentration of ammonia has a higher effect on the poisoning effect than the actual exposure time. From their observations, they also suggested that the presence of ammonia affects primarily the anode side. However, they did not put together a detailed hypothesis in regards to the poisoning mechanism. The transient effect of exposure of a PEM fuel cell to ammonia has also been investigated by Halseid et al. [133] experimentally. Using their experimental apparatus, they looked into the effects of levels of contamination, exposure time, operating temperature, and the catalyst used. Their study was tailored toward formulating a comprehensive understanding of the effects of ammonia. They argued that ammonia can diffuse rapidly through the Nafion membrane, thus affecting the cathode catalyst layer, which is a contradiction to the conclusions by Uribe et al. [131] and Soto et al. [132]. In order to support this argument, Halseid et al. [133] examined the ohmic resistance as well as the hydrogen oxidation reaction (HOR) and oxygen reduction reaction (ORR). They concluded that an increase in the bulk membrane resistance can only explain parts of the total loss and a lower ionic conductivity in both catalyst layers can affect the fuel cell performance. The presence of ammonia affects the adsorption process of hydrogen on Pt, while desorption of hydrogen is not affected [134]. The effects of ammonia on the oxygen reduction reaction are not as straightforward and not as clear. They explained that the effect could be caused by two main mechanisms. Firstly, the formation of adsorbed NH_3 originating from the electrochemical oxidation of ammonium blocks the active sites on the cathode. Secondly, a mixed potential on the cathode due to the ORR and simultaneous oxidation of ammonium can influence the overall performance. Further, their investigation of the temperature effect on the poisoning rate led to the speculation that higher temperatures can lead to higher recovery times. In general, they found that full recovery of ammonia poisoning can take up to days. They also suggested that the presence of even 1 ppm of ammonia in the fuel stream could lead to catastrophic losses in the performance of the cell.

   The discussion given above raises the issue for a better understanding of ammonia poisoning. Due to its applicability and reliability, mathematical modeling is of great importance to this knowledge. However, literature is lacking mathematical models of ammonia poisoning. In this section, some of the experimental work on the poisoning of ammonia is presented. As seen earlier, in order to be able to numerically simulate any poisoning phenomenon, a full understanding of the phenomenon is needed. The mathematical formulation of ammonia poisoning is similar to that of the poisoning of any other species. Experimental work can be used to formulate the reaction rates of these phenomena as well as offer a direction for the numerical investigations.

7.5.2 Poisoning of Mixture Contaminants

In reality, various contaminants can be present simultaneously in the fuel stream due to the reforming process. Hence, it is necessary to understand the effect of these contaminants as mixtures on the performance of the cell. The authors of this chapter found that literature is lacking in both experimental and numerical studies in regards to this area. One experimental study, which takes into account the presence of various contaminants in the fuel stream, was found. The purpose of this chapter is to give the reader an idea about the effects of a mixture of contaminants on the cell. The experimental measurements presented here can be used as the basis for studying numerically the poisoning phenomenon of a mixture of contaminants.

7.5.2.1 Mixture of CO and H₂S

The presence of carbon monoxide and hydrogen sulfide as a mixture in the fuel stream of PEM fuel cells could be a devastating combination. This is because they are both competing for platinum sites. As seen earlier, a small amount of either gas in the fuel stream can result in severe losses in the performance of the cell. Hence, it is crucial to understand the effects of a mixture of CO and H_2S on the overall performance of the cell. Rockward et al. [135] observed cyclic voltammetry measurements for an anode exposed to 500 ppm CO and 50 ppm H_2S at 2, 4, 6, 8, 10, and 12 min, as shown in Figure 7.36. Investigating the results of their measurements, a decrease in charge associated with CO oxidation and a corresponding increase in charge associated with S oxidation with time is observed. Further, they noticed that by minute 12, the CO stripping peak disappeared leaving only S oxidation. They suggest that in the presence of H_2S, CO is eventually displaced from the surface and S species continue to accumulate. They finally conclude that the responses of the CVs suggest that there may be some interesting multicomponent phenomenon occurring. They, however, do not go into more detail because this was beyond the scope of their investigation. The need for more investigation is apparent. Numerical simulation of this phenomenon could be very helpful in that many different parameters could be studied simultaneously.

7.5.2.2 Other Binary Mixtures

Understanding the effects of other binary mixtures on the fuel cell is also very essential. Some of these mixtures are given as:

- CO and CO_2
- CO and NH_3
- NH_3 and H_2S

The literature is lacking of experimental and numerical studies of these binary mixtures. It is, however, crucial to investigate performance losses due to these mixtures. Experimental investigations can be used to outline the magnitude of their effects. The experimental measurements can then be used to outline a more comprehensive study for tertiary or higher order mixtures. Further, they are used to formulate the reaction rates for the purpose of developing a numerical model.

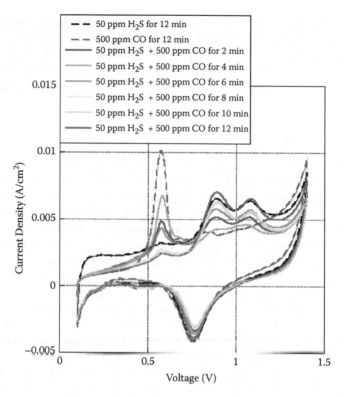

FIGURE 7.36
A family of cyclic voltammetry taken after simultaneously injecting the anode with 500 ppm CO and 50 ppm H_2S at 2, 4, 6, 8, 10, and 12 min. Neat H_2 and individually 500 ppm CO and 50 ppm H_2S after 12 min exposure are shown for reference. Arrows indicate direction of increasing exposure time for the CO and H_2S mixture. (From Rockward, T. et al., 2007. *ECS Transactions*, **11**:821–829. With permission.)

7.5.3 Tertiary or Higher Order Mixtures

It is clear by now that many impurities can be present simultaneously in the fuel stream. These contaminants can be present as a mixture of three or more gases. Following the discussion presented so far, some tertiary mixtures include:

- CO, CO_2, and H_2S
- CO, CO_2, and NH_3
- CO, NH_3, and H_2S
- CO_2, H_2S, and NH_3

Unfortunately, the literature is lacking both experimental and numerical investigations of the effects of these mixtures. In order to better understand the cell behavior with a fuel containing these mixtures, experiments should first be performed. The measured data will reveal the contribution of each gas separately and as a mixture on the overall magnitude of the losses. Hence, the overall rates of reaction could be obtained. Further, a better understanding of the reaction kinetics could be established.

The contamination of higher order mixtures has been studied by Rockward et al. [135]. They used a fuel mixture as specified by the U.S. FreedomCAR Tech Team. The mixture contained more than 99.9% hydrogen, 10 ppb H_2S, 0.1 ppm CO, 5 ppm CO_2, 1 ppm NH_4, and 50 ppm C_2H_4. The performance was then compared to the performance of the cells operated with pure hydrogen and their findings are shown in Figure 7.37.

Figure 7.37 is a visual illustration of the resistance change with time for a cell run on a contaminated fuel mixture and an identical cell running on ultrahigh purity (UHP) hydrogen. A significant increase in the resistance is observed in the contaminated cell. The cell resistance increased over 30% from 0.097 to 0.128 $\Omega.cm^2$ during a 1000 h test. This suggests that a potential loss of 25 mV due to ohmic effects is experienced. Rockward et al. [135] believed that many of these losses are a result of the presence of ammonia in

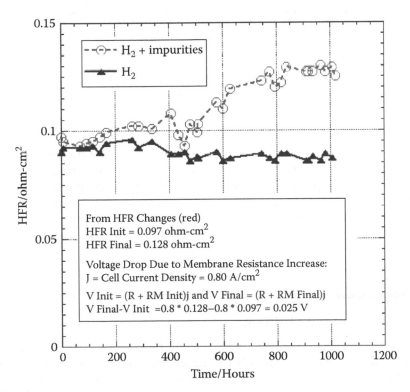

FIGURE 7.37
HFR versus time for the baseline test cell and 0.2 mg Pt/cm^2. Fuel cells operated on pure H_2 and on the FreedomCAR fuel mixture. (From Rockward, T. et al., 2007. *ECS Transactions*, 11:821–829. With permission.)

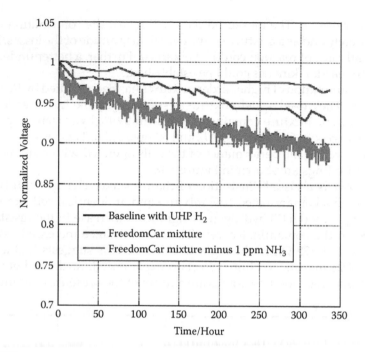

FIGURE 7.38
The voltage response at 0.8A/cm² for ultrahigh purity hydrogen, FreedomCAR fuel mixture, and ammonia-free FreedomCAR mixture. (From Rockward, T. et al., 2007. *ECS Transactions*, 11:821–829. With permission.)

the fuel stream. In order to test this hypothesis, they ran a fuel cell using the fuel specification mixture without ammonia (Figure 7.38). After 300 h, they observed that the performance loss of the cell run without ammonia to be much smaller than that run with 1 ppm ammonia. The voltage loss over 300 h in the presence of ammonia is around 20 mV; hence, the presence of ammonia results in very severe losses in comparison to other contaminants.

7.6 Summary and Outlook

The aim of this chapter was to outline a comprehensive method to numerically simulate the many poisoning phenomena that occur in PEM fuel cells and their mitigation methods. A full mathematical formulation for CO, CO_2, and H_2S poisoning of the anode of PEM fuel cells was put forward. Further, the mathematical formulation for the mitigation of CO effects using oxygen bleeding and CO-tolerant catalysts was also discussed.

A comprehensive literature review of the most up-to-date numerical work in the area of anode contamination showed that most of the research efforts are tailored toward the understanding of CO poisoning. Two types of models (empirical and numerical models), which focus on the understanding of CO poisoning, are found in literature. Empirical models are used to gain knowledge on the overall voltage loss due to the presence of CO in the fuel stream. Numerical models, however, give a more comprehensive understanding of the phenomenon. The investigation of the transient, as well as the steady-state effects on the anode due to CO presence, is abundant in literature. It is shown that CO poisoning is a transient effect. Further, studying the polarization curve of a contaminated cell, it is obvious that CO electrooxidation occurs at high anode overpotential, thus increasing the overall voltage losses of the cell. Efforts are also put toward understanding the effect of oxygen bleeding on the mitigation of CO. It is found that even with a small concentration of oxygen, the effects of CO contamination on the cell can be greatly lessened. The numerical models are also used to carry out parametric studies to understand the effects of varying various operating and physical parameters on CO poisoning.

Numerical models of the contamination of the anode by CO_2 and H_2S are not as abundant as their counterparts for CO poisoning. The numerical models found in literature, which take into account the presence of carbon dioxide, are focused on understanding the reverse water-gas shift reaction (WGSR) and its effects on the performance of the cell. The effect of hydrogen sulfide presence on the polarization curve and liquid water distribution in the cell is also studied numerically. It is found that with trace amounts of H_2S, the cell performance can be dramatically decreased. The effects of H_2S poisoning can be sometimes very severe and cannot be reversed, unlike their counterparts of CO poisoning.

The last section of the chapter deals with understanding the poisoning of the anode by other contaminants, such as ammonia and mixtures of contamination. A comprehensive literature review of the contamination of the anode of PEM fuel cells revealed that numerical models that consider other contaminants are nonexistent in literature. Therefore, much work is needed in understanding these types of contaminations numerically.

The discussion presented in this chapter revealed that many areas in numerical modeling of contamination of the anode of PEM fuel cells still need much work. The next section is designed to highlight these main areas.

7.6.1 Outlook

Many of the numerical models available, which look into the many poisoning phenomena in PEM fuel cells, are one-dimensional. Neglecting the dependency of the performance on dimensionality limits the knowledge that could be acquired about these phenomena. In reality, the distribution of the current

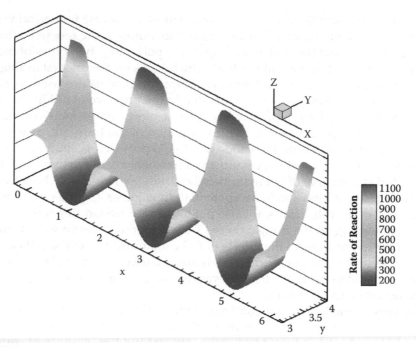

FIGURE 7.39
Two-dimensional distribution of the rate of electrooxidation reaction in the catalyst layer (taken from the work done for the studies in Zamel and Li [136] and Zamel and Li [137]). The modeling domain used to generate this result is given in Figure 7.40. (From Zamel, N., X. Li, 2008. *International Journal of Energy Research*, **32**:698–721; Zamel, N., X. Li, 2009. Forthcoming, *International Journal of Energy Research*.)

density in the catalyst layer is highly nonuniform. Hence, the rate of the electrooxidation reaction of species is highly nonuniform (see Figure 7.39 and Figure 7.40). It is crucial to understand the effects of this nonuniformity on the poisoning phenomena. The results of this investigation will facilitate in designing the catalyst layer.

The same argument can be used for the lack of nonisothermal models in literature. As it has been made clear in this chapter, the temperature plays a vital role in determining the overall magnitude of poisoning. In reality, PEM fuel cells run at a nonisothermal condition. The temperature, especially, increases in the catalyst layer due to the heat generation during the electrooxidation reaction. Consequently, more work is needed to understand the effects of the nonuniformity of temperature on contamination of PEM fuel cells.

Another important aspect, on which more attention should be paid, is more detailed kinetic models for different phenomena. A more detailed kinetic model for platinum alloys as the catalysts is needed. There has been much experimental work to investigate all of these alloys; however, the reaction kinetics for all the alloys have not been fully identified. This is also the case for ammonia poisoning as well as contamination by mixtures. More

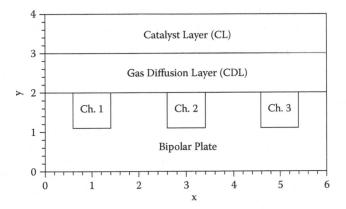

FIGURE 7.40
Two-dimensional modeling domain used to generate the results given in Figure 7.39 — x and y axes have been normalized—x axis is normalized by the height of the flow channel that is 1 mm, here x = 6 denotes a distance of 6 mm; the section ($0 \le y \le 2$) is normalized by the channel height that is 1 mm and y = 2 denotes a distance of 2 mm. The section ($2 \le y \le 3$) is normalized by the gas diffusion layer thickness that is 150 μm and the section ($3 \le y \le 4$) is normalized by the catalyst layer thickness that is 5 μm. The flow direction is in the y-direction. (From Zamel, N., X. Li, 2008. *International Journal of Energy Research*, **32**:698–721; Zamel, N., X. Li, 2009. Forthcoming, *International Journal of Energy Research*.)

work also should be focused on determining possible mitigation methods of ammonia and mixture poisoning. Literature is lacking both experimental and numerical studies in this area.

Finally, contamination of the cell by low concentration of impurities must be investigated. As discussed earlier in the introduction section, PEM fuel cells have low tolerances toward impurities; however, numerical studies found in literature do not examine the effects of these low concentrations. At these low concentrations, the reaction kinetics of these contaminants can be altered and, hence, should be investigated.

References

1. T. Ralph, G. Hards, J. Keating, S. Campbell, D. Wilkinson, M. Davis, J. St-Pierre, M. Johnson, 1997. Low cost electrodes for proton exchange membrane fuel cells: Performing in single cells and Ballard stacks. *Journal of the Electrochemical Society*, **144**:3845–3857.
2. G. Voecks, DOE consultant and coordinator of the contamination group, The development of hydrogen fuel quality specification, *Canada—USA PEM Network Research Workshop*, Feb. 16-17, 2009, Vancouver, British Columbia.
3. J. Bentley, B. Barnett, S. Hynek, Multifuel reformer and on-board hydrogen storages, *1992 Fuel Cell Seminar*, 1992, 455–460.

4. P. Loftus, J. Thijssen, J. Bentley, J. Bowman, Development of a multi-fuel partial oxidation reformer for transportation applications, *1994 Fuel Cell Seminar*, 1994, 487–490.

5. J. Divisek, H.-F. Oetjen, V. Peinecke, V.M. Schmidt, U. Stimming, 1998. Components for PEM fuel cell systems using hydrogen and CO containing fuels, *Electrochemica Acta*, **43**:3811–3815.

6. M.S. Wilson, C. Derouin, J. Valerio, S. Gottesfedl, Electrocatalysis issues in polymer electrolyte fuel cells, *Proceedings for the 28th IECEC*, 1993, 1.1203–1.208.

7. X. Cheng, Z. Shi, N. Glass, L. Zhang, J. Zhang, D. Song, Z.-S. Liu, H. Wang, J. Shen, 2007. A review of PEM hydrogen fuel cell contamination: Impacts, mechanisms and mitigation. *Journal of Power Sources*, **165**:739–756.

8. J. Kim, S.M. Lee, S. Srinivasan, C. Chamberlin, 1995. Modeling the proton exchange membrane fuel cell performance with an empirical equation. *Journal of the Electrochemical Society*, **142**:2670–2674.

9. J. Amphlett, R. Baumert, R. Mann, B. Peppley, P. Roberge, T. Harris, 1995. Performance modeling of the Ballard Mark IV solid polymer electrolyte fuel cell I. Mechanistic model development. *Journal of the Electrochemical Society*, **142**:1–8.

10. J. Amphlett, R. Baumert, R. Mann, B. Peppley, P. Roberge, T. Harris, 1995. Performance modeling of the Ballard Mark IV solid polymer electrolyte fuel cell II. Empirical model development. *Journal of the Electrochemical Society*, **142**:9–15.

11. A. Rodrigues, J. Amphlett, R. Mann, B. Peppley, P. Roberge, Carbon monoxide poisoning of proton-exchange membrane fuel cells, *Proceedings of the 32 Intersociety Energy Conversion Engineering Conference*, Honolulu, HI, 27 July–1 August 1997, 768–773.

12. R. Bellows, E. Marucchi-Soos, D. Buckley, 1996. Analysis of reaction kinetics for carbon monoxide and carbon dioxide on polycrystalline platinum relative to fuel cell operation. *Industrial and Engineering Chemistry Research*, **35**:1235–1242.

13. D. Chu, R. Jiang, C. Walker, 2000. Analysis of PEM fuel cell stacks using an empirical current voltage equation. *Journal of Applied Electrochemistry*, **30**:365–370.

14. D. Bernardi, M. Vergrugge, 1992. A mathematical model of the solid-polymer-electrolyte fuel cell. *Journal of Electrochemical Society*, **139**:2477–2491.

15. T. Springer, M. Wilson, S. Gottesfeld, 1993. Modeling and experimental diagnostics in polymer electrolyte fuel cells. *Journal of the Electrochemical Society* **140**:3513–3526.

16. T. Springer, T. Zawodzinski, S. Gottesfeld,. 1991. Polymer electrolyte fuel cell model. *Journal of the Electrochemical Society*, **138**:2334–2342.

17. J.-T. Wang, R. Savinell, 1992. Simulation studies on the fuel electrode of a H_2-O_2 polymer electrolyte fuel cell. *Electrochimica Acta*, **37**:2737–2745.

18. H. Dhar, L. Christner, A. Kush, 1987. Nature of CO adsorption during H_2 oxidation in relation to modeling for CO poisoning of a fuel cell anode. *Journal of the Electrochemical Society*, **134**:3021–3026.

19. T. Springer, T. Zawadzinski, S. Gottesfeld, 1997. Modeling of polymer electrolyte fuel cell performance with reformate feed streams: effects of low levels of CO in hydrogen. In S. Srinivasan, J. McBreen, A. Khandkar, and V. Tilak, eds., *Electrode Materials and Processes for Energy Conversion and Storage IV*, pp. 15–24, Pennington, NJ: The Electrochemical Society.

20. T. Springer, T. Rockward, T. Zawodzinski, S. Gottesfeld, 2001. Model for polymer electrolyte fuel cell operation on reformate feed: Effects of CO, H_2 dilution, and high fuel utilization. *Journal of the Electrochemical Society*, **148**:A11–A23.

21. J.J. Baschuk, X. Li, 2003. Modeling CO poisoning and O_2 bleeding in a PEM fuel cell Anode. *International Journal of Energy Research*, **27**:1095–1116.

22. T. Zhou, H. Liu, 2004. A 3-D model for PEM fuel cells operated on reformate. *Journal of Power Sources*, **138**:101–110.

23. H.S. Chu, C.P. Wang, W.C. Liao, W.M. Yan, 2006. Transient behavior of CO poisoning of the anode catalyst layer of a PEM fuel cell. *Journal of Power Sources*, **159**:1071–1077.

24. C.P. Wang, H.S. Chu, 2006. Transient analysis of multicomponent transport with carbon monoxide poisoning effect of a PEM fuel cell. *Journal of Power Sources*, **159**:1025–1033.

25. N. Zamel, X. Li, 2007. Transient analysis of carbon monoxide poisoning and oxygen bleeding in a PEM fuel cell anode catalyst layer. *International Journal of Hydrogen Energy*, **33**:1335–1344.

26. F. Barbir, B. Balasubrumanian, J. Neutzler. Trade-off design analysis of operating pressure and temperature in PEM fuel cell systems. *In Proceedings of the ASME Advanced Energy Systems Division*, pp. 305–315, New York, 1999. ASME Advanced Energy Systems Division, The American Society of Mechanical Engineers.

27. P. Stonehart, P. Ross, 1975. The commonality of surface processes in electrocatalysis and gas-phase heterogeneous catalysis. *Catalysis Reviews-Science and Engineering*, **12**:1–35.

28. S. Gilman, 1964. The mechanism of electrochemical oxidation of carbon monoxide and methanol on platinum II: the "reactant pair" mechanism for electrochemical oxidation of carbon monoxide and methanol. *Journal of the Electrochemical Society*, **68**:70–80.

29. W. Vogel, J. Lundquist, P. Ross, P. Stonehart, 1975. Reaction pathways and poisons-II: The rate controlling step for electrochemical oxidation of hydrogen on Pt in acid and poisoning of the reaction by CO. *Electrochimica* Acta, **20**:79–93.

30. J.J. Baschuk, X. Li, 2001. Carbon monoxide poisoning of proton exchange membrane fuel cells. *International Journal of Energy Research*, **25**:695–715.

31. S. Gottesfeld, J. Pafford, 1988. A new approach to the problem of carbon monoxide poisoning in fuel cells operating at low temperatures. *Journal of the Electrochemical Society*, **135**:2651–2652.

32. R.P.H. Gasser, 1985. *An introduction to chemisorption and catalysis by metals*. Clarendon Press, Oxford, U.K., chap. 9.

33. T. Engel, G. Ertle, 1982. Oxidation of carbon monoxide. In *The Chemical Physics of Solid Substances and Heterogeneous Catalysis*, Vol. 4: *Fundamental Studies of Heterogeneous Catalysis*, King, D., Woodruff, D. (eds). Elsevier: Amsterdam, 73–93.

34. P. Norton, 1982. *The Chemical Physics of Solid Substances and Heterogeneous Catalysis*, Vol. 4: *Fundamental Studies of Heterogeneous Catalysis*, King, D., Woodruff, D. (eds). Elsevier: Amsterdam, 27–72.

35. E.I. Santiago, V.A. Paganin, M. do Carmo, E.R. Gonzalez, E.A. Ticianelli, 2005. Studies of CO tolerance on modified gas diffusion electrodes containing ruthenium dispersed on carbon. *Journal of Electroanalytical Chemistry*, **575**:53–60.

36. M. do Carmo, V.A. Paganin, J.M. Rosolenand, E.R. Gonzalez, 2005. Alternative support for the preparation of catalysts for low-temperature fuel cells: the use of carbon nanotubes. *Journal of Power Sources*, **142**:169–176.

37. A. Chambers, N.M. Rodriguez, K. Baker, R. Terry, 1995. Catalytic engineering of carbon nanostructures. *Langmuir*, **11**:3862–3866.

38. T. Kyotani, L. Tsai, A. Tomita, 1997. Formation of nanorods and nanoparticles in uniform carbon nanotubes prepared by a template carbonization method. *Chemical Communications*, **7**:701–702.

39. S. Lee, S. Mukerjee, E. Ticianelli, J. McBreen, 1999. Electrocatalysis of CO tolerance in hydrogen oxidation reaction in PEM fuel cells. *Electrochimica Acta*, **44**:3283–3293.

40. K. Wang, H.A. Gasteiger, N.M. Markovic, P.N. Ross Jr., 1996. On the reaction pathway for methanol and carbon monoxide electrooxidation on Pt-Sn alloy versus Pt-Ru alloy surfaces. *Electrochimica Acta*, **41**:2587–2593.

41. R. Srinivasan, R.J. De Angeles, B.H. Davis, 1987. Alloy formation in Pt-Sn alumina catalysts: *In situ* x-ray diffraction study. *Journal of Catalysis*, **106**:449–457.

42. R. Srinivasan, R.J. De Angeles, B.H. Davis, 1990. Structural studies of Pt-Sn catalysts on high and low surface area alumina supports. *Catalysis Letters*, **4**:303–308.

43. R. Srinivasan, L.A. Rice, B.H. Davis, 1991. Electron microdiffraction study of Pt-Sn alumina reforming catalysts. *Journal of Catalysis*, **129**:257–268.

44. R. Srinivasan, B.H. Davis. 1992. X-ray diffraction and electron microscopy studies of platinum-tin-silica catalysts. *Applied Catalysis A*, **87**:45–67.

45. H.A. Gasteiger, N.M. Markovic, P.N. Ross, 1995. Electro-oxidation of CO and H_2/CO mixtures on a well characterized Pt_3SN electrode surface. *Journal of Physical Chemistry*, **99**:8945–8949.

46. V.I. Kuznestov, A.S. Belyi, E.N. Yuchenko, M.D. Smolikov, M.T. Protasova, E.V. Zatolokina, V.K. Duplayakin, 1986. Mössbauer spectroscopic and chemical analysis of the composition of Sn-containing components of Pt-Sn/Al_2O_3(Cl) reforming catalyst. *Journal of Catalysis*, **99**:159–170.

47. E.M. Crabb, R. Marshall, T. David, 2000. Carbon monoxide electro-oxidation properties of carbon supported PtSn catalysts prepared using surface organometallic chemistry. *Journal of the Electrochemical Society*, **147**:4440–4447.

48. T.J. Schmidt, Z. Jusys, H.A. Gasteiger, R.J. Behm, U. Endruschat, H. Boennemann. 2001. *Journal of Electroanalytical Chemistry*, **501**:132–140.

49. D.D. Eley, P.B. Moore, 1981. The adsorption and reaction of CO and O_2 on PdAu alloy wires. *Surface Sciences*, **111**:325–343.

50. T.J. Schmidt, M. Noeske, H.A. Gasteiger, R.J. Behm, P. Britz, H. Bönnemann, 1998. PtRu alloy colloids as precursors for fuel cell catalysts. *Journal of the Electrochemical Society*, **145**:925–931.

51. D.C. Papageorgopoulos, M. Keijzer, F.A. de Bruijn, 2002. The inclusion of Mo, Nb and Ta in Pt and PtRu carbon supported electrocatalysts in the quest for improved CO tolerant PEMFC anodes. *Electrochimica Acta*, **48**:197–204.

52. S. Mukerjee, S.J. Lee, E.A. Ticianelli, J. McBeen, B.N. Grgur, N.M. Markovic, P.N. Ross, J.R. Giallombardo, E.S. De Castroc, 1999. Investigation of enhanced CO tolerance in proton exchange membrane fuel cells by carbon supported PtMo alloy catalyst. *Electrochemical and Solid State Letters*, **2**:12–15.

53. B.N. Grgur, N.M. Markovic, P.N. Ross, 1999. The electro-oxidation of H_2 and H_2/ CO mixtures on carbon supported Pt_xMo_y. *Journal of the Electrochemical Society*, 146:1613–1619.
54. H. Zhang, Y. Wang, E.R. Fachini, C.R. Cabrera, 1999. Electrochemically codeposited PtMo oxide electrode for catalytic oxidation of methanol in acid solution. *Electrochemical and Solid State Letters*, 2:437–439.
55. R.C. Urian, A.F. Gullá, S. Mukerjee, 2003. Electrocatalysis of reformate tolerance in proton exchange membrane fuel cells: Part 1. *Journal of Electroanalytical Chemistry*, 554,555:307–324.
56. A. Lima, C. Coutanceau, J.M. Léger, C. Lamy, 2001. Investigation of ternary catalysts for methanol electro-oxidation. *Journal of Applied Electrochemistry*, 31:379–386.
57. M. Götz, H. Wendt, 1998. *Electrochimica Acta*, 43:3637–3644.
58. K.Y. Chen, P.K. Shen, A.C.C. Tseung, 1995. CO oxidation on Pt-Ru/WO_3 electrodes. *Journal of the Electrochemical Society*, 142:L85–L86.
59. A.C.C. Tseung, K.Y. Chen, 1997. Hydrogen spill-over effect on Pt/WO_3 anode catalysts. *Catalysis Today*, 38:439–443.
60. K.Y. Chen, Z. Sun, A.C.C. Tseung, 2000. Preparation and characterization of high performance PtRu/WO_3/C anode catalysts for the oxidation of impure hydrogen. *Electrochemical and Solid State Letters*, 3:10–12.
61. C. Roth, M. Goetz, H. Fuess, 2001. Synthesis and characterization of carbon supported Pt-Ru-WO_x catalysts by spectroscopic and diffraction methods. *Journal of Applied Electrochemistry*, 31:793–798.
62. Z. Hou, B. Yi, H. Yu, Z. Lin, H. Zhang, 2003. CO tolerance electrocatalyst of PtRu-H_xMeO_3/C (Me=W, Mo) made by composite support method. *Journal of Power Sources*, 123:116–125.
63. M. Watanabe, S. Motoo, 1975. Electrocatalysis by AD-atoms: Part II. Enhancement of the oxidation of methanol on platinum by ruthenium AD-atoms. *Electroanalytical Chemistry and Interfacial Electrochemistry*, 60:267–273.
64. S. Enbäck, G. Lindbergh, 2005. Experimentally validated model for CO oxidation on PtRu/C in a porous PEFC electrode. *Journal of the Electrochemical Society*, 152:A23–A31.
65. E.I. Santiago, G.A. Camara, E.A. Ticianelli, 2003. CO tolerance on PtMo/C electrocatalysts prepared by the formic acid method. *Electrochimica Acta*, 48:3527–3534.
66. E.I. Santiago, M.S. Batista, E.M. Assaf, E.A. Ticianelli, 2004. Mechanism of CO tolerance on molybdenum-based electrocatalysts for PEMFC. *Journal of the Electrochemical Society*, 151:A944–A949.
67. E. Christoffersen, P. Liu, A. Ruban, H.L. Skriver, J.K. Nørskov, 2001. Anode materials for low-temperature fuel cells: A density functional theory study. *Journal of Catalysts*, 199:123–131.
68. J.-H. Wee, K.-Y. Lee, 2006. Overview of the development of CO-tolerant anode electrocatalysts for proton exchange membrane fuel cells. *Journal of Power Sources*, 157:128–135.
69. T.V. Choudhary, D.W. Goodman, 1999. Stepwise methane steam reforming: A route to CO-free hydrogen. *Catalysis Letters*, 59:93–94.
70. S.H. Lee, J. Han, K.-Y. Lee, 2002. Development of PROX (preferential oxidation of CO) system for 1kW$_e$ PEMFC. *Korean Journal of Chemical Engineering*, 19:431–433.

71. S.H. Lee, J. Han, K.-Y. Lee, 2002. Development of 10 kW$_e$ preferential oxidation system for fuel cell vehicle. *Journal of Power Sources*, **109**:394–402.
72. W.H.J. Hogarth, J.C. Diniz Da Costa, G.Q. Lu, 2005. Solid acid membranes for high temperature (>140°C) proton exchange membrane fuel cells. *Journal of Power Sources*, **142**:223–237.
73. S. Wasmus, A. Valeriu, G.D. Mateescu, D.A. Tryk, R.F. Savinell, 1995. Characterization of H$_3$PO$_4$ equilibrated Nafion 117 membranes using ^{1}H and ^{31}P NMR spectroscopy. *Solid State Ionics*, **80**:87–92.
74. O. Savadogo, 2004. Emerging membranes for electrochemical systems: Part II. High temperature composite membranes for polymer electrolyte fuel cell (PEFC) applications. *Journal of Power Sources*, **127**:135–161.
75. H. Yu, Z. Hou, B. Yi, Z. Lin, 2002. Composite anode for CO tolerance proton exchange membrane fuel cells. *Journal of Power Sources*, **105**:52–57.
76. W. Shi, M. Hou, Z. Shao, J. Hu, Z. Hou, P. Ming, B. Yi, 2007. A novel proton exchange membrane fuel cell anode for enhancing CO tolerance. *Journal of Power Sources*, **174**:164–169.
77. C.-H. Wan, Q.-H. Zhuang, 2007. Novel layer wise anode structure with improved CO-tolerance capability for PEM fuel cell. *Electrochimica Acta*, **52**:4111–4123.
78. E. Gileadi, E. Kirowa-Eisner, J. Penciner, 1975. *Interfacial Electrochemisty: An Experimental Approach*. Addison-Wesley: Reading, PA.
79. W. Vogel, J. Lundquist, P. Ross, P. Stonehart, 1975. Reaction pathways and poisons-II: The rate controlling step for electrochemical oxidation of hydrogen on Pt in acid and poisoning of the reaction by CO. *Electrochimica* Acta, **20**:79–93.
80. H. Wu, X. Li, P. Berg, 2007. Numerical analysis of dynamic processes in fully humidified PEM fuel cells. *International Journal of Hydrogen Energy*, **32**:2022–2031.
81. C. Marr, X. Li, 1999. Composition and performance modeling of catalyst layer in proton exchange membrane fuel cell. *Journal of Power Sources*, **77**:17–27.
82. J. Dean, 1999. *Lange's Handbook of Chemistry*. McGraw-Hill: New York.
83. E-TEK. *Gas Diffusion Electrodes and Catalyst Materials* catalog, 1995.
84. J. Newman, 1991. *Electrochemical Systems*, 2nd ed., Prentice Hall International Series in the Physical and Chemical Engineering Sciences. Prentice-Hall: Upper Saddle River, NJ.
85. S. Um, C.Y. Wang, 2004. Three-dimensional analysis of transport and electrochemical reactions in polymer electrolyte fuel cells. *Journal of Power Sources*, **125**:40–51.
86. G. Hu, J. Fan, S. Chen, Y. Liu, K. Cen, 2004. Three-dimensional numerical analysis of proton exchange membrane fuel cells (PEMFCs) with conventional and interdigitated flow fields. *Journal of Power Sources*, **136**:1–9.
87. G. Mohan, B.P. Rao, S.K. Das, S. Pandiyan, N. Rajalakshmi, K.S. Dhathathreyan, 2004. Analysis of flow maldistribution of fuel and oxidant in a PEMFC. *Transactions of the ASME*, **126**:262–270.
88. W.-M. Yan, C.-H. Yang, C.-Y. Soong, F. Chen, S.-C. Mei, 2006. Experimental studies on optimal operating conditions for different flow field designs of PEM fuel cells. *Journal of Power Sources*, **160**:284–292.
89. T. Kanezaki, X. Li, J.J. Baschuk, 2006. Cross-leakage flow between adjacent flow channels in PEM fuel cells. *Journal of Power Sources*, **162**:415–425.
90. X. Li, I. Sabir, J. Park, 2007. A flow channel design procedure for PEM fuel cells with effective water removal. *Journal of Power Sources*, **163**:933–942.

91. G. Wallis, 1969. One-dimensional two-phase flow. McGraw-Hill: New York.
92. I.E. Idelchik, 1994. *Handbook of Hydraulic Resistance.* 3rd ed., CRC Press: Boca Ration, FL.
93. R. Reid, J. Prausnitz, T. Sherwood, 1997. *The Properties of Gases and Liquids.* McGraw-Hill: New York.
94. F. Incropera, D. Dewitt, 1996. *Fundamentals of Heat and Mass Transfer.* John Wiley & Sons: New York.
95. R. Perry, C. Chilton, 1973. *Chemical Engineers' Handbook,* 5th ed. McGraw-Hill: New York.
96. S. Lee, S. Mukerjee, E. Ticianelli, J. McBreen, 1999. Electrocatalysis of CO tolerance in hydrogen oxidation reaction in PEM fuel cells. *Electrochimica Acta,* **44**:3283–3293.
97. T. Zawodzinski, C. Karuppaiah, R. Uribe, S. Gottesfeld, 1997. Aspects of CO tolerance in polymer electrolyte fuel cells: some experimental findings. In *Electrode Materials and Processes for Energy Conversion and Storage IV,* S. Srinivasan, J. McBreen, A. Khandkar, V. Tilak (eds). The Electrochemical Society: Pennington, NJ, 139–146.
98. T. Zawodzinski, J. Bauman, T. Rockward, P. Haridos, F. Uribe, S. Gottesfeld, 1999. Enhanced CO tolerance in polymer electrolyte fuel cells with Pt-Mo anodes. *Proceedings of the 2nd International Symposium on Proton Conducting Membrane Fuel Cells,* pp. 200–208. The Electrochemical Society Proceedings Series: Pennington, NJ.
99. J. Giner, 1963. Electrochemical reduction of CO2 on platinum electrodes in acid solutions. *Electrochimica Acta,* 8:857–865.
100. P. Giunta, N. Amadeo, M. Laborde, 2006. Simulation of a low temperature water gas shift reactor using the heterogeneous model/application to a PEM fuel cell. *Journal of Power Sources,* **156**:489–496.
101. R. Radhakrishnan, R.R. Willigan, Z. Dardas, T.H. Vanderspurt, 2006. *AIChE Journal,* **52**:1888–1894.
102. M.S. Batista, E.I. Santiago, E.M. Assaf, E.A. Ticianelli, 2005. Evaluation of water-gas shift and CO methanation processes for purification of reformate gases and the coupling to a PEM fuel cell system. *Journal of Power Sources,* **145**:50–54.
103. G. Germani, P. Alphonse, M. Courty, Y. Schuurman, C. Mirodatos, 2005. Platinum/ceria/alumina catalysts on microstructures for carbon monoxide conversion. *Catalysis Today,* **110**:114–120.
104. R. Radhakrishnan, R.R. Willigan, Z. Dardas, T.H. Vanderspurt, 2006. Water gas shift activity and kinetics of Pt/Re catalysts supported on ceria-zirconia oxides. *Applied Catalysis B: Environmental,* **66**:23–28.
105. F.A. de Bruijn, D.C. Papageorgopoulos, E.F. Sitters, G.J.M. Janssen, 2002. The influence of carbon dioxide on PEM fuel cell anodes. *Journal of Power Sources,* **110**:117–124.
106. G.J.M. Janssen, 2004. Modeling study of CO_2 poisoning on PEMFC anodes. *Journal of Power Sources,* **136**:45–54.
107. M. Minutillo, A. Perna, 2008. Behaviour modeling of a PEMFC operating on diluted hydrogen feed. *International Journal of Energy Research,* **32**:1297–1308.
108. S. Ball, A. Hodgkinson, G. Hoogers, S. Maniguet, D. Thompsett, B. Wong, 2002. The proton exchange membrane fuel cell performance of a carbon supported PtMo catalyst operating on reformate. *Electrochemical and Solid-State Letters,* 5:A31–A34.

109. G. Karimi, X. Li, 2006. Analysis and modeling of PEM fuel cell stack performance: Effect of *in situ* reverse gas shift reaction and oxygen bleeding. *Journal of Power Sources*, **159**:943–950.

110. G.C. Chinchen, P.J. Denny, J.R. Jennings, M.S. Spencer, K.C. Waugh, 1988. Synthesis of methanol: Part 1: Catalysts and kinetics. *Applied Catalysis*: **36**:1–65.

111. A. Heinzel, W. Benz, F. Mahlendorf, O. Niemzig, J. Roes, 2002. Extended abstract, in *Proceedings of the Fuel Cell Seminar*, Palm Springs, CA.

112. R. Mohtadi, W. Lee, S. Cowan, J. van Zee, M. Murthy, 2003. Effects of hydrogen sulfide on the performance of a PEMFC. *Electrochemistry and Solid State Letters*, **6**:A272–A274.

113. S. Knights, N. Jia, C. Chuy, J. Zhang, 2005. Fuel cell progress, challenges and markets. *Proceedings of the 2005 Fuel Cell Seminar*, Palm Springs, CA.

114. R. Mohtadi, W.-K. Lee, J.W. Van Zee, 2005. The effect of temperature on the adsorption rate of hydrogen sulfide on Pt anodes in a PEMFC. *Applied Catalysis B: Environmental*, **56**: 37–42.

115. J. Kim, S.M. Lee, S. Srinivasan, C. Chamberlin, 1995. Modeling the proton exchange membrane fuel cell performance with an empirical equation. *Journal of the Electrochemical Society*, **142**:2670–2674.

116. J. Amphlett, R. Baumert, R. Mann, B. Peppley, P. Roberge, T. Harris, 1995. Performance modeling of the Ballard Mark IV solid polymer electrolyte fuel cell: I. Mechanistic model development. *Journal of the Electrochemical Society*, **142**:1–8.

117. S. Srinivasan, E.A. Ticianelli, C.R. Derouin, A. Redondo, 1988. Advances in solid polymer electrolyte fuel cell technology with low platinum loading electrodes. *Journal of Power Sources*, **22**:359–375.

118. Z. Shi, D. Song, J. Zhang, Z.-S. Liu, S. Knights, R. Vohra, N.Y. Jia, D. Harvey, 2007. Transient analysis of hydrogen sulfide contamination on the performance of a PEM fuel cell. *Journal of the Electrochemical Society*, **154**:B609–B615.

119. A.A. Shah, F.C. Walsh, 2008. A model for hydrogen sulfide poisoning in proton exchange membrane fuel cells. *Journal of Power Sources*, **185**:287–301.

120. T. Loučka, 1971. Adsorption and oxidation of sulfur and of sulfur dioxide at the platinum electrode. *Journal of the Electroanalytical Chemistry*, **31**:319–322.

121. T. Loučka, 1972. The adsorption of sulfur and simple organic substances on platinum electrodes. *Journal of the Electroanalytical Chemistry*, **36**:369–381.

122. E. Najdeker, E. Bishop, 1973. The formation and behavior of platinum sulfide on platinum electrodes. *Journal of Electroanalytical Chemistry*, **41**:79–87.

123. R. Jayaram, A.Q. Contractor, H. Lal, 1978. Formic acid oxidation at platinized platinum electrodes. *Journal of the Electroanalytical Chemistry*, **87**:225–237.

124. A. Contractor, H. Lal, 1979. Two forms of chemisorbed sulfur on platinum and related studies. *Journal of Electroanalytical Chemistry*, **96**:175–181.

125. D. Imamura, Y. Hashimasa, 2007. Effect of sulfur containing compounds on fuel cell performance. *ECS Transactions*, **11**:853–862.

126. M. Mathieu, M. Primet, 1984. Sulfurization and regeneration of platinum. *Applied Catalysis*, **9**:361–370.

127. T. Loučka, 1973. The formation and reduction of an oxide layer on a platinum electrode partially covered with adsorbed sulfur. *Journal of Electroanalytical Chemistry*, **44**: 221–227.

128. Y. Wang, H. Yan, E. Wang, 2001. The electrochemical oxidation and the quantitative determination of hydrogen sulfide on a solid polymer electrolyte-based system. *Journal of Electroanalytical Chemistry*, **497**:163–167.
129. R. Borup, M. Inbody, J. Tafoya, T. Semelsberger, L. Perry, 2002. *Durability studies: Gasoline/reformate durability*, from 2002 National Laboratory R&D Meeting DOE Fuel Cells for Transportation Program: http://www.eere.energy.gov/hydrogenandfuelcells/pdfs/nn0123ba.pdf
130. H.Y. Zhu, 1996. Room temperature catalytic ammonia synthesis over an AB_5-type intermetallic hydride. *Journal of Alloys and Compounds*, **240**:L1–L3.
131. F.A. Uribe, S. Gottesfeld, T.A. Zawodzinksi, 2002. Effect of ammonia as potential fuel impurity on proton exchange membrane fuel cell performance. *Journal of the Electrochemical Society*, **149**:A293–A296.
132. H.J. Soto, W.-K. Lee, J.W.V. Zee, M. Murthy, 2003. Effect of transient ammonia concentrations on PEMFC performance. *Electrochemical and Solid-State Letters*, **6**:A133–A135.
133. R. Halseid, P.J.S. Vie, R. Tunold, 2006. Effect of ammonia on the performance of polymer electrolyte membrane fuel cells. *Journal of Power Sources*, **154**:343–350.
134. R. Halseid, J.S. Wainright, R.F. Savinell, R. Tunold, 2007. Oxidation of ammonium on platinum in acidic solutions. *Journal of the Electrochemical Society*, **154**:B263–B270.
135. T. Rockward, I.G. Urdampilleta, F.A. Uribe, E.L. Brosha, B.S. Pivovar, F.H. Garzon, 2007. The effects of multiple contaminants on polymer electrolyte fuel cells. *ECS Transactions*, **11**:821–829.
136. N. Zamel, X. Li, 2008. A parametric study of multi-phase and multi-species transport in the cathode of PEM fuel cells. *International Journal of Energy Research*, **32**:698–721.
137. N. Zamel, X. Li, 2009. Non-isothermal, multi-phase modeling of PEM fuel cell cathode. *International Journal of Energy Research*. DOI: 10.1002/er.1572.

8

Membrane Contamination Modeling

Thomas E. Springer and Brian Kienitz

CONTENTS

8.1 Introduction

The scientific community has made great progress in increasing the durability of polymer electrolyte membrane fuel cell (PEMFC) systems, but durability must further increase before we can consider fuel cells economically viable [1]. As durability increases, new modes of fuel cell contamination and failure are exposed. We expect state of the art PEMFC systems to run for thousands of hours. This means that each sulfonate group in typical perfluorosulfonic acid (PFSA) membranes used in today's PEMFC systems will associate with several million protons over the lifetime of the systems. Even if other cations replace only a small fraction of the protons entering the electrolyte membrane, these contaminant cations can build up in the system and degrade the fuel cell system performance over time.

Cationic contamination in PEMFC systems occurs when another cation replaces the proton associated with a sulfonate group in the polymer electrolyte. These contaminant cations then block protons from further associating with that site. The degree of cationic contamination is dependent on the specific cationic contaminant, the concentration of that contaminant, the exposure of that contaminant to the fuel cell, and the specific conditions of operation. Investigators have conducted a relatively small amount of research on the problem of cationic contamination of PEMFC systems as compared to other modes of fuel cell degradation. This is likely because of the slow nature of cationic contamination. In general, cationic contaminants will enter the polymer electrolyte slowly over time. In most settings, the effects of cationic contamination will not be significant until the fuel cell has run for many hours, unless specific probing enhances these effects. Interest in the effects of cationic contamination has intensified as the durability of PEMFC systems has increased.

Cationic contaminants may emanate from many sources. Metals, such as iron and copper, in system components may ionize due to corrosion exchange with protons in the membrane. Metallic salts, such as sodium and calcium, may enter the fuel cell from coastal water or from deicing agents. The most likely source of cationic contaminants is from the fuel line. Hydrogen from reformed hydrocarbons usually contain parts per million (ppm) of ammonia. This ammonia can be oxidized to ammonium ions and enter the polymer electrolyte.

Cationic contaminants tend to build up in the polymer electrolyte. This is because the sulfonate sites have a higher affinity for most other cations than protons and because most other cations do not partake in a suitable reaction to exit the polymer electrolyte phase [2,3]. In the case of ammonia, there is a suitable reaction at the cathode to remove ammonium ions from the system, but this reaction is likely slower than proton reduction. Some other metal ions, such as copper and cobalt, are electrochemically active in the fuel cell potential window and tend to "plate out" of the system. In general, once a cationic contaminant is in the polymer electrolyte phase it tends to stay there until the membrane has an acid treatment.

The properties of polymer electrolyte membranes change with the cation associated with the sulfonic acid sites. The conductivity of the membrane decreases when larger cations replace the protons. Other cations are at least four times less mobile than protons. The mobility of multivalent cations is lower than monovalent cations at similar molecular weight, but this is somewhat offset by the fact that multivalent ions carry more than one charge per molecule. The cationic composition also affects other properties of the polymer electrolyte membrane. Nonproton forms of the membrane do not hold as much water as the protonic form. In addition, the electroosmotic drag of water by other cations is greater than that by protons. Parameters for membranes equilibrated with several cationic species and mixtures of species have been measured and reported in the literature [2-4].

Although research has been limited, several studies conducted show the effects of cationic contamination on PEMFC performance. These studies, in general, have been of two types. In the first, researchers introduced cationic contaminant to the feed stream and measured fuel cell performance over time. In the second, they precontaminated the fuel cell to known levels to understand how contamination level directly relates to the percentage of protons displaced. A brief summary of the experiments and results from these studies follow below.

Time-dependent measurements collected when a cationic contaminant is included in the fuel cell fuel feed stream show a decrease in performance over time [5-10]. The most studied of these feed stream cationic contaminants is ammonia. Fuel cell performance quickly degrades when traces of ammonia are present in the hydrogen feed stream. The rate of degradation increases as a function of ammonia concentration or exposure. Uribe et al. [9] measured the performance of two fuel cells run potentiostatically at 0.5 V with different concentrations of ammonia in the hydrogen feed stream for 1 h and the subsequent recovery with zero ammonia concentration in the feed stream. When they added 13 ppm NH_3 to the hydrogen feed stream, the current density at 0.5 V was reduced by 25%. When they added 130 ppm NH_3 to the feed stream, the current density was reduced by 50% after 1 h of exposure. In the first case, where they added 13 ppm NH_3, the performance recovery was almost complete after an additional 12 h of operation on neat hydrogen. Performance did not fully recover for fuel cells exposed to higher concentrations of NH_3, although the period of the experiment may not have been long enough to reach equilibrium. In addition, the high frequency resistance increased after the addition of ammonia. A similar study showed that even though the high frequency resistance of the membrane increased, this increase only accounted for roughly 15% of total performance losses [11].

Other studies reveal similar results. They show that ammonia levels as low as 0.3 ppm in the hydrogen stream negatively affect the fuel cell over time [8]. Recovery seems possible in the case of ammonia poisoning. In general, performance increases when the fuel cell is switched to pure hydrogen feed streams. This is likely because ammonium ions have a mechanism to leave the membrane by converting back to ammonia. Other cations, such as sodium and iron, seem to stick in the membrane. In these cases, one needs a proper counter ion for the contaminant to leave the system.

Hydrogen pump cells, where hydrogen oxidizes at the anode and reduces back to hydrogen at the cathode, were also utilized to better understand the nature of performance loss due to cationic contamination. The results of the hydrogen pump also showed a limiting current under contaminated conditions, a phenomenon not known for fully protonated systems [11]. This indicates that a mechanism of proton limitation exists in contaminated cases.

Researchers have also confirmed the change in limiting current in systems that had protons quantifiably exchanged with other cations. They varied the cesium ion exchange fraction using mixed solutions of cesium salts and hydrochloric acid. They then confirmed the fraction of sites exchanged using x-ray fluorescence and measured polarization curves as a function of extent of contamination. These experiments showed that the limiting current and maximum power decreased as contamination increased. These effects were noticeable in systems with as little as 5% of the protons exchanged with cesium [12].

We infer three main effects of cationic contamination on PEMFC systems from these experimental results. First, the system performance seems to drop all across the polarization spectrum. Second, the high frequency resistance of the fuel cell increases as contamination increases, but this increase does not fully account for the decrease in system performance. Third, the limiting current of PEMFC systems decreases as contamination increases. We can explain these effects through physical modeling of the fuel cell system.

The goal of this chapter is to lay down the groundwork for explaining and predicting the effects of cationic contamination with macroscopic physics-based mathematical models. The chapter progresses from simple steady state dilute solution models to time dependents concentrated solution theory. As the focus of this chapter is the polymer electrolyte membrane and as modern fuel cell electrodes contain an electrolyte phase, we briefly discuss electrode modeling. To sharpen the focus on the membrane, we also restrict the modeled system to hydrogen pump rather than fuel cell mode. Although this chapter is comprehensive, it does not try to be definitive. We try to point out the limitations of the models and the data on which we based our empirical relationships.

8.2 Membrane Model: Single Cation Issues

8.2.1 Water Content

Before describing the modeling of cation contamination in the membrane, we should describe how we intend to treat the membrane itself as a medium for transport. Membranes, such as Nafion®, are perfluorinated ionomers composed of long chained polymers with side chains ending in a SO_3^- radical attached to a positive cation. We can picture the morphology of these membranes as a structure of hydrophilic pores formed along the charge sites and hydrophobic pores along the region of the Teflon®-like CF_2 backbone. There are two phases of water in the membrane: (1) that associated with external water in the vapor phase and (2) that associated with the additional water added when external water is in the liquid phase. Most hydrogen–air fuel cell membranes operate in the vapor phase. Liquid water at the boundaries would flood and block the reactant gases. Thus, the vapor phase is most relevant. People have treated the membrane as a single homogeneous phase with diffusion of water, migration of protons, and electric osmotic interaction between them [13,14]; as a porous structure with a pore distribution function of hydrophilic and hydrophobic pores [15,16]; and as a combined model, which uses the homogeneous structure for the vapor phase and a hydrophobic porous distribution for the liquid phase [17,18].

For this chapter, we will restrict ourselves to modeling the vapor phase of the membrane in which the membrane is in equilibrium with water vapor at its boundaries. This is the case in normal operation in hydrogen–air fuel cells. We will treat the membrane as a homogeneous phase with measurable properties like ion conductivity, water diffusion, and electroosmotic drag coefficients. We could model membranes as a porous structure of capillary tubes with a pore size distribution, which gives a physical picture relating the membrane water content to the external relative humidity. In the vapor region, such models provide a more physical and perhaps more satisfying picture of what is happening in the membrane, but they do not provide any more help in predicting performance of cation contamination than by using concentrated solution theory with measured parameters.

The homogeneous phase model relies on thermodynamic principles to structure the model. Experimental data determine the constitutive properties needed to model transport of water and ions in the membrane. The region near the charge sites is highly hydrophilic, capable of taking up many molecules of water for each charge site. This ratio, denoted as $\lambda = C_{H_2O} / C_{SO_3^-}$, is the water to sulfonic acid site concentration ratio. The weight of a membrane from the completely dry state to the fully hydrated with liquid water state can determine λ. It is often observed for Nafion of equivalent weight 1100, when the membrane is fully in contact with liquid water, $\lambda = 22$ and when it is fully in contact with saturated water vapor, $\lambda = 14$. This phenomenon, known as Schroeder's paradox may disappear if the state of the membrane includes not just its current situation, but also its previous thermal and humidification history [19]. We need to relate the water content to a thermodynamic state of the membrane. The structure and free energy of the system are altered when water is introduced into the membrane. The electrochemical potential μ_i is defined as the change in Gibbs free energy G per mole of water added at constant temperature, pressure, and other species. The term electrochemical potential used for μ_i includes electrical potential energy $\mu_i = \mu_{chem} + Z_iF\phi$, for a species i with charge Z_i and potential ϕ.

It is often more convenient to use water activity a_w, which is defined to be unity at a selected reference state of μ_{ref}.

$$\mu_w = \frac{\partial G}{\partial n_w}\bigg]_{T,P,n_j} , \quad a_w = e^{\frac{\mu_w - \mu_{ref}}{RT}} \tag{8.1}$$

If the membrane is in equilibrium with external water vapor, the water activity and electrochemical potential in each phase will be equal. Water vapor near atmospheric pressure well approximates an ideal gas. Therefore, the activity of water vapor is the ratio of the partial pressure of water p_w to the saturation pressure p_{sat} that would exist if the water were in equilibrium with liquid water at the surrounding temperature. We can also express activity in terms of the mole fraction x_w and the total pressure p in bars.

$$a_w = \frac{p_w}{p_{sat}} = x_w \frac{p}{p_{sat}} \tag{8.2}$$

The saturation partial pressure of water P_{sat} as a function of temperature is given by the Antoine equation in atm as

$$P_{sat} = 10^{\left(5.1905 - \frac{1730.63}{233.426 + T_c}\right)} \tag{8.3}$$

Equilibrating membranes over salt solutions that set the vapor pressure have provided membrane water activity versus water content data $a_w(\lambda)$. In particular, for Nafion with an equivalent weight EW of 1100 at 30°C, λ has been measured as a function of activity with the fitted function [20]

$$\lambda = .043 + 17.81a_w - 39.85a_w^2 + 36.0a_w^3 \tag{8.4}$$

At higher temperatures Weber and Newman [18] discuss that the water uptake at saturated vapor temperature decreases to around $\lambda = 9$ at 60°C and rises to 10 at 80°C. In addition to weight, thickness measurements can yield data on swelling volume with λ. If the fuel cell constrains the membrane's in-plane dimensions, swelling will mostly occur in the thickness or z direction.

As water swells the membrane, we assume the partial molar volume of water plus the partial molar volume of the dry membrane adds to the total membrane volume per mole. If M_{EW} is the equivalent weight of the dry membrane, M_w (18 gm/mol) is the mass per mole of water, ρ_{dry} is the dry membrane density, the mass per equivalent of the swelled membrane is $M_{EW} + \lambda M_w$, and the volume per equivalent is $(M_{EW}/\rho_{dry}) + (\lambda M_w/\rho_w)$. If ρ_w is 1 gm/cm², the wet membrane density is

$$\rho_{wet} = \rho_{dry} \frac{M_{EW} + \lambda M_w}{M_{EW} + \lambda M_w \rho_{dry}} \tag{8.5}$$

and the charge site concentration is

$$C_t = \frac{\rho_{dry}}{M_{EW} + \lambda M_w \rho_{dry}} \text{mol/cm3} \tag{8.6}$$

8.2.2 Water Diffusion Flux

In the absence of current flow through the membrane, water will flow down a gradient in electrochemical potential with a velocity v given by

$$v = -\frac{D'_w}{RT} \frac{\partial \mu_w}{\partial z} = -D'_w \frac{\partial \ln a_w}{\partial z} = -D'_w \frac{1}{a_w} \frac{da_w}{dz} \tag{8.7}$$

The water flux N_w expressed in terms of λ is

$$N_w = -D'_w C_t \frac{\lambda}{a_w} \frac{da_w}{d\lambda} \frac{\partial \lambda}{\partial x} \tag{8.8}$$

For 1100 EW Nafion from equation (8.4), we see that

$$\frac{da_w}{d\lambda} = \frac{1}{17.81 - 79.7a_w + 108a_w^2}. \tag{8.9}$$

D'_w was measured using spin echo nuclear magnetic experiments in membranes at various λ values by Zawodzinski [21]. The fitted equation is

$$D'_w = 10^{-6} \exp\left[2416\left(\frac{1}{303} - \frac{1}{273.15 + T_c}\right)\right](2.563 + .0264\lambda^2 - 0.000671\lambda^3) \tag{8.10}$$

Figure 8.1 to Figure 8.3 show the water diffusion coefficient, membrane activity, and the correction factor $(1/a_w)(da_w/d\lambda)$ needed to use $\nabla\lambda$ for the water driving force instead of $\nabla\mu_w$. Figure 8.4 shows the variation of water saturation with temperature.

8.2.3 Conductivity

In normal fuel cell operation, protons are the only cation species in the membrane. In the absence of a water-activity gradient, H^+ or H_3O^+ ions move solely by migration. The potential gradient in the membrane is related to the current density by

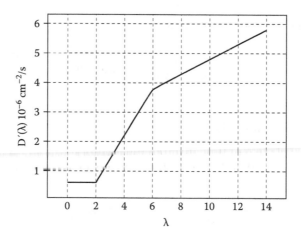

FIGURE 8.1
Diffusion coefficient for water in Nafion® 1100 EW in vapor phase as measured by Zawodzinski [21].

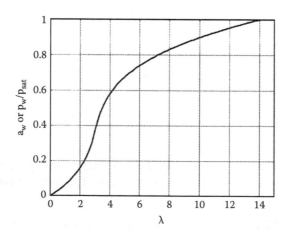

FIGURE 8.2
Water activity at 30°C in Nafion EW 1100 versus water uptake λ as measured by Zawodzinski [20].

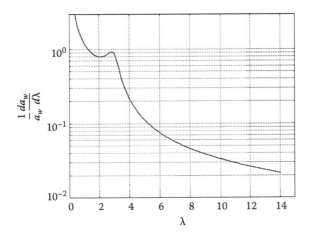

FIGURE 8.3
Conversion factor from electrochemical potential to water content gradient as driving force for water flux.

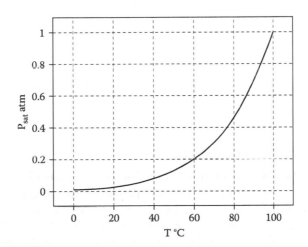

FIGURE 8.4
Saturated vapor pressure of water.

$$J = -\sigma \frac{\partial \phi}{\partial z}$$ (8.11)

where σ is measured in siemans per centimeter (S/cm). The conductivity depends on type of membrane, water content, temperature, and cation species present. A different cation in the membrane will change the water content at a given temperature and will change the conductivity. For proton conductivity used in this chapter's models, we used fitted data taken by Zawodzinski for 1100 EW Nafion, which is shown in Figure 8.5 [22].

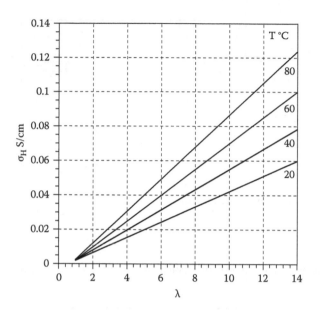

FIGURE 8.5
Proton conductivity dependence on water content and temperature for 1100 EW Nafion.

$$\sigma_H = \exp\left(1268\left(\frac{1}{303} - \frac{1}{T_K}\right)\right)(0.005139\lambda - .00326) \qquad (8.12)$$

8.2.4 Water Electroosmotic Drag

While this section has covered nonionic issues in the membrane model, the movement of protons through the membrane is an ionic issue. However, their interaction with water can be considered a nonionic issue. Protons moving through the membrane attach to water. The drag coefficient ξ_H is the ratio of the moles of water per mole of protons in the absence of a water activity gradient. In the vapor phase, the drag coefficient ξ_H appears to be about unity for λ above 4.5 [23]. When saturated with liquid water that fills the hydrophobic regions of the membrane, ξ_H goes to 2.5. If there is another cation B^+ present and moving, it too moves water so the total water flux, if λ is uniform, is

$$N_w = \xi_H N_H + \xi_B N_B \qquad (8.13)$$

Hydrogen pump experiments measuring the water lost on the anode and gained on the cathode can be used to measure ξ_H [24]. In liquid, hydrogen-charged palladium electrodes pressed against the membrane can be used to move water back and forth while measuring water height in capillary tubes on either side. By using the alpha to beta phase transition of Pd (palladium) to store H, no gas bubbles distort the measurement.

8.3 Membrane Model: Multiple Cation Issues

In this chapter, where possible, we will use the definitions contained in *Electrochemical Systems*, 3rd ed. (Newman and Thomas-Alyea [25]). We will discuss membrane contaminant models in one space dimension and time, which is sufficient to highlight all the major effects, but is readily extendable to two or three dimensions. In general, positive ionic current density, denoted by J in A/cm², will flow from left to right. We denote distance by z in absolute units (cm) and by x in scaled units. If L is the membrane thickness, $x = z/L$.

The total charge concentration in the membrane is C_T. If the membrane contains a blocking cation B with charge Z_B, we will let y_H represent the ion fraction of H⁺ cations to sulfonate sites. This is also its charge fraction. The ion fraction of B^{+Z} is

$$y_B = \frac{1 - y_H}{Z_B} \tag{8.14}$$

and the charge fraction of B^{+Z} is $1 - y_H$. The total charge flux through the membrane is

$$N_T = N_H + Z_B N_B \tag{8.15}$$

Note that the charge fraction of the two species always adds to 1, but the sum of the ion fractions is less than 1 unless Z_B is 1.

Section 8.2 discussed property values for a_w, D'_w, σ, and ξ as functions of λ and temperature for Nafion in protonated form. We need to obtain these property values when other cations are present. In general, we need data for the vapor region, whereas there has been more experimental data with cations in liquid saturated membranes. Likewise, we need data as a function of the ion fraction of protons. Okada et al. [3,26-28] have studied the effects of metal cation impurities on water transport properties in polymer electrolyte membranes. Much of this data is in the liquid region and there are uncertainties on what data one should use. Our purpose in this chapter is to discuss how one might model cation impurity effects and to show general behavior one might expect by modeling steady state, transient, and frequency response in the membrane and modeling observable metal potentials through connected electrodes with kinetic overpotentials and Nernst corrections. As this is a relatively emerging field, specific data used should be left to the modeler.

8.3.1 Ammonium Ion Contamination Example

Let us see what occurs when we introduce a current of 0.25 A/cm^2 into a 180 μm-thick Nafion membrane containing 50% ammonium ions NH$_4^+$ and 50% hydrogen ions H$^+$. Since we are only concerned with the cations in the membrane, for this example we can assume imaginary reversible point electrodes at that anode and cathode. The NH$_4^+$ ions cannot enter or leave the membrane on a short time scale. With no current in the membrane, the ions will be uniformly distributed throughout the fractional length x. After we introduce current entering the membrane at the anode side at x = 0 and leaving the cathode side at x = 1, the proton ion fraction y_H distribution profile will shift over time to a new distribution.

In Figure 8.6, we see the how the distribution shifts at various time intervals from 0.05 s to 100 s. Table 8.1 lists base case parameters for this example and others. At the anode boundary, the proton concentration will initially increase while the NH$_4^+$ concentration will correspondingly decrease. Initially both protons and blocking ions will move toward the cathode by migration down the electric field. No blocking cations can enter or leave through the boundary electrodes, thus, at the cathode, the exiting hydrogen ions will be supplied mostly from the H$^+$ migration flux with the drop in H$^+$ concentration supplying the rest of the current. The NH$_4^+$ flux blocked at the

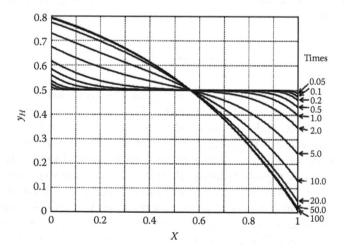

FIGURE 8.6

Proton distribution of an initial 50% H$^+$, 50% NH$_4^+$ fraction following introduction of 0.25 A/cm^2 current density into Nafion® 1100, 180-μm thickness at indicated time intervals from .05 to 100 s. Anode at left (x = 0) and cathode at right (x = 1).

TABLE 8.1

Some Base Case Parameters

Variable	Value
α	0.143
D_H	$4.58 \cdot 10^{-5}$ cm^2/s
D_B	$2.34 \cdot 10^{-6}$ cm^2/s
σ_H	0.1S /cm
Z	1
C_t	0.00182 mol/cm^3
λ	14
D'_w	$5.80 \cdot 10^{-6}$ cm^2/s
b	0.0287 V
L	0.018 cm

cathode will increase its concentration corresponding to the decrease in H⁺ concentration because electroneutrality must hold.

As time progresses equilibrium is established. At this time at every position across the membrane, the proton migration and diffusion fluxes will add to supply the total current density, while the NH₄⁺ migration flux is everywhere opposite to its diffusion flux and no net NH₄⁺ current will flow. While the current is flowing, water in the membrane will also move to the cathode side by electroosmotic drag from the ionic current. If this results in a gradient in water content in the membrane, there may be a back diffusion of water opposing the electroosmotic drag. Below is a description of why these effects occur, how to model the effects, and some of the resulting predictions and verifications.

8.3.2 Migration

Assume for now we can apply an ionic potential across the membrane (later we will discuss the anode and cathode boundaries needed to convert to electronic potential). Charged ions move down the gradient of electrochemical potential, which includes both concentration and electric potential gradients. The charge site concentration C_t in the membrane is fixed because the sulfonate sites are immobile (as long as the water content is uniform so that swelling effects can be neglected). The net charge will be zero everywhere. If only a single cation species is present, as is H⁺ in a normal fuel cell, the current can be related to the electric field in terms of any of three related membrane properties, mobility u_i, diffusion coefficient D_i, or ionic conductivity:

$$J_i = v_i Z_i F \frac{C_t}{Z_i} = -u_i F^2 C_t \frac{d\phi}{dz} = -\frac{D_i}{RT} Z_i F^2 C_t \frac{d\phi}{dz} = -\sigma_i \frac{d\phi}{dz} \qquad (8.16)$$

Equation (8.16) uses the fact that mobility is related to the diffusion coefficient by the Nernst Einstein relation $u_i = D_i / RT$ and to the average ion velocity by $v_i = -u_i Z_i F \partial \phi / \partial z$. R is the gas constant 8.3144 J/mol/K, T is the temperature in K, F is the Faraday constant 96485 J/mol.

In general, the conductivity σ_i for species i is that measured when it is the only cation present.

$$\sigma_i = \frac{C_t Z_i F^2 D_i}{RT} = \frac{Z_i F C_t D_i}{b} \text{ where } b = \frac{RT}{F} \tag{8.17}$$

8.3.3 Ion Diffusion

When two ionic species are present and the water content is uniform, the H$^+$ flux is

$$N_H = -D_H C_t \frac{\partial y_H}{\partial z} - \frac{D_H F}{RT} C_t y_H \frac{\partial \phi}{\partial z}$$

$$= -\frac{D_H C_T}{RT} \left(RT \frac{\partial y_H}{\partial z} + y_H F \frac{\partial \phi}{\partial z} \right) = -\frac{D_H C_T}{RT} \frac{\partial \mu_H}{\partial z} \tag{8.18}$$

It is convenient to let

$$\alpha = \frac{D_B}{D_H} = \frac{\sigma_B}{Z_B \sigma_H}. \quad \sigma_B = \alpha Z_B \sigma_H \tag{8.19}$$

The B$^+$ flux is

$$N_B = -\alpha D_H C_t \frac{\partial y_B}{\partial z} - \frac{\alpha D_H C_t F Z_B}{RT} y_B \frac{\partial \phi}{\partial z}$$

$$= -\alpha \frac{D_H C_t}{RT} \left(\frac{\partial y_B}{\partial z} + y_B F Z_B \frac{\partial \phi}{\partial z} \right) = -\alpha \frac{D_H C_t}{RT} \frac{\partial \mu_B}{\partial z} \tag{8.20}$$

Here we see that the diffusion and migration terms in the flux equations combine into the gradient of electrochemical potential of each species. We now need to see how to modify these fluxes when a gradient in water electrochemical potential exists.

8.3.4 Combined Ion and Water Fluxes

In dilute solutions in which multiple solutes do not interfere with each other, each species separately interacts only with the solvent. In that case, the flux of species i is given by the Nernst–Planck equation and charge conservation

$$N_i = -D_i \frac{\partial C_i}{\partial z} - Z_i u_i F C_i \frac{\partial \phi}{\partial z} + C_i v \quad \text{and} \quad \sum_i C_i Z_i = 0 \qquad (8.21)$$

whose three terms give the diffusion, migration, and convection fluxes, respectively, in terms of only properties associated with that species. The summation of charge concentration includes both positive and negative charges. However, when the species interact with each other, this is no longer the case.

Concentrated solution theory allows the treatment of all species interacting with each other. We discuss this in more detail in the appendix to this chapter where we derive equation (8.22). There we show why the matrix relating electrochemical potential to flux is symmetric. If we know how the water flux is related to the proton flux, we can determine how the proton flux is affected by the water flux. In this equation, the C_i are the concentrations of species H, B, and w. The L_{ij} are unknown, only their symmetry property that $C_i C_j L_{ij} = C_j C_i L_{ji}$ is known.

$$\begin{bmatrix} N_H \\ N_B \\ N_w \end{bmatrix} = \begin{bmatrix} C_H v_H \\ C_B v_B \\ C_w v_w \end{bmatrix} = \begin{bmatrix} C_H C_H L_{HH} & C_H C_B L_{HB} & C_H C_w L_{Hw} \\ C_B C_H L_{BH} & C_B C_B L_{BB} & C_B C_w L_{Bw} \\ C_w C_H L_{wH} & C_w C_B L_{wB} & C_w C_w L_{ww} \end{bmatrix} \begin{bmatrix} \nabla \mu_H \\ \nabla \mu_B \\ \nabla \mu_w \end{bmatrix} ,$$

$$C_i C_j L_{ij} = C_j C_i L_{ji} \qquad (8.22)$$

We know some of the coefficients in the first two rows of equation (8.22) from equation (8.18) and equation (8.20), but not the coefficients for μ_w. For this model, we will assume N_H and N_B are each respectively independent of μ_B and μ_H. Then $L_{HB} = L_{BH} = 0$. In the third row, for N_w, we know the coefficient for $\nabla \mu_w$ when $\nabla \mu_H$ and $\nabla \mu_B$ are zero and we know that $N_w = \xi_H N_H + \xi_B N_B$ when $\nabla \mu_w = 0$. If we substitute the known coefficients in equation (8.22), we get

$$N_H = -\frac{D_H C_t}{RT} \nabla \mu_H + 0 \cdot \nabla \mu_B + C_H C_w L_{Hw} \nabla \mu_w$$

$$N_B = 0 \cdot \nabla \mu_H - \alpha \frac{D_H C_t}{RT} \nabla \mu_B + C_B C_w L_{Bw} \nabla \mu_w$$

$$N_w = -\frac{D_H C_t}{RT} \xi_H \nabla \mu_H - \alpha \frac{D_H C_t}{RT} \xi_B \nabla \mu_B \qquad (8.23)$$

$$+ \left(\xi_H C_H C_w L_{Hw} + \xi_B C_B C_w L_{Bw} - C_t \lambda \frac{D'_w}{RT} \right) \nabla \mu_w$$

From the symmetry relations, we know

$$C_H C_w L_{Hw} = -\frac{D_H C_t}{RT}\xi_H \text{ and } C_B C_w L_{Bw} = -\alpha\frac{D_H C_t}{RT}\xi_B$$

Equation (8.24) gives the fluxes in terms of the electrochemical potential gradients.

$$\begin{bmatrix} N_H \\ N_B \\ N_w \end{bmatrix} = \frac{D_H C_t}{RT}\begin{bmatrix} -1 & 0 & -\xi_H \\ 0 & -\alpha & -\alpha\xi_B \\ -\xi_H & -\alpha\xi_B & -\xi_H\xi_H - \alpha\xi_B\xi_B - \lambda D_w' / D_H \end{bmatrix}\begin{bmatrix} \nabla\mu_H \\ \nabla\mu_B \\ \nabla\mu_w \end{bmatrix}$$

$$(8.24)$$

Now that we have determined all the coefficients in terms of measured quantities, we can reform the equations for solving.

8.4 Transport Equations

8.4.1 Continuity Equation

The basic equation needed to model what happens in the membrane when its boundaries become subjected to electric potential, proton flux, or water activity/pressure is the continuity equation. Because we have no sources or sinks inside the membrane, only at the boundaries, continuity requires that

$$\frac{\partial C_i}{\partial t} + \frac{\partial N_i}{\partial z} = 0 \qquad (8.25)$$

The state variables that we will use to describe the instantaneous state of the membrane will be the proton ion fraction, ion potential, and water content (y_H, ϕ, and λ). The three fluxes we will use are the hydrogen flux N_H, the total charge flux $N_T = N_H + Z_B N_B$, and the water flux N_w. The concentration associated with the total flux is just C_t, which is constant in time. In terms of the scaled position variable, equation (8.25) becomes

$$C_t L\frac{\partial y_H}{\partial t} + \frac{\partial N_H}{\partial x} = 0, \quad \frac{\partial N_T}{\partial x} = 0, \quad C_t L\frac{\partial \lambda}{\partial t} + \frac{\partial N_w}{\partial x} = 0 \qquad (8.26)$$

The potential ϕ has no partial time derivative. This is because the total charge is always fixed. If the current changes, the potential responds instantaneously and only shifts as the y_H distribution shifts.

8.4.2 Flux Equations in Scaled Solution Variables

To complete putting equation (8.26) in a form amenable to solution, we need to express the electrochemical potentials in terms of our state variables.

$$\nabla \mu_H = RT \frac{\partial y_H}{\partial z} + F y_H \frac{\partial \phi}{\partial z};$$

$$\nabla \mu_B = -\frac{RT}{Z_B} \frac{\partial y_H}{\partial z} + Z_B F \frac{(1 - y_H)}{Z_B} \frac{\partial \phi}{\partial z};$$

$$\nabla \mu_W = RT \nabla \ln a_w = RT d_{a\lambda} \frac{\partial \lambda}{\partial z} \text{ where } d_{a\lambda} = \frac{d \ln a_w}{d\lambda}$$

(8.27)

If we replace the $\nabla \mu_i$ in equation (8.24) with gradients in the state variables of equation (8.27), we get the fluxes needed in equation (8.28) through equation (8.31) to model transport in the membrane. We also convert the position variable z to the scaled distance $x = z / L$.

$$N_H = -\frac{D_H C_t}{L} \left(\frac{\partial y_H}{\partial x} + \frac{y_H}{b} \frac{\partial \phi}{\partial x} + \xi_H d_{a\lambda} \frac{\partial \lambda}{\partial x} \right)$$

(8.28)

$$N_B = \frac{\alpha D_H C_t}{L} \left(\frac{1}{Z_B} \frac{\partial y_H}{\partial x} - \frac{(1 - y_H)}{b} \frac{\partial \phi}{\partial x} - \xi_B d_{a\lambda} \frac{\partial \lambda}{\partial x} \right)$$

(8.29)

$$N_t = N_H + Z_B N_B = \frac{C_t D_H}{L} \left((\alpha - 1) \frac{\partial y_H}{\partial x} + \frac{(\alpha Z_B - 1) y_H - \alpha Z_B}{b} \frac{\partial \phi}{\partial x} \right.$$

$$\left. -(\xi_H + \alpha Z_B \xi_B) d_{a\lambda} \frac{\partial \lambda}{\partial x} \right)$$

(8.30)

$$N_w = \frac{C_t D_H}{L} \left(\left(\frac{\alpha \xi_B}{Z_B} - \xi_H \right) \frac{\partial y_H}{\partial x} + \frac{((\alpha \xi_B - \xi_H) y_H - \alpha \xi_B)}{b} \frac{\partial \phi}{\partial z} \right.$$

$$\left. -(\xi_H^2 + \alpha \xi_B^2) d_{a\lambda} \frac{\partial \lambda}{\partial x} \right) - \frac{C_t D_w' \lambda}{L} d_{a\lambda} \frac{\partial \lambda}{\partial x}$$

(8.31)

8.4.3 Boundary Conditions

The partial differential equations defined in the previous two sections must be supplied boundary conditions. In general, there are two types of boundary conditions. Neumann conditions specify a flux entering the region and Dirichlet conditions place a constraint on a state variable at the boundary. In the examples in this chapter, we will specify the current density J at $t > 0$, fix the potential at the anode side, and fix the water content at both anode and cathode sides. We will specify the initial conditions at time $t = 0$ that would exist if the current were zero. There, y_H will have a uniform value of y_{in}. The potential everywhere is zero. We arbitrarily let λ vary linearly across the membrane. These conditions are formulized in equation (8.32):

$$\text{At } x = 0, \quad N_H = \frac{J}{F}, \quad \phi = 0, \quad \lambda = \lambda_A$$

$$\text{At } x = 1, \quad N_H = -\frac{J}{F}, \quad N_T = -\frac{J}{F}, \lambda = \lambda_C \tag{8.32}$$

$$\text{At } t = 0, \quad y_H(x) = y_{in}, \phi(x) = 0, \lambda(x) = \lambda_A(1-x) + \lambda_C x$$

In the solution of the membrane equations, it is convenient to set the anode ion potential to zero. Later, we will add overpotential and Nernst potential shifts to the ion potential distribution calculated with the 0 anode boundary potential. Thus, $\phi(1) = \phi_{CA} = \phi_C - \phi_A$ when an offset is applied.

8.5 Solution Methods and Predictions

We let the subscripts A and C indicate the value of the subscripted variable at the anode or cathode boundary, respectively. Thus, $y_A = y_H(0)$, $y_C = y_H(1)$, $\lambda_A = \lambda(0)$, and $\lambda_C = \lambda(1)$. If we fix λ_A and λ_C to the same value, in steady state, $\partial\lambda/\partial x$ will be zero everywhere. In this case, we can get an analytic solution. Otherwise, we must solve the equations numerically.

8.5.1 Steady State Analytic Solution

To get an analytic, steady state solution with constant water content, we use equation (8.28) and equation (8.29) without the $\partial\lambda/\partial x$ terms. Equation

(8.17) and equation (8.19) are used to convert diffusion coefficients to proton conductivity.

$$J = FN_H = -F\frac{D_H C_t}{L}\left(\frac{\partial y_H}{\partial x} + \frac{y_H}{b}\frac{\partial \phi}{\partial x}\right) = -\frac{\sigma_H b}{L}\left(\frac{\partial y_H}{\partial x} + \frac{y_H}{b}\frac{\partial \phi}{\partial x}\right) \tag{8.33}$$

$$0 = \frac{\alpha D_H C_t}{L}\left(\frac{\partial y_H}{\partial x} - \frac{Z_B(1-y_H)}{b}\frac{\partial \phi}{\partial x}\right) = \frac{\alpha \sigma_H b}{L}\left(\frac{\partial y_H}{\partial x} - \frac{Z_B(1-y_H)}{b}\frac{\partial \phi}{\partial x}\right) \tag{8.34}$$

We see immediately from equation (8.34) that the solution is independent of α or blocking conductivity, but is modified by the blocking ion charge. We can immediately determine the potential through the membrane in terms of the values of ϕ and y_H at the anode side.

$$\frac{\partial \phi}{\partial x} = \frac{b}{Z_B(1-y_H)}\frac{\partial y_H}{\partial x}, \ \phi(y_H) = \phi_A + \frac{b}{Z_B}\int_{y_A}^{y_H}\frac{dy_H}{1-y_H} = \frac{b}{Z_B}\ln\left(\frac{1-y_A}{1-y_H}\right) \tag{8.35}$$

Substituting $\partial \phi/\partial x$ into equation (8.33), we get

$$J = -\frac{\sigma_H b}{L}\left(\frac{Z_B - (Z_B - 1)y_H}{Z_B(1-y_H)}\right)\frac{\partial y_H}{\partial x} \tag{8.36}$$

This yields an implicit function for $y_H(x)$.

$$\frac{Z_B JL}{\sigma_H b}x + (Z_B - 1)(y_H - y_A) + \ln\left(\frac{1-y_H}{1-y_A}\right) = 0 \tag{8.37}$$

In particular, equation (8.37) has an explicit solution when $Z_B = 1$.

$$y_H(x) = 1 - (1-y_A)\exp\left(\frac{JL}{\sigma_H b}x\right) \tag{8.38}$$

This expresses y_H in terms of y_A, which we need to determine in terms of the known initial y_{in}. We know that the average value of y_H across the membrane will always be equal to y_{in}.

$$\int_0^1 y_H(x)dx = y_{in}$$

(8.39)

Integrating equation (8.39), we get y_A in terms of y_{in} when $Z_B = 1$.

$$y_A = \frac{1-\exp\left(\dfrac{JL}{\sigma_H b}\right)+\dfrac{JL}{\sigma_H b}(1-y_{in})}{1-\exp\left(\dfrac{JL}{\sigma_H b}\right)}$$

(8.40)

At the cathode side

$$x = 1, \quad y_C = 1-(1-y_A)\exp\left(\frac{JL}{\sigma_H b}\right)$$

(8.41)

The current will reach a limiting value when y_C drops to zero. This occurs because migration of ions is proportional to the ion concentration. When y_C drops to zero at the cathode, migration flux is also zero there. At the limiting current, all current at the cathode is carried by diffusion. When y_C is zero in equation (8.41), we see that the limiting current J_{lim} becomes the solution of the implicit equation

$$1-(1-y_A(J_{lim}))\exp\left(\frac{J_{lim}L}{\sigma_H b}\right)=0$$

(8.42)

Figure 8.7 shows the limiting current density as a function of membrane thickness for various values of y_{in}. The previous analytic solution gives an accurate picture of the concentration and potential distribution for steady state conditions with uniform membrane water content. However, for time-dependent solutions and later frequency response solutions, we need numerical computation.

8.5.2 Predictions from Analytic Steady State Solution without Water

The analytic equations can give us the general behavior of the membrane with cation contamination. Our base case thickness L is 180 μm, proton

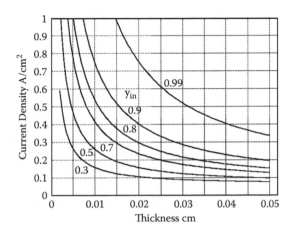

FIGURE 8.7
Limiting current density as a function of membrane thickness for various initial H⁺ fractions.

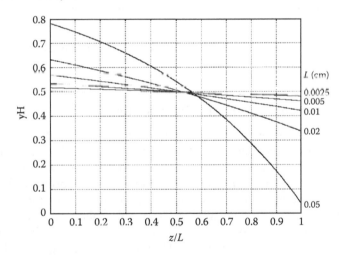

FIGURE 8.8
Proton distribution in membrane at a fixed current density of 0.09 A/cm² for various thicknesses, with initial 50% initial proton ion fraction. The distribution would be the same if we fixed the thickness and varied the current density.

conductivity σ_H is 0.1 S/cm, initial H⁺ fraction y_{in} is 0.5, and diffusion coefficient ratio α is 0.143. Figure 8.7 shows that thinner membranes will have higher limiting currents. Figure 8.8 shows that, for the same current density, the distribution becomes more pronounced with thicker membranes. Note that all the distributions cross each other at about the same location until we approach the limiting distribution. Figure 8.9 shows the current components carried by diffusion and migration near limiting current density. Note that the B⁺ migration and diffusion fluxes everywhere cancel each other and that diffusion of H⁺ increases toward the cathode side. At

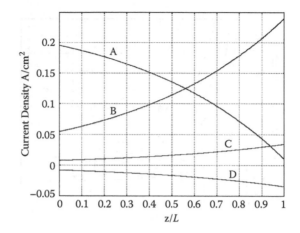

FIGURE 8.9
Components of a 0.25 A/cm² current density for the base case carried by: A. H⁺ migration, B. H⁺ diffusion, C. B⁺ migration, and B⁺ diffusion. No water activity gradient: $\lambda = 14$.

limiting current, all the proton current moves by diffusion at the cathode boundary.

8.5.3 Time Domain Numerical Solutions with Water

Equation (8.25) represents a set of partial differential equations (PDEs) that must be solved in one space dimension and propagated in time. Typically, one discretizes the space variable $0 \le x \le 1$ into n intervals. If the equations are linear, one can write difference equations at each point and solve the resulting matrix equation. The solution is progressed in time by also discretizing the time derivative with a Crank–Nicholson-like scheme. If the equations are nonlinear, the situation is more complicated. The appendix of *Electrochemical Systems* (Newman and Thomas-Alyea [25]) gives a discussion of numerical solution of partial differential equations.

Fortunately there are software programs these days that can be given the PDEs, boundary conditions, and input parameters, and will set up and solve the required equations. One such program, used for the examples in this chapter, is Comsol Multiphysics® (Stockholm, Sweden).

If $\mathbf{u}$ is the vector of our state variables (y_H, ϕ, λ), the equations we will be solving are of the form

$$d_a \frac{\partial \mathbf{u}}{\partial t} + \nabla \cdot \left(-c\nabla \mathbf{u} - a\mathbf{u} + \gamma \right) + a\mathbf{u} + \beta \cdot \nabla \mathbf{u} = f \qquad (8.43)$$

Here the matrix c can have nonlinear terms containing some of the state variables, u_i. In particular, for the time solution, most of this equation is not used. In order to solve equation (8.28), equation (8.30), and equation (8.31), we only need the form

$$d_a \frac{\partial \mathbf{u}}{\partial t} + \nabla \cdot (-c\nabla \mathbf{u}) = 0 \tag{8.44}$$

$d_a = (C_t, 0, C_t)$, $\mathbf{u} = \begin{bmatrix} y_h \\ \phi \\ \lambda \end{bmatrix}$. Let $\beta = \dfrac{D_H C_t}{L}$. From equation (8.28) to (8.31),

c becomes

$$c = \beta \begin{bmatrix} 1 & \dfrac{y_H}{b} & \xi_H d_{a\lambda} \\ 1-\alpha & \dfrac{\alpha Z_B - (\alpha Z_B - 1)y_H}{b} & (\xi_H + \alpha Z_B \xi_B)d_{a\lambda} \\ \left(\xi_H - \dfrac{\alpha \xi_B}{Z_B}\right) & \dfrac{\alpha \xi_B - (\alpha \xi_B - \xi_H)y_H}{b} & \left(\xi_H^2 + \alpha \xi_B^2 + \dfrac{C_t D_w' \lambda}{\beta L}\right)d_{a\lambda} \end{bmatrix}$$

$$\tag{8.45}$$

The boundary conditions were given in equation (8.32).

The form of the PDEs we will use to calculate linearized frequency response is

$$\nabla \cdot (-c\nabla \mathbf{u} - a\mathbf{u}) + a\mathbf{u} = 0 \tag{8.46}$$

8.5.4 Predictions from Time Domain Numerical Solution with Water

Previously, we saw how the diffusion and migration fluxes varied across the membrane in the absence of water effects. If we include water effects but maintain uniform λ across the membrane, the water will have little effect. Now, we impose a water activity gradient by setting λ_A to 11 and λ_C to 14. In the absence of current, this will cause water to flow toward the anode. Let us look at the same case as in Figure 8.9 at 0.25 A/cm², shown in Figure 8.10. The potential at the cathode was −45 mV when λ was constant, but was −66 mV when λ_A was 11. Because the forward flow of protons was impeded by the backflowing water, its migration flux is decreased while its diffusion flux is increased. The ion fraction y_H distribution is the same in both cases.

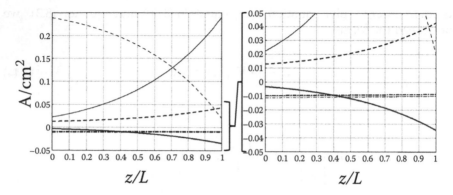

FIGURE 8.10
Diffusion (solid), migration (dashed), and osmotic H^+ (thin line) and B^+ (thick line) components of current density at 0.25 A/cm² and $\lambda_A = 11$. See Figure 8.9 for $\lambda_A = \lambda_C = 14$ case.

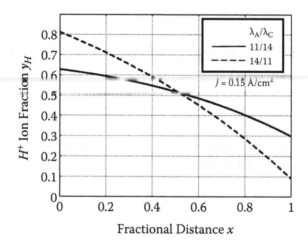

FIGURE 8.11
Equilibrium proton distribution shift with opposite water activity gradients with yin = 0.5 at a current density of 0.15 A/cm2 and including electroosmotic drag.

When there is an activity gradient across the fuel cell, the proton distribution shifts toward the direction of additional water flow. This is shown in Figure 8.11.

8.5.5 Dynamic Resistance Effects

When current is stepped from one value to another, the concentration profiles do not change instantaneously. Thus, only the migration flux will change immediately. If the cathode potential steps to a lower potential, the current will experience a step increase followed by a slower additional increase as the

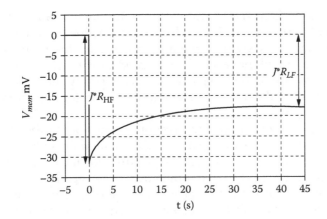

FIGURE 8.12
Potential response to 0.1 A/cm² current density step with $y_{in} = 0.5$, $L = .018$ cm, $\lambda = 14/14$.

new y_H distribution forms. If we step the current, as in Figure 8.12, the potential drops and then rises some. The initial effective conductivity will be

$$\sigma_{eff} = y_H\sigma_H + y_R\sigma_R = y_H\sigma_H + \frac{1-y_H}{Z_B}\alpha Z_R\sigma_H = \sigma_H\left(y_H + \alpha\left(1-y_H\right)\right)$$

$$(8.47)$$

High frequency resistance measurements, usually done to measure the membrane resistance for direct current, will see this effective conductivity. At low frequency and DC current, only protons flow with the conductivity σ_H. The effective conductivity is only uniform across the membrane when no current is flowing and λ is uniform. The high and low frequency resistances are

$$R_{HF} = \frac{L}{\sigma_H}\int_0^1 \frac{dx}{y_H(x)+\alpha\left(1-y_H(x)\right)}; \quad R_{LF} = \frac{L}{\sigma_H}; \quad (8.48)$$

In our base case example of $y_{in} = 0.5$, $L = .018$ cm, R_{HF} is 0.315 Ω cm² at .01 A/cm² but increases to 0.364 Ω cm² at .025 A/cm².

For a step of 0.1 A/cm² in a 180 μm membrane, we see the 32 mV initial drop corresponds to a high frequency resistance of 0.32 Ωcm² and a subsequent low frequency resistance of 0.18 Ωcm².

8.5.6 Frequency Domain Equations

AC impedance spectroscopy is a useful probing tool for investigating membrane cation contamination. If the current through the membrane is varied slightly by a sinusoidal perturbation, the ionic potential everywhere will vary at the same frequency with some amplitude and phase shift relative to the current perturbation. The hydrogen ion fraction y_H and water content λ will also vary everywhere. Generally, one applies the perturbation after the system has settled into a steady state mode. Likewise, when the current initially has a perturbation at a particular frequency applied, the initial transient adjustment to the new frequency must pass before measuring voltage amplitude and phase shift.

To model the system frequency response, we rewrite the PDEs, represented by equation (8.43), by replacing $\mathbf{u}(x,t)$ with $\mathbf{u}_o(x)+\mathbf{u}'(x,t)$, the steady state solution plus a small time varying perturbation $\mathbf{u}'$. The steady state and the perturbation were set by the input current density $J(t) = J_o + J'(t)$. Note we could have perturbed the other inputs λ_A and λ_C, but this is not experimentally practical. We substitute $\mathbf{u}_o + \mathbf{u}'$ into equation (8.43) and express the c matrix as $c_1\mathbf{u} + c_2$. This is to cover the nonlinear situation in the migration terms containing $y_H \, \partial\phi/\partial x$ and $y_H \, \partial\lambda/\partial x$ in equation (8.45).

$$d_a\left(\frac{\partial \mathbf{u}_o}{\partial t} + \frac{\partial \mathbf{u}'}{\partial t}\right) + \nabla \cdot \left(c_1\mathbf{u}_o + c_1\mathbf{u}' + c_2\right)\left(\nabla \mathbf{u}_o + \nabla \mathbf{u}'\right) = 0$$

If we subtract the steady state solution $\partial\mathbf{u}_o/\partial t = 0 = \nabla \cdot \left(c_1\mathbf{u}_o + c_2\right)\nabla \mathbf{u}_o$, there remains a new set of PDEs in the perturbed variables. We know all the steady state variables.

$$\frac{\partial \mathbf{u}'}{\partial t} + \nabla \cdot \left(c_1\mathbf{u}_o + c_2\right)\nabla \mathbf{u}' + \nabla \mathbf{u}_o c_1 \mathbf{u}' = 0 \tag{8.49}$$

Now, we Laplace transform the primed variables and their time derivatives from the time domain t to the frequency domain ω

$$\bar{\mathbf{u}}(x,\omega) = L\left[\mathbf{u}'(x,t)\right], \quad L\left[\frac{d\mathbf{u}'}{dt}(x,t)\right] = j\omega\bar{\mathbf{u}}(x,\omega) \tag{8.50}$$

The new PDEs in the frequency domain are still functions of x and are now in the form of equation (8.46).

$$\nabla \cdot \left(c_1\mathbf{u}_o + c_2\right)\nabla \bar{\mathbf{u}} + \nabla \mathbf{u}_o c_1 \bar{\mathbf{u}} + j\omega d_a \bar{\mathbf{u}} = 0 \tag{8.51}$$

The perturbed state variables are complex and the coefficients use the steady state solution as functions of x. Instead of a time solution, the PDE solver now has a series of steady state solutions with ω as a varying parameter.

The problem variables are complex numbers. Because this problem is linear, if we set the perturbed transformed current density $\bar{J}(\omega)$ to unity, all the transformed state variables will be so-called transfer functions of that variable with respect to current density.

The three PDEs that are solved in the frequency domain are

$$\nabla \cdot -\beta \left[\frac{\partial \bar{y}}{\partial x} + \frac{y_o}{b} \frac{\partial \bar{\phi}}{\partial x} + \xi_H d_{a\lambda} \frac{\partial \bar{\lambda}}{\partial x} + \frac{1}{b} \frac{\partial \phi_o}{\partial x} \bar{y} \right] + C_t L j \omega \bar{y} = 0 \tag{8.52}$$

$$\nabla \cdot \beta \left[(\alpha - 1) \frac{\partial \bar{y}}{\partial x} + \frac{((\alpha Z_B - 1)y_o - \alpha Z_B)}{b} \frac{\partial \bar{\phi}}{\partial x} - (\xi_H + \alpha Z_B \xi_B) d_{a\lambda} \frac{\partial \bar{\lambda}}{\partial x} \right.$$
$$\left. + \frac{(\alpha Z_B - 1)}{b} \frac{\partial \phi_o}{\partial x} \bar{y} \right] = 0 \tag{8.53}$$

$$\nabla \cdot \beta \left[\left(\frac{\alpha \xi_B}{Z_B} - \xi_H \right) \frac{\partial \bar{y}}{\partial x} + \frac{((\alpha \xi_B - \xi_H)y_o - \alpha \xi_B)}{b} \frac{\partial \bar{\phi}}{\partial z} \right.$$
$$\left. - \left(\xi_{II}^2 + \alpha \xi_B^2 + \frac{C_t D_w' \lambda_o}{\beta L} \right) d_{u\lambda} \frac{\partial \bar{\lambda}}{\partial x} + \right. \tag{8.54}$$
$$\left. \left(\frac{\alpha \xi_B - \xi_H}{b} \frac{\partial \phi_o}{\partial z} \right) \bar{y} - \frac{C_t D_w'}{\beta L} d_{a\lambda} \frac{\partial \lambda_o}{\partial x} \bar{\lambda} \right] + C_t L j \omega \bar{\lambda} = 0$$

8.5.7 Impedance Predictions

The current step in Figure 8.12 has its analog in the frequency domain. The membrane impedance is shown in Figure 8.13, calculated at a steady current density of one half the step size. We see the low and high frequency

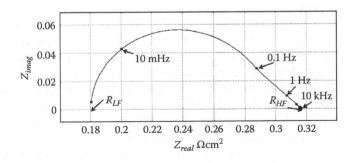

FIGURE 8.13
Membrane impedance of Figure 8.12 membrane at .05 A/cm² steady current density, which is the average of the initial and final current density in Figure 8.12.

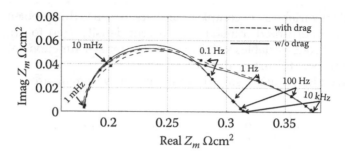

FIGURE 8.14
Membrane impedance for base case with and without water drag and at low and high current densities. The proton distribution for the two current densities is shown in Figure 8.15.

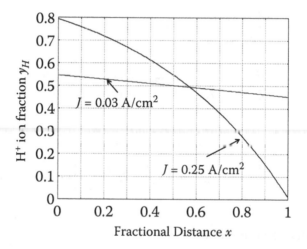

FIGURE 8.15
Proton distribution for the two current densities of the impedance plots in Figure 8.14. The distributions are the same with or without water drag, but the potential distributions are different.

resistance directly. Also, note that the imaginary part is positive in the membrane, although the Nernst potential will make this difficult to measure. The 45° slope at high frequency is characteristic of diffusion of the ion species at higher frequencies. In Figure 8.14, we see the effects on the impedance by water drag and by high steady current density. When the water activity is the same at anode and cathode, the driving force for the water flux must be supplied by an increased potential for the same current flow. Thus, the impedance is increased. Likewise, at higher current density section 8.5.5 indicates that the high frequency resistance will increase.

When the anode and cathode humidities are different, we saw in Figure 8.11 that the proton distribution is shifted in the direction of additional water flow toward the cathode. Also in Figure 8.16, the impedance increases at high frequency when the additional water flow is toward the cathode. However,

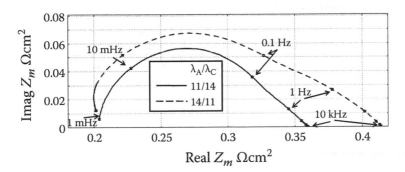

FIGURE 8.16
Membrane impedance for the two cases in Figure 8.11.

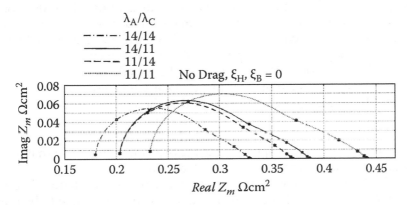

FIGURE 8.17
By turning off electroosmotic drag, we see the effect of lower membrane conductivity when λ is lowered. The frequencies can be seen in Figure 8.16.

the impedance decreases at low frequency for this case because more current is carried by diffusion, which is driven by the steeper gradient seen in Figure 8.11. Some of the shift in impedance results from the electroosmotic water drag and some results from the shift in conductivity caused by the lower value of λ necessary to impose a water activity gradient. Figure 8.17 shows calculated impedance with no water drag for four combinations of anode/cathode humidification levels.

When we have an activity gradient, the DC resistance as defined as the voltage across the membrane divided by the current density no longer has meaning because of the potential generated by the water flux. Table 8.2 shows the open circuit potential ϕ_C and the H^+ ion fraction at anode and cathode for differences in λ_A and λ_C for base case parameters. However, when we vary the potential or current density for AC impedance calculations, λ_A and λ_C are not allowed to vary from their steady values. In practice, this may be difficult to achieve.

TABLE 8.2

Open Circuit Potential and H⁺ Ion Fraction Shift
from Difference in Boundary Humidification

λ_A	λ_C	$\phi_C - \phi_A$	y_A	y_C
11	14	−7.598 mV	0.450	0.544
14	11	+7.598 mV	0.544	0.450

8.6 Electrodes and Nernst Potential

The preceding sections have discussed what occurs inside the membrane. The potential has been that in the ionomer. Our goal here is to develop an analytic model of the potential from the metal on each side into and through the ionomer. As this chapter is concerned with membrane modeling, we will only consider a hydrogen pump cell with hydrogen electrodes on the anode and cathode sides. We will treat point electrodes on the boundaries of the one-dimensional membrane. The point electrode model will supply the necessary interface equations to use in a distributed electrode model, which one would need in order to expect the model to match experimental data. However, this chapter focuses on the membrane, not the electrode. We will assume we can maintain equal H_2 gas pressure on each side so that the gas contribution to any Nernst potential will be zero.

Let C_{H+}^{ref} be the concentration of hydrogen ions for the standard state of $H_2 \Leftrightarrow 2H^+$ with zero voltage shift. We take our reference state to be the H⁺ concentration in the membrane when it is fully protonated and $y_{in} = 1$. We will assume a sufficiently fast Tafel process, the dissociative adsorption or desorption of hydrogen on the Pt catalyst, such that the coverage fraction does not change. The kinetic reaction will then be driven solely by the Volmer process, whereby an electron exchange occurs on the catalyst surface and $H_{ads} \Leftrightarrow H^+ + e^-$. The anode and cathode overpotentials are what drive the kinetic reaction that generates current density. At the anode, the ion potential ϕ_A is related to the metal potential V_A by

$$\phi_A = V_A - \eta_A + b\ln\left(\frac{C_{H+}^A}{C_{H+}^{ref}}\right) = V_A - \eta_A + b\ln y_A \tag{8.55}$$

The metal potential difference across the cell may be obtained by summing the potential drop from the overpotentials, the membrane potential difference, and the Nernst potentials.

$$V_C - V_A = \eta_C - \eta_A + \phi_C - \phi_A + b\ln y_C - b\ln y_A \tag{8.56}$$

From equation (8.55) and equation (8.56), the ion potential ϕ_C at the cathode is

$$\phi_C = V_C - \eta_C - b\log y_C + 2b\log y_A \qquad (8.57)$$

When the current is zero, V_C and η_C are zero, y_C equals y_A, and we can see that $\phi_C = b\log y_A$, the same as ϕ_A. If we have 50% contaminant in a membrane with no current flowing, $\phi_A = \phi_C = b\ln 0.5 = -30.4\,\text{mV}$ at 80 C. We will normally take V_A to be zero.

The forward anodic process has no ion concentration dependence, but the backward cathodic process is proportional to the cathode H⁺ concentration. For both η_A and η_C to be zero when $J = 0$, the kinetic expression for current density is

$$J = \pm i_o A_r \left[e^{\frac{\eta_i}{2b}} - \frac{y_i}{y_{in}} e^{-\frac{\eta_i}{2b}} \right] + \text{for anode}, -\text{for cathode.} \qquad (8.58)$$

The subscript i represents A or C at the respective electrodes. The equilibrium of anode, cathode, and membrane are all coupled. The sign is negative for the cathode where the current flows from electrolyte to metal.

Figure 8.18 and Figure 8.19 show the variation of cell voltage, the over potentials, the electrode ion potential, and the Nernst potentials with current density. Figure 8.18 has 50% H⁺ and Figure 8.19 has 99.9% H⁺. We have

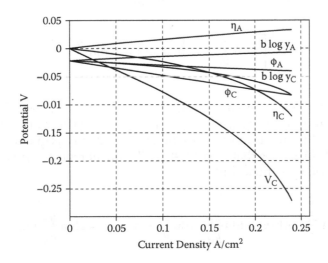

FIGURE 8.18
Various anode and cathode potentials as a function of current density for $y_{in} = 0.5$ and $i_o A_r = 0.3$ A/cm². The anode metal potential $V_A = 0$.

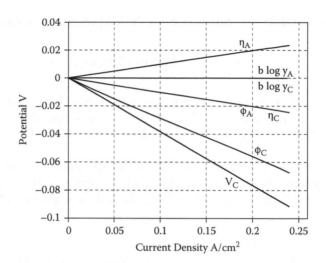

FIGURE 8.19
Same as Figure 8.18, but $y_{in} = 0.99$.

selected a low total exchange current density for the electrodes, 0.3 A/cm².
The Nernst potential shift in the membrane in Figure 8.18 disappears in
Figure 8.19. We see η_c drops increasingly in the former case, but is just equal
to $-\eta_A$ in the latter.

Let us see what happens to the overpotential at each electrode if we let the
total exchange current density $i_o A_r$ become extremely large compared to the
actual current density. From equation (8.58)

$$\left[e^{\frac{\eta_i}{2b}} - \frac{y_i}{y_{in}} e^{-\frac{\eta_i}{2b}} \right] \approx 0 \text{ and } \eta_i = b \ln \frac{y_i}{y_{in}} \tag{8.59}$$

The overpotential of each electrode is now determined solely by the concentration and not by $i_o A_r$. Let V_A be our ground potential of zero. After
substituting the anode and cathode overpotentials from equation (8.59) into
equation (8.56), we see that V_C consists of the membrane potential drop plus
a concentration potential shift.

$$V_C = b \ln \frac{y_C}{y_A} - b \ln y_A + b \ln \frac{y_C}{y_{in}} - b \ln \frac{y_A}{y_{in}} + \phi_C - \phi_A = 2b \ln \frac{y_C}{y_A} + \phi_C - \phi_A \tag{8.60}$$

Figure 8.20 shows the same situation as Figure 8.18, but with an increase of
1000 in $i_o A_r$. V_C is now the same as in equation (8.60).

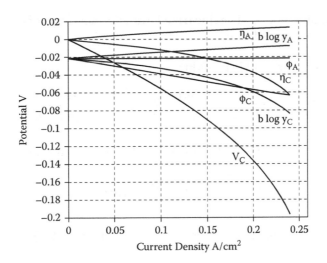

FIGURE 8.20
Same as Figure 8.18, but $i_oA_r = 300$ A/cm².

8.6.1 Electrode Impedance

Let us now examine what the effect of the electrode is dynamically. To the electron exchange current in equation (8.58) must be added the displacement current through the double layer capacitance at the interface.

$$j = \pm\left(i_oA_r\left[e^{\frac{\eta_i}{2b}} - \frac{y_i}{y_{in}}e^{-\frac{\eta_i}{2b}}\right] + C_{Di}\frac{d\eta_i}{dt}\right)$$ (8.61)

As we did in the membrane, again, we linearize and Laplace transform equation (8.61) to get the electrode impedance.

$$\bar{J}(\omega) = \pm i_oA_r\left[\frac{1}{2b}e^{\frac{\eta_0}{2b}}\bar{\eta}(\omega) + \left(\frac{1}{2b}\frac{y_0}{y_{in}}\bar{\eta}(\omega) - \frac{\bar{y}(\omega)}{y_{in}}\right)e^{-\frac{\eta_0}{2b}}\right] \pm j\omega C_D\bar{\eta}(\omega)$$ (8.62)

If we set $\bar{J}(\omega) = 1$ in equation (8.62), then $\bar{\eta}(\omega)$ becomes the electrode impedance.

$$Z_{electrode} = \frac{\bar{\eta}(\omega)}{\bar{J}(\omega)} = \bar{\eta}(\omega) = \pm\frac{1 \pm i_oA_r\frac{\bar{y}(\omega)}{y_{in}}e^{-\frac{\eta_0}{2b}}}{\frac{i_oA_r}{2b}\left(e^{\frac{\eta_0}{2b}} + \frac{y_0}{y_{in}}e^{-\frac{\eta_0}{2b}}\right) + j\omega C_D}$$ (8.63)

The electrode impedance equation has positive signs for the anode impedance and negative for the cathode impedance. Equation (8.56) was the time domain equation for V_C. The linearized, transformed form of equation (8.56) gives us the cell impedance.

$$V_C = \eta_C - \eta_A + \phi_{CA} + b\ln y_C - b\ln y_A$$

$$Z_{tot} = -\frac{\bar{V}_C(\omega)}{\bar{J}(\omega)} = -\frac{\bar{\eta}_C(\omega)}{\bar{J}(\omega)} + \frac{\bar{\eta}_A(\omega)}{\bar{J}(\omega)} - \frac{\bar{\phi}_{CA}(\omega)}{\bar{J}(\omega)} - \frac{b}{y_C}\frac{\bar{y}_C(\omega)}{\bar{J}(\omega)} + \frac{b}{y_A}\frac{\bar{y}_A(\omega)}{\bar{J}(\omega)} \qquad (8.64)$$

When the exchange current i_o gets very large, we saw in equation (8.59) that

$$e^{\frac{\eta_i}{2b}} = \sqrt{\frac{y_i}{y_{in}}} \;,\; e^{-\frac{\eta_i}{2b}} = \sqrt{\frac{y_{in}}{y_i}} \text{ when } i_o \to \infty.$$

Here, subscript i refers to anode or cathode. Equation (8.63) reduces to

$$\bar{\eta}_i(\omega) = \pm\frac{1 \pm i_0 A_r \dfrac{\bar{y}(\omega)}{y_{in}}e^{-\frac{\eta_0}{2b}}}{\dfrac{i_0 A_r}{2b}\left(e^{\frac{\eta_0}{2b}} + \dfrac{y_i}{y_{in}}e^{-\frac{\eta_0}{2b}}\right) + j\omega C_D} \to \frac{2b\dfrac{\bar{y}(\omega)}{y_{in}}e^{-\frac{\eta_0}{2b}}}{\left(e^{\frac{\eta_0}{2b}} + \dfrac{y_i}{y_{in}}e^{-\frac{\eta_0}{2b}}\right)} \to b\frac{\bar{y}(\omega)}{y_i}$$

$$(8.65)$$

$$Z_{tot} = -\bar{\phi}_{CA}(\omega) - 2b\frac{\bar{y}_C(\omega)}{y_C} + 2b\frac{\bar{y}_A(\omega)}{y_A} \qquad (8.66)$$

Let us see what we can understand from this equation, which is the impedance we would expect if we had perfect electrodes with no potential losses. Because impedance should be positive and our model has positive current with negative cathode voltage, the signs in equation 8.66 are reversed. The first term is the membrane impedance, which as we saw in Figure 8.13 has a positive imaginary part. The latter two terms include the Nernst potential fluctuation effects at the cathode and anode each with negative imaginary contribution to the total impedance. The resulting impedance as seen in Figure 8.21 has negative imaginary part. The impedance was calculated for current densities of .01, 0.13 and 0.2 A/cm² using the base case diffusion coefficients with this large value of electrode exchange current density.

8.6.2 Model Comparison to Experiment

At the time of this writing, there is very little data available to support the model equations we have developed in this chapter. Los Alamos researchers have supplied some preliminary hydrogen pump data that provides support for the model, with the caveat that they have encountered some observations they do not yet fully understand.

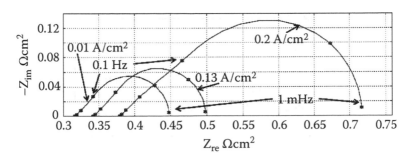

FIGURE 8.21
Calculated impedance for 50% Na^+ contaminated Nafion 117 membrane for high electrode exchange current $i_oA_r = 300$ A/cm².

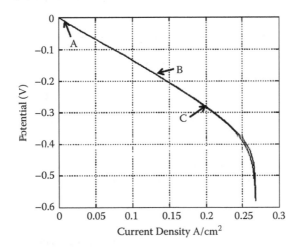

FIGURE 8.22
Polarization curve for 50% Na^+ hydrogen pump impedance experiments. Points A, B, and C indicate where impedance measurements were made.

We will show hydrogen pump impedance data from two 5-cm² Nafion 117 membrane electrode assemblies. Both 5-cm² cells were operated at 80°C in hydrogen pump mode with saturated water vapor and 2 atm hydrogen on both electrodes. They measured the AC impedance with a Parstat 2273 analyzer. The first membrane was prepared with 50% Na^+ cation and had 0.2 mg/cm² Pt on carbon electrodes with E-tek double-sided carbon cloth gas diffusion electrodes. The anode and cathode humidifier and cell temperatures were 94/92/80°C, respectively, and both H_2 pressures were 30 psig and the flows were 200/100 sccm (standard cubic centimeters per minute). Figure 8.22 shows the polarization curve. If the swept cathode voltage did not go much below −0.6 V, they obtained repeatable results. When extended below −0.9 V, they did observe temporary changes cell resistance.

Figure 8.23 to Figure 8.25 show two full-range impedance plots and the high frequency part of all three. We note that the high frequency impedance

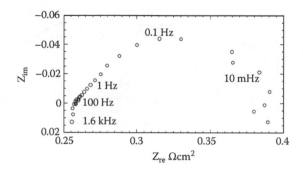

FIGURE 8.23
Measured impedance with 50% Na^+ at point A in Figure 8.22, 0.01 A/cm^2.

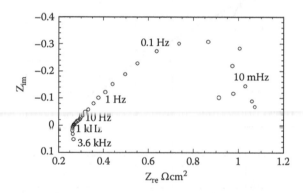

FIGURE 8.24
Measured impedance with 50% Na^+ at point C in Figure 8.22, 0.2 A/cm^2.

of the measured data is lower than the modeled impedance, indicating higher conductivity than the base case. The measured high frequency resistance difference between low and high current density is less than that predicted by the model. On the other hand, the measured low frequency resistance is higher than that predicted for a high exchange current electrode. We will not speculate on these differences with this preliminary data, but will point out that the general trends of the data and the model are the same.

In the second 5-cm^2 membrane electrode assembly, Los Alamos researchers prepared Nafion 117 with 50% Cs and 50% H^+ and 6 mg/cm^2 Pt black electrodes. The top graph of Figure 8.26 shows the measured impedance. We must warn the reader that this data has not been repeated yet and is highly questionable. We speculate that the cesium presence increased the electrode overpotential enabling a double loop in the impedance to be observed. We show it to illustrate the electrode contribution to model prediction. The bottom graph of Figure 8.26 shows a calculated impedance. The point electrode model is incapable of predicting distributed ionic resistance. The 45° angle

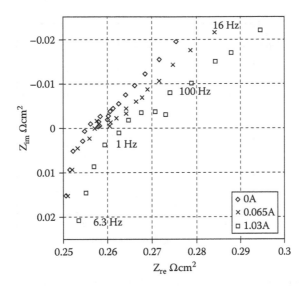

FIGURE 8.25
The high frequency part of impedances measured at 0, 0.13, and 0.2 A/cm² in the 5 cm² cell with 50% Na⁺.

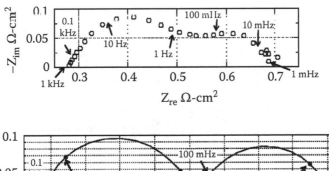

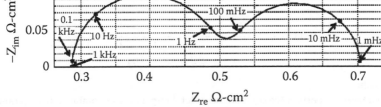

FIGURE 8.26
Top graph shows measured impedance with 50% Cs⁺ in Nafion 117 at 0.01 A/cm²and 6 mg/cm² Pt black electrodes. Bottom graph is point electrode model approximation: $C_D = 0.2$ F/cm², $i_o A_r = 0.3$ A/cm².

of the measured impedance seen at high frequency apparently indicates distributed ionic resistance in the electrode. The lower frequency loop would be expected to have the 45° angle associated with cation diffusion in the membrane. This was seen at high frequency in the previous Na⁺ data, where the

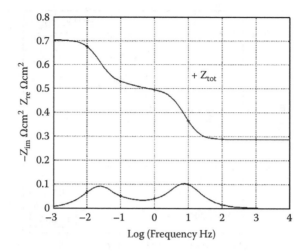

FIGURE 8.27
Bode plot of the modeled impedance in Figure 8.26 showing the real and imaginary parts as a function of log frequency. Figure 8.28 shows the components that add to this impedance.

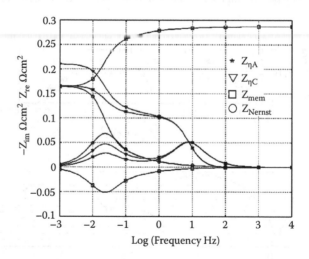

FIGURE 8.28
Real and imaginary parts of the components of total impedance as combined in equation (8.64).

electrode effects were less pronounced. The major features of the model are still apparent in the data.

To understand better what is occurring, let us examine in more detail the modeled impedance in Figure 8.26. Figure 8.27 shows the same impedance in a Bode plot of real and imaginary parts plotted as a function of frequency. We saw in equation (8.64) that there are four major components to the total measurable impedance Z_{tot}. The components are the anode and cathode overpotentials, the membrane ionic potential difference and the Nernst

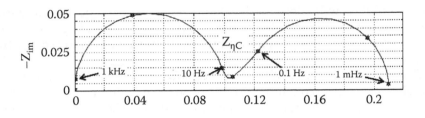

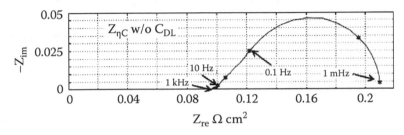

FIGURE 8.29
Cathode overpotential impedance with and without the double layer capacitance.

correction caused by the different proton concentrations between the anode and cathode sides. We have plotted these separate components in Figure 8.28. We note that the double loop occurs only in the anode and cathode overpotentials. Note these impedances are slightly different because we set the DC current density at 0.01 A/cm² instead of zero, precisely to illustrate the point. At high frequency, the double layer capacitance shorts the overpotential, contributing to one loop. At lower frequency, the shifting proton distribution with current variation changes the impedance in the lower loop. Figure 8.29 shows a Nyquist plot of the cathode impedance with and without the double layer capacitance. The high frequency loop disappears when the capacitance is removed.

8.7 Conclusions

Mathematical modeling is a useful tool for exploring the mechanisms of cationic contamination. One can use this new insight to help understand the effects of cationic contamination on fuel cell performance. Ultimately, the complexity of the needed model depends on its purpose. For many applications, simple models can qualitatively explain the mechanisms of cationic contamination on PEMFC systems. Models that are more complex help to quantify experimental data and make specific predictions. Understanding exactly how and to what extent cationic contamination affects PEMFCs will allow for development of hydrogen purity standards, understanding of acceptable corrosion rates, and for creating new modes for diagnosis, new component design, and better methods to recover fuel cell performance.

This chapter has provided the basic equations to understand and to model the effects of cation contamination in polymer membranes. This is a relatively new field with much left to be understood. The effects of different cations on water uptake and membrane structure have not specifically been addressed, but the modeler can apply these effects using the equations we have discussed. We have not specifically discussed operation in the fuel cell mode, but have concentrated on the hydrogen pump. The differences are in the electrodes, not in the membrane.

8.8 Appendix: Concentrated Solution Theory

We will treat the membrane as a liquid with various species moving in it with different velocities relative to the other species. It will be an isotropic medium and we will restrict the membrane species to have zero velocity. Effects on the nanoscale or on variations in a random porous structure of the membrane are ignored. We will use macroscopic transport properties whose values may be functions of temperature, water content, and ion fraction. Because the membrane thickness is small compared to its in-plane dimensions, a one-dimensional model will suffice. Our medium will contain four species: membrane, water, protons, and blocking cations that will be denoted with subscripts m, w, H, and B, respectively, and subscripts i and j can represent any of these. We will pick a geometric reference system fixed to the membrane. If we are treating swelling, this reference system should be the dry membrane, but we will ignore this for now. For one species i to move with velocity v_i, it will require a driving force to move it through species j proportional to the difference of their mean velocities and proportional to the mole fraction C_j / C_T of species j in the medium. C_T is the total concentration. The total driving force will be the sum of the interaction with all other species. By definition, this driving force must be the gradient of the electrochemical potential of species i. Each species j contributes to total velocity v_i. The diffusion coefficient D_{ij} is the proportionality constant between that contribution and $\nabla \mu_i$. A key point to note here is that since the force on i from j must be opposite to the force of j on i, it follows that $D_{ij} = D_{ji}$. This is the Onsager reciprocal relation. To express the driving force on a unit volume basis, we must multiply by concentration.

$$C_i \nabla \mu_i = RT \sum_j \frac{C_i C_j}{C_T D_{ij}} (v_j - v_i) = \sum_j K_{ij} (v_j - v_i) \text{ where } K_{ij} = \frac{C_i C_j RT}{C_T D_{ij}} \qquad (8.67)$$

Equation (8.67) is the driving force per unit volume acting on species i. We can reduce this equation to two well-known examples. Consider a single dilute species i in a solvent j. Then the mean velocity of the solvent is zero and its concentration is essentially the total concentration C_T. In dilute solutions, the activity approaches the concentration:

$$\frac{C_i}{RT_i} \nabla \mu_i = C_i \nabla \ln a_i = \nabla C_i = -\frac{C_i v_i}{D_i} \qquad (8.68)$$

and we get Fick's law $N_i = -D \nabla C_i$. If we use multicomponent ideal gases still letting activity reduce to concentration, we have the Stefan–Maxwell equation:

$$\frac{C_i}{RT} \nabla \mu_i = C_T \nabla x_i = \sum_{j \neq i} \frac{C_i C_j}{D_{ij} C_T} (v_j - v_i)$$

$$\nabla x_i = \sum_{j \neq i} \frac{C_i C_j v_j - C_i C_j v_i}{C_T^2 D_{ij}} = \sum_{j \neq i} \frac{x_i N_j - x_j N_i}{C_T D_{ij}} \qquad (8.69)$$

In order to model transport in the membrane, we need the fluxes of each species in terms of the driving forces. At this point, equation (8.67) gives us the driving forces (i.e., gradients in concentration, activity, potential, pressure, etc.) in terms of the fluxes. We need the inverse equations. We defined the friction coefficients K_{ij} in equation (8.67) to use their symmetric property that $K_{ij} = K_{ji}$. We illustrate with a three-component system.

$$
\begin{aligned}
C_1 \nabla \mu_1 &= K_{12}(v_2 - v_1) + K_{13}(v_3 - v_1) \\
C_2 \nabla \mu_2 &= K_{21}(v_1 - v_2) + K_{23}(v_3 - v_2) = \\
C_3 \nabla \mu_3 &= K_{31}(v_1 - v_3) + K_{32}(v_2 - v_3)
\end{aligned}
\begin{bmatrix}
-K_{12} - K_{13} & K_{12} & K_{13} \\
K_{12} & -K_{12} - K_{23} & K_{23} \\
K_{13} & K_{23} & -K_{13} - K_{23}
\end{bmatrix}
\cdot
\begin{bmatrix}
v_1 \\
v_2 \\
v_3
\end{bmatrix}
$$

$$(8.70)$$

If we add the three equations in equation (8.70), we see that the Gibbs–Duhem equation holds: $\sum C_i \nabla \mu_i = 0$. Let us fix species 3, which might represent the membrane, with velocity $v_3 = 0$. The third equation in (8.70) is superfluous, as $K_{ij} = K_{ji}$. The remaining 2×2 matrix equation can be inverted to determine the velocities and thus the fluxes of species 1 and 2.

$$\begin{bmatrix} v_1 \\ v_2 \end{bmatrix} = \begin{bmatrix} -K_{12} - K_{13} & K_{12} \\ K_{12} & -K_{12} - K_{23} \end{bmatrix}^{-1} \begin{bmatrix} C_1 \nabla \mu_1 \\ C_2 \nabla \mu_2 \end{bmatrix}$$

$$\begin{bmatrix} v_1 \\ v_2 \end{bmatrix} = \begin{bmatrix} \dfrac{-K_{12} - K_{23}}{d_{en}} & \dfrac{-K_{12}}{d_{en}} \\ \dfrac{-K_{12}}{d_{en}} & \dfrac{-K_{12} - K_{13}}{d_{en}} \end{bmatrix} \begin{bmatrix} C_1 \nabla \mu_1 \\ C_2 \nabla \mu_2 \end{bmatrix} = \begin{bmatrix} L_{11} & L_{12} \\ L_{21} & L_{22} \end{bmatrix} \begin{bmatrix} C_1 \nabla \mu_1 \\ C_2 \nabla \mu_2 \end{bmatrix} \quad (8.71)$$

where $d_{en} = K_{12}K_{13} + K_{12}K_{23} + K_{13}K_{23}$ and $L_{21} = L_{12}$.

We see that the inverted L matrix is symmetric. Extend this example to our 4 species system.

If we multiply each of the velocities by their corresponding concentrations and put the concentrations inside the matrix, we have the fluxes expressed as a symmetric matrix times the gradients in electrochemical potential. Substitute H, B, w for 1, 2, 3.

$$\begin{bmatrix} N_H \\ N_B \\ N_w \end{bmatrix} = \begin{bmatrix} C_H v_H \\ C_B v_B \\ C_w v_w \end{bmatrix} = \begin{bmatrix} C_{HH}C_H L_{HH} & C_H C_B L_{HB} & C_H C_w L_{Hw} \\ C_B C_H L_{BH} & C_B C_B L_{BB} & C_B C_w L_{Bw} \\ C_w C_H L_{wH} & C_w C_B L_{wB} & C_w C_w L_{ww} \end{bmatrix} \begin{bmatrix} \nabla \mu_H \\ \nabla \mu_B \\ \nabla \mu_w \end{bmatrix},$$

$$C_i C_j L_{ij} = C_j C_i L_{ji} \quad (8.22)$$

The purpose of tracking the symmetric matrices through from the velocity differences in equation (8.67) to those relating fluxes to driving forces was to obtain three extra relations. In general, if we have n fluxes and n driving forces, we have n^2 coefficients of which $n(n-1)/2$ are eliminated and only $n(n+1)/2$ coefficients need to be obtained from experiment or more *ab initio* methods. In our case, we need six properties to determine the system because we have three fluxes, N_H, N_B, and N_w.

Acknowledgments

The authors are indebted to Dr. Fernando Garzon, Dr. Rangachary Mukundan, Dr. Bryan Pivovar, Dr. Thomas A. Zawodzinski, Jr., Dr. Adam Weber, and Thiago Lopes for many helpful discussions and for providing data to support the modeling.

Nomenclature

Variable	Definition	Units
α	D_B/D_H	
a_w	activity of water	
A_r	catalyst to geometric area ratio	
b	RT/F	V
C	concentration	mol/cm^3
C_t	SO_3^- concentration	mol/cm^3
C_T	total concentration of all species	mol/cm^3
d_a	time scaling factor, equation (8.43)	
$d_{a\lambda}$	$d\ln a_w/d\lambda$, equation (8.27)	
D_i	diffusion coefficient of ion i	cm^2/s
$\mathcal{D}_{ij}$	generalized concentrated solution dif. coef.	
D'_w	diffusion coefficient of water in membrane	cm^2/s
f	frequency	Hz
F	Faraday constant	96485 C/mol
ϕ	electric potential	V
G	Gibbs free energy	J
η	overpotential	V
$\bar{\eta}(\omega)$	Laplace transformed overpotential	V-s
i_o	exchange current density	A/cm^2
J	current density	A/cm^2
$\bar{J}(\omega)$	Laplace transformed current density	$A\text{-}s/cm^2$
K_{ij}	friction coefficients in equation (8.68)	$J\text{-}s/cm^5$
λ	membrane H_2O to SO_3^- ratio	
L	membrane thickness	cm
L_{ij}	inverted friction coefficients	$cm^5/J\text{-}s$
M_{EW}	equivalent weight of dry membrane per SO_3^-	gm/equivalent
μ_i	electrochemical potential of species i	J/mol
M_w	mass per mole of water	18 gm/mol
μ_w	electrochemical potential of water	J/mol
N_i	molar flux	$mol/cm^2/s$
p	total pressure	atm
p_{sat}	saturated vapor pressure	atm
p_w	partial pressure of water vapor	atm
R	gas constant	8.3144 J/mol/K
ρ_{dry}	dry membrane density	2.1 gm/mol
ρ_w	density of liquid water	1 gm/cm^3
σ_i	conductivity of species i	S/cm
T	temperature	K
T_c	temperature	Celsius

(Continued)

(Continued)

Variable	Definition	Units
u_i	mobility of species i D/RT	cm-mol/J-s
v	velocity	cm/s
ω	angular frequency	radian/s
x	scaled distance z/L	
ξ_i	electroosmotic drag coefficient species i	
y_B	blocking ion fraction	
y_H	hydrogen ion fraction	
y_{in}	initial proton ion fraction distribution	
z	distance	cm
Z_B	charge on blocking cation	

Subscripts	Species or Location
A	anode end
C	cathode end
i, j	generalized subscripts
B	blocking ion
H	hydrogen ion
w	water

References

1. Borup, R., et al. 2007. Scientific aspects of polymer electrolyte fuel cell durability and degradation. *Chemical reviews* 107 (10): 3904–3951.
2. Eisenberg, A., Yeager, H.L. 1982. *Perfluorinated ionomer membranes.* ACS: Washington, D.C.
3. Okada, T. et al. 2002. Ion and water transport characteristics of perfluorosulfonated ionomer membranes with H^+ and alkali metal cations. *Journal of Physical Chemistry B* 106 (6): 1267–1273.
4. Halseid, R., Vie, P.J.S., Tunold, R. 2004. Influence of ammonium on conductivity and water content of Nafion 117 membranes. *Journal of the Electrochemical Society* 151 (3): A381–8.
5. Mikkola, M.S., et al. 2007. The effect of NaCl in the cathode air stream on PEMFC performance. *Fuel cells* 7 (2): 153–8.
6. Rajalakshmi, N., Jayanth, T., Dhathathreyan, K. 2003. Effect of carbon dioxide and ammonia on polymer electrolyte membrane fuel cell stack performance. *Fuel Cells* 3 (4): 177.
7. Rockward, T., et al. 2007. The effects of multiple contaminants on polymer electrolyte fuel cells. *ECS Transactions* 11 (1): 821–829.
8. Soto, H.J., et al. 2003. Effect of transient ammonia concentrations on PEMFC performance. *Electrochemical and Solid-State Letters* 6 (7): A133–5.

9. Uribe, F.A., Gottesfeld, S., Zawodzinski, T.A. 2002. Effect of ammonia as potential fuel impurity on proton exchange membrane fuel cell performance. *Journal of the Electrochemical Society* 149 (3): A293–6.
10. Uribe, F.A., et al. 2004. Effects of some air impurities on PEM fuel cell performance. *Meeting Abstracts*: 332–332.
11. Halseid, R., Vie, P.J.S., Tunold, R. 2006. Effect of ammonia on the performance of polymer electrolyte membrane fuel cells. *Journal of Power Sources* 154 (2): 343–50.
12. Kienitz, B. 2009. The effects of cationic contamination on polymer electrolyte membrane fuel cells. PhD disser., Case Western Reserve University, Cleveland, OH.
13. Janssen, G.J.M. 2001. A phenomenological model of water transport in a proton exchange membrane fuel cell. *Journal of the Electrochemical Society* 148 (12): A1313–A1323.
14. Springer, T.E., Zawodzinski, T.A., Gottesfeld, S. 1991. Polymer electrolyte fuel cell model. *Journal of the Electrochemical Society* 138 (8): 2334–42.
15. Bernardi, D.M., Verbrugge, M.W. 1992. A mathematical model of the solid-polymer-electrolyte fuel cell. *Journal of the Electrochemical Society* 139 (9): 2477–2491.
16. Eikerling, M., et al. 1998. Phenomenological theory of electro-osmotic effect and water management in polymer electrolyte proton-conducting membranes. *Journal of the Electrochemical Society* 145 (8): 2684-2699.
17. Weber, A.Z., Newman, J. 2003. Transport in polymer-electrolyte membranes. *Journal of the Electrochemical Society* 150 (7): A1008 A1015.
18. Weber, A.Z., Newman, J. 2004. Transport in polymer-electrolyte membranes. *Journal of the Electrochemical Society* 151 (2): A311–A325.
19. Onishi, L.M., Prausnitz, J.M., Newman, J. 2007. Water & Nafion equilibria. Absence of Schroeder's paradox. *The Journal of Physical Chemistry B* 111 (34): 10166–10173.
20. Zawodzinski, T.A., et al. 1993. Water uptake by and transport through Nafion 117 membranes. *Journal of the Electrochemical Society* 140 (4): 1041–1047.
21. Zawodzinski, T.A., et al. 1991. Determination of water diffusion-coefficients in perfluorosulfonate ionomeric membranes *Journal of Physical Chemistry* 95 (15): 6040–6044.
22. Zawodzinski, T.A., et al. 1993. Characterization of polymer electrolytes for fuel cell applications. *Solid State Ionics* 60 (1–3): 199–211.
23. Zawodzinski, T.A., et al. 1995. The water-content dependence of electroosmotic drag in proton-conductin polymer electrolytes *Electrochimica Acta* 40 (3): 297–302.
24. Vanderborgh, N., Springer, T. 1983. Water-flow through solid polymer electrolytes (SPE) during fuel cell operation. Paper presented at the *164th Meeting of the Electrochemical Society*, October, Washington, D.C.
25. Newman, J.S., Thomas-Alyea, K.E. 2004. *Electrochemical Systems*, 3rd ed., Wiley-Interscience: Hoboken, N.J., pp. 647.
26. Okada, T., et al. 2005. Membrane transport characteristics of binary cation systems with Li^+ and alkali metal cations in perfluorosulfonated ionomer. *Electrochimica Acta* 50 (16-17): 3569–3575.

27. Okada, T., et al. 1999. The effect of impurity cations on the transport characteristics of perfluorosulfonated ionomer membranes. *Journal of Physical Chemistry B* 103 (17): 3315–3322.
28. Okada, T., et al. 1998. Transport and equilibrium properties of Nafion membranes with H^+ and Na^+ ions. *Journal of Electroanalytical Chemistry* 442 (1–2): 137–145.

9

Impurity Mitigation Strategies

Cunping Huang, Xinyu Huang, and Marianne Rodgers

CONTENTS

9.1 Introduction

Understanding of impurities of hydrogen fuel streams in terms of their sources and properties is the first step in the development of alleviation strategies. Impurities in hydrogen may originate from either hydrogen production feedstock or from production processes. They may also derive from onboard hydrogen storage materials. It is self-evident that if an impurity can be removed at its source there will be no need for a pretreatment process in the fuel system of a vehicle. The impurity can also be eliminated during the pretreatment step. In either case, the requirements for proton exchange membrane (PEM) fuel cell catalysts will be lowered.

This chapter consists of three sections: (1) Hydrogen Production Technologies and Hydrogen Purification (section 9.2 by Cunping Huang), (2) System-Level Mitigation Strategies (section 9.3 by Xinyu Huang), and (3) Onboard Fuel Cell Impurity Mitigation Strategies (section 9.4 by Marianne Rodgers).

9.2 Hydrogen Production Technologies and Hydrogen Purification

9.2.1 Overview of Hydrogen Production Technologies

Based on raw materials, hydrogen production can be classified as hydrogen generation from carbon-based feedstocks and hydrogen generation from water. Carbon-based feedstocks include fossil fuels (natural gas, hydrocarbons, and coal) and renewable biomass with higher efficiency, and all are widely used today. Water-based feedstock enables carbon neutral hydrogen production because, as a clean energy carrier when its chemical energy is converted into electrical energy via a fuel cell, hydrogen is oxidized by oxygen and water is the only product. Although difficult to process, water is considered to be a renewable hydrogen production resource with no greenhouse gas emission issues. Hydrogen production technologies have been extensively reviewed in literature. The objective of this section is to briefly review these technologies based on fundamentals of chemistry.

From chemical reaction viewpoints, hydrogen production technologies can be divided into three categories: hydrogen production via reforming reactions, decomposition reactions, and partial oxidation reactions. Both reforming and decomposition reactions are highly endothermic processes that require thermal heat energy input to the systems. A partial oxidation process requires less external thermal energy because oxidizing a fraction of the needed fuel from a feedstock provides some energy for the main process. If heat released from oxidation is equal to or slightly greater than that needed for hydrogen production, the process is defined as *autothermal*. A water reduction process uses a reducing agent (e.g., carbon monoxide) to react with water for the production of hydrogen. In addition to a water gas shift reaction ($CO + H_2O \rightarrow H_2 + CO_2$) at high temperatures, some metals or metal oxides can reduce water to produce hydrogen. Table 9.1 lists the major hydrogen production processes accomplished via chemical reactions.

TABLE 9.1

Summary of Hydrogen Production Technologies

Process	Reaction	ΔH°_{298} (kJ/mol)
Reforming Processes		
Steam reforming of natural gas	$CH_4 + H_2O \rightarrow CO + 3H_2$	206
Steam reforming of coal	$CH_xO_y + (1-y)H_2O \rightarrow (0.5x + 1 - y)H_2 + CO$	
Steam reforming of biomass	$C_nH_mO_z + 2(n-z)H_2O \rightarrow nCO + (n + 0.5m - z)H_2$	
Steam reforming of hydrocarbons	$C_nH_m + nH_2O \rightarrow nCO + 0.5(m+2n)H_2$	49
Steam reforming of oxygenated hydrocarbon	$CH_3OH + H_2O \rightarrow CO_2 + 3H_2$	49.4
Carbon dioxide reforming of methane	$CH_4 + CO_2 \rightarrow 2CO + 2H_2$	247.3
Hydrogen sulfide reforming of methane	$CH_4 + 2H_2S \rightarrow CS_2 + 4H_2$	233.0
Decomposition processes		
Methane decomposition (pyrolysis)	$CH_4 \rightarrow C + 2H_2$	75.6
Water decomposition	$H_2O \rightarrow 0.5O_2 + H_2$	285.8
Hydrogen sulfide decomposition	$H_2S \rightarrow 0.5S_2 + H_2$	79.9
Partial oxidation & autothermal processes		
Methane partial oxidation	$CH_4 + 0.5O_2 \rightarrow CO + 2H_2$	−36
Autothermal steam reforming of methane	$CH_4 + xO_2 + yH_2O \rightarrow (0.5x + y) CO + 3H_2$	−192
Methanol partial oxidation	$CH_3OH + 0.5O_2 \rightarrow CO_2 + 2H_2$	−192.2
Coal gasification	$C + xO_2 \rightarrow CO$	
Biomass gasification	$C_nH_mO_z + 0.5(n-z)O_2 \rightarrow nCO + 0.5mH_2$	
Biomass autothermal reforming	$C_nH_mO_z + yO_2 + 2(n-y-0.5z)H_2O \rightarrow nCO_2 + 2(n-y-0.5z+0.25m)H_2$	
Water reduction processes		
Water gas shift reaction	$CO + H_2O \rightarrow CO_2 + H_2$	−41.15
Metal-based water reduction	$Fe + H_2O \rightarrow FeO + H_2$	
Metal oxide-based water reduction	$FeO + H_2O \rightarrow Fe_2O_3 + H_2$	

9.2.1.1 Hydrogen Production from Natural Gas and Hydrocarbons

Carbon-based feedstocks include natural gas, light hydrocarbons, coal, and biomass. Basically, carbon-based hydrogen production technologies are thermal catalytic processes in which part of the feedstock is burned to provide the thermal energy needed for the hydrogen production. A brief review of these basic hydrogen production technologies follows.

9.2.1.1.1 Hydrogen Production via Steam Reforming of Natural Gas and Hydrocarbons

Natural gas is a naturally occurring gas mixture consisting primarily of methane. It is found associated with fossil fuels, e.g., in coal beds. Before natural gas can be used as a fuel or feedstock for the production of hydrogen, it must be pretreated to reduce the concentrations of higher molecular weight hydrocarbons, such as ethane, propane, etc. The other major objective of natural gas pretreatment is the removal of any sulfur component, primarily H_2S, because it can deactivate thermal catalysts in hydrogen production processes. When used as a fuel, the burning of H_2S-contaminated natural gas releases SO_2 gas into the atmosphere, resulting in acid rain and its associated damage to the environment. Typical compositions of distributed natural gases are given in Table 9.2.

The average sulfur content in natural gas is about 5.5 mg/m³, which includes the 4.9 mg/m³ of sulfur in the odorant (mercaptan) added to gases for safety reasons.

TABLE 9.2

Typical Natural Gas Compositions

Component	Typical Analysis (mol%)	Range (mol%)
Methane (CH_4)	95.2	87.0 – 96.0
Ethane (C_2H_6)	2.5	1.5 – 5.1
Propane (C_3H_8)	0.2	0.1 – 1.5
iso-Butane (i-C_4H_{10})	0.03	0.01 – 0.3
normal-Butane (C_4H_{10})	0.03	0.01 – 0.3
iso-Pentane (i-C_5H_{12})	0.01	Trace – 0.14
normal-Pentane (C_5H_{12})	0.01	Trace – 0.04
Hexanes plus (C_6H_{14})	0.01	Trace – 0.06
Nitrogen (N_2)	1.3	0.7 – 5.6
Carbon dioxide (CO_2)	0.7	0.1 – 1.0
Oxygen (O_2)	0.02	0.01 – 0.1
Hydrogen (H_2)	Trace	Trace – 0.02
Specific gravity	0.58	0.57 – 0.62
Gross heating value (MJ/m³), dry basis	37.8	36.0 – 40.2

Source: Chemical composition of natural gas: http://www.uniongas.com/aboutus/aboutng/composition.asp (accessed May 19, 2009). With permission.

Steam reforming of distributed natural gas and light hydrocarbons is an efficient, economical, and widely used process for near-term hydrogen production. Steam reforming can be defined as a high temperature thermochemical process converting hydrocarbon feedstocks in natural gas, mainly methane, into hydrogen and carbon monoxide. The reactions associated with steam reforming reactions include [2]:

Steam reforming of methane:

$$CH_4 + H_2O \rightarrow CO + 3H_2 \qquad \Delta H^\circ = + 206.16 \text{ kJ/mol } CH_4 \qquad (9.1)$$

Steam reforming of propane:

$$C_3H_8 + 3H_2O \rightarrow 3CO + 7H_2 \qquad \Delta H^\circ = + 498.6 \text{ kJ/mol } C_3H_8 \qquad (9.2)$$

Steam reforming of methanol:

$$CH_3OH + H_2O \rightarrow CO_2 + 3H_2 \qquad \Delta H^\circ = 49 \text{ kJ/mol } CH_3OH \qquad (9.3)$$

Steam reforming of ethanol:

$$C_2H_5OH + H_2O \rightarrow 2CO + 4H_2 \qquad \Delta H^\circ = 254.77 \text{ kJ/mol } C_2H_5OH \qquad (9.4)$$

Steam reforming of gasoline (using iso-octane and toluene as example compounds):

$$C_8H_{18} + 8H_2O \rightarrow 8CO + 17H_2 \qquad \Delta H^\circ > 0 \qquad (9.5)$$

$$C_7H_8 + 7H_2O \rightarrow 7CO + 11H_2 \qquad \Delta H^\circ > 0 \qquad (9.6)$$

In summary, steam reforming of hydrocarbons can be written as:

$$C_nH_m + nH_2O \rightarrow nCO + 0.5(m + 2n)H_2$$

Economically, steam reforming of methane is operated at temperatures between 700 and 850°C and pressures between 3 and 25 atm in the presence of a nickel-based catalyst. In order to prevent coking, or carbon buildup, on the catalysts, typically the mass ratio of steam-to-carbon is about 3 or higher [3]. The energy efficiency of steam reforming is defined as: high heating value of hydrogen produced/high heating value of energy input. In most cases, the external heat needed to drive the reaction is provided by the combustion of a fraction of the natural gas feedstock (up to 25%) and waste gas. The thermodynamic heat of combustion of hydrogen (its high heating value (HHV)) equates to the standard heat of formation (ΔH_f°) of the product water, i.e.,

$$H_2(g) + 0.5O_2(g) \rightarrow H_2O(l) \qquad \Delta H_f^\circ = -285.83 \text{ kJ/mol } H_2 \qquad (9.7)$$

The HHV of hydrogen is the maximum amount of heat that can be derived from the combustion of H_2 when the product water is condensed to 25°C. The energy conversion efficiency for large-scale steam methane reformers is in the range of 70 to 80%, among the highest of current commercially available hydrogen production methods [2].

The steam reforming reaction is a highly endothermic process producing a gas mixture of CO and H_2. The CO produced in the reforming further reacts with steam over a catalyst to produce H_2 and carbon dioxide (CO_2). This process is called a water gas shift (WGS) reaction. WGS reaction occurs in two stages: a high temperature shift (HTS) at 350 to 475°C and a low temperature shift (LTS) at 200 to 250°C.

$$CO + H_2O \rightarrow CO_2 + H_2 \qquad \Delta H^\circ = -41.15 \text{ kJ/mol CO} \qquad (9.8)$$

As indicated by the overall steam methane reformation reaction, reaction (9.9), two moles of hydrogen produced from steam reforming are from methane and the other two moles are from water splitting.

$$CH_4 + H_2O \rightarrow CO_2 + 4H_2 \qquad \Delta H^\circ = +206.16 \text{ kJ/mol } CH_4 \qquad (9.9)$$

The exiting gas from WGS processes contains mostly H_2 (70 to 80%), and CO_2, CH_4, H_2O, and small quantities of CO. Depending on the applications, the H_2-rich gas mixture requires a purification process to meet a purity level.

One major challenge of steam reforming for the production of hydrogen is the emission of carbon dioxide. Although the steam reforming reaction produces four times as much hydrogen as carbon dioxide in moles (reaction (9.9)), the amount of CO_2 produced in mass is twice as much as hydrogen in a centralized plant, which can impose a serious environmental concern. To avoid emission of CO_2 into the atmosphere, it must be concentrated, captured, and sequestered.

9.2.1.1.2 Hydrogen Production via Partial Oxidation of Natural Gas and Hydrocarbons

In a partial oxidation process, methane and other hydrocarbons in natural gas react exothermically with less than stoichiometric oxygen to produce carbon monoxide and hydrogen. Typically, the partial oxidation of natural gas is much faster than steam reforming and, therefore, requires smaller reactors. Partial oxidation reactions are given as follows [1]:

Partial oxidation of methane (POM):

$$CH_4 + 0.5O_2 \rightarrow CO + 2H_2 \qquad \Delta H^\circ = -35.7 \text{ kJ/mol } CH_4 \qquad (9.10)$$

Partial oxidation of propane:

$$C_3H_8 + 1.5O_2 \rightarrow 3CO + 4H_2 \qquad \Delta H^\circ = -226.89 \text{ kJ/mol } C_3H_8 \qquad (9.11)$$

Partial oxidation of methanol:

$$CH_3OH + 0.5O_2 \rightarrow CO_2 + 2H_2 \qquad \Delta H^\circ = -192.2 \text{ kJ/mol } CH_3OH \qquad (9.12)$$

Partial oxidation of ethanol:

$$C_2H_5OH + 0.5O_2 \rightarrow 2CO + 3H_2 \qquad \Delta H^\circ = 12.94 \text{ kJ/mol } C_2H_5OH \qquad (9.13)$$

Partial oxidation of gasoline (using iso-octane and toluene as example compounds):

$$C_8H_{18} + 4O_4 \rightarrow 8CO + 9H_2 \qquad \Delta H^\circ < 0 \qquad (9.14)$$

$$C_7H_8 + 3.5O_2 \rightarrow 7CO + 4H_2 \qquad \Delta H^\circ < 0 \qquad (9.15)$$

Partial oxidation of hydrocarbons produces gas mixtures consisting of hydrogen and carbon monoxide and a small amount of carbon dioxide. Subsequently, a WGS reaction (reaction (9.8)) is needed to convert carbon monoxide to produce additional hydrogen. Taking methane partial oxidation as an example, the overall reaction can be written as:

$$CH_4 + 0.5O_2 + H_2O = CO_2 + 3H_2 \qquad (9.16)$$

Partial oxidation of methane (reaction (9.16)) produces one mole less of hydrogen than that of steam reforming of methane (reaction (9.9)). Like steam reforming of methane, partial oxidation of methane is also a commercially available process. The thermodynamic advantages of partial oxidation of methane over a steaming reforming process can be summarized as [4]:

1. Partial oxidation is, typically, much faster than steam reforming of methane and requires a smaller reactor vessel.
2. The H_2/CO ratio (= 2) produced in a stoichiometric partial oxidation is ideal for the synthesis of liquid fuels.
3. A partial oxidation process produces a low concentration of carbon dioxide.
4. Partial oxidation is mildly exothermic, thus avoiding the need for large amounts of expensive superheated steam.

The disadvantage of the process, as in most exothermic reactions, is that the overheat spot (hot spot) can occur in the catalyst bed causing catalyst deactivation. Additionally, partial oxidation of methane produces less hydrogen per unit of input hydrocarbon than that obtained by steam reforming of the same hydrocarbon. Although the efficiency of the partial oxidation process is relatively high (70 to 80%), partial oxidation systems are typically less energy efficient than steam reforming of methane because the higher temperatures involved result in high heat losses [3]. Besides, a partial oxidation reactor is less expensive than a steam reforming reactor, while the downstream WGS reaction and hydrogen purification systems are likely to be more expensive [5]. It should be pointed out that the ratio of carbon to oxygen atoms (C/O) in both steam reforming and partial oxidation processes is always C/O = 1.

9.2.1.1.3 Steam Autothermal Reforming of Natural Gas and Hydrocarbons

Recently, catalytic autothermal reforming has received much research attention as a viable process for hydrogen production for fuel cell applications [6]. Autothermal steam reforming, sometimes referred to as oxy-steam reforming, of hydrocarbons, is a combination of the advantages of steam reforming and partial oxidation processes and both steam reforming and partial oxidation occur in one reactor. In this process, a hydrocarbon feedstock reacts with both steam and oxygen to produce a hydrogen-rich gas mixture. The ratio of fuel (e.g., methane), oxygen, and steam is adjusted based on the heat generated from the partial oxidation reaction, which supplies the heat needed for the catalytic steam reforming reaction. The catalyst and two ratios, O_2/C and H_2/C, affect the temperature and the reaction pathways of the autothermal reforming. Controlling relative rates of partial oxidation and steam reforming can lead to a lower temperature process than partial oxidation and steam reforming [6]. In this sense, unlike the steam reforming reactor, the autothermal reformer does not require an external heat source and indirect heat exchangers, which makes the reactor simpler and more compact than steam reforming reactors reducing the capital cost of hydrogen production. Thus, autothermal reforming offers higher system efficiency than a partial oxidation process [3]. Additionally, as compared to the steam reforming process, autothermal reforming can be operated in smaller and lighter units, and with lower operating temperatures [7,8]. Autothermal reforming is a complicated process that involves oxidation reactions and steam reforming processes. The reactions involved in autothermal process can be expressed as follows [6,9,10]:

Partial oxidations:

$$CH_4 + 2O_2 \rightarrow CO_2 + 2H_2O \qquad \Delta H_{298} = -890.3 \text{ kJ/mol} \qquad (9.17)$$

$$CH_4 + O_2 \rightarrow CO_2 + 2H_2 \qquad \Delta H_{298} = -322.2 \text{ kJ/mol} \qquad (9.18)$$

$$CH_4 + 1.5O_2 \rightarrow CO + 2H_2O \qquad \Delta H_{298} = -519.6 \text{ kJ/mol} \qquad (9.19)$$

$$CH_4 + 0.5O_2 \rightarrow CO_2 + 2H_2 \qquad \Delta H_{298} = -35.7 \text{ kJ/mol} \qquad (9.20)$$

$$CO + 0.5O_2 \rightarrow CO_2 \qquad \Delta H_{298} = -283.0 \text{ kJ/mol} \qquad (9.21)$$

$$H_2 + 0.5O_2 \rightarrow H_2O \qquad \Delta H_{298} = -253.0 \text{ kJ/mol} \qquad (9.22)$$

Steam reforming reactions:

$$CH_4 + H_2O \rightarrow CO + 3H_2 \qquad \Delta H_{298} = 250.1 \text{ kJ/mol} \qquad (9.23)$$

$$CH_4 + 2H_2O \rightarrow CO_2 + 4H_2 \qquad \Delta H_{298} = 163.2 \text{ kJ/mol} \qquad (9.24)$$

$$CO + H_2O \rightarrow CO_2 + H_2 \qquad \Delta H_{298} = -41.8 \text{ kJ/mol} \qquad (9.25)$$

Decomposition reactions:

$$2CO \rightarrow C + CO_2 \quad \Delta H_{298} = -171.5 \text{ kJ/mol (Boudouard reaction)} \quad (9.26)$$

$$CH_4 \rightarrow C + 2H_2 \quad \Delta H_{298} = 75.3 \text{ kJ/mol (Methane dissociation)} \quad (9.27)$$

The formation of solid carbon (coke) is undesirable due to its adverse effect of deactivating the catalysts. Thermodynamic analysis indicates that high temperature and high concentration of O_2 and H_2O can suppress the formation of coke [6] and the ideal autothermal reforming of methane may be expressed as [6]:

$$CH_4 + x(O_2 + 3.76N_2) + (2 - 2x)H_2O(l) \rightarrow CO_2 + (4 - 2x)H_2 + 3.76 \times N_2 \quad (9.28)$$

where x is the O_2/C molar ratio that determines:

1. The minimum amount of water, $(2 + 2x)$, theoretically required to convert the carbon in CH_4 to CO_2.
2. The maximum hydrogen yield, $(4 - 2x)$, in the product.
3. The heat of reaction, $\Delta H°_{f, CO2} - \Delta H°_{f, CH4} - (2 - 2x) \Delta H°_{f, H2O}$, which determines the adiabatic reaction temperature.

9.2.1.1.4 Hydrogen Production from CO_2 Reforming and Oxy-CO_2 Reforming of Methane

Catalytic reforming of methane with carbon dioxide, also known as dry reforming, has recently attracted considerable attention due to the simultaneous utilization and reduction of two types of greenhouse gases, CO_2 and CH_4, for the production of synthesis gas. The reactions of CO_2 reforming and oxy-CO_2 reforming are as follows:

$$CH_4 + CO_2 \rightarrow 2CO + 2H_2 \qquad \Delta H°_{298} = 247 \text{ kJ/mol } CH_4 \qquad (9.29)$$

$$CH_4 + 0.5O_2 \rightarrow 2CO + 2H_2 \qquad \Delta H°_{298} = -38 \text{ kJ/mol } CH_4 \qquad (9.30)$$

$$CH_4 + xCO_2 + (1 - x)/2O_2 \rightarrow (1 + x) CO + 2H_2$$
$$\Delta H° = (285x - 38)\text{kJ/mol } (0 < x < 1) \qquad (9.31)$$

It should be pointed out that CO_2 reforming is mainly used for the production of synthesis gas that can then be converted to paraffinic liquid fuels through Fischer–Tropsch reactions [11] on Fe, Co, and Ru metal-based catalysts:

$$nCO + 2nH_2 \rightarrow (-CH_2-)_n + nH_2O \qquad (9.32)$$

Synthesis gas can also be used to synthesize methanol over Cu/ZnO catalyst [11]:

$$CO + 2H_2 \rightarrow CH_3OH \qquad \Delta H^\circ = -91.1 \text{ kJ/mol} \qquad (9.33)$$

In terms of hydrogen production, CO generated from CO_2 reforming undergoes a WGS reaction to produce H_2 and CO_2. The overall reaction is the same as for steam reforming of methane:

$$CH_4 + CO_2 \rightarrow 2CO + 2H_2 \qquad \text{(CO}_2 \text{ reforming of methane)}$$

$$2CO + 2H_2O \rightarrow 2CO_2 + 2H_2 \qquad \text{(Water gas shift reaction)}$$

$$\text{Overall: } CH_4 + 2H_2O \rightarrow CO_2 + 4H_2 \qquad \text{(Steam reforming of methane)}$$

Therefore, there is no energy efficiency gain or greenhouse reduction effect using CO_2 reforming of methane for the production of hydrogen.

9.2.1.1.5 Hydrogen Production via Decomposition and Oxy-Decomposition of Natural Gas and Hydrocarbons

Methane decomposition (or pyrolysis, dissociation, cracking) is a CO_2-free process for the production of hydrogen and carbon [12,13]:

$$CH_4 \rightarrow C + 2H_2 \qquad \Delta H^\circ_{298} = 75.6 \text{ kJ/mol CH}_4 \qquad (9.34)$$

$$C_nH_{2n+2} \rightarrow nC + (n + 1)H_2 \qquad \Delta H^\circ_{298} = \text{hydrocarbon dependent} \quad (9.35)$$

The thermal energy required for methane decomposition is less than that for steam reforming of methane, 37.8 versus 63 kJ/mol H_2. Theoretically, no CO_2 will be produced via methane decomposition. However, in an oxy-methane decomposition process, a small portion of methane is used as fuel to provide thermal energy for the remaining methane decomposition:

$$xCH_4 + xO_2 \rightarrow xCO_2 + 2xH_2 \qquad (9.36)$$

$$(1 + x)CH_4 + xO_2 \rightarrow C + xCO_2 + (2x + 1)H_2 \qquad (9.37)$$

Then, the CO_2 emission could potentially be as low as 0.05 mol/mol H_2, compared to 0.43 mol CO_2/mol H_2 from a steam reforming process.

When the hydrogen produced is used for fuel cell applications, hydrocarbon pyrolysis can produce CO-free hydrogen. In addition to high hydrogen purity and minimized CO_2 emission, depending on the catalysts employed, hydrocarbon decomposition can generate value-added carbon products, such as carbon filaments, carbon nanotubes, turbostratic carbon, and graphitic carbon. In most cases, methane decomposition uses Ni-based catalysts that have been widely studied due to their high activity and the capability of producing high value carbon filaments and carbon nanotubes at moderate temperatures (500 to 700°C) [14]. Fe-based catalysts operate at a higher temperature range than that of Ni-based catalysts and are also capable of generating carbon nanotubes [15]. It occurs when the active sites of the metallic particles are covered by nonreactive graphitic layers. Additionally, methane decomposition by metal-based catalysts can also produce other forms of carbon, including graphitic carbon, turbostratic, and carbidic carbon (Table 9.3) [13] at elevated temperature, and the catalysts are rapidly deactivated due to carbon deposition.

However, there are two major issues related to the application of metal-based catalysts: (1) catalyst deactivation and (2) separation of carbon products from catalysts. Carbon-based catalysts offer certain advantages over metal catalysts because of their availability, durability, and low cost [13]. The other important features of carbon-based catalysts are that they are sulfur tolerant and can withstand higher temperatures. The benefit of using catalytically active carbon particles for methane decomposition is that carbon formed from the reaction can deposit on the surface of the original carbon catalysts. These carbon particles serve as a new carbon catalyst to continue the decomposition reaction. No new external catalysts are needed. Additionally, no separation of carbon products from catalysts is needed because the catalysts and products are the same types of carbon. Many types of carbon-based catalysts have been used in hydrogen production via methane decomposition.

TABLE 9.3

Methane Decomposition Catalysts and Carbon Products

Catalyst	Reaction Temperature	Products
Ni-based catalysts	500 – 700°C	Carbon filaments or carbon nanotubes
Fe-based catalysts	680 – 970°C	Carbon filaments or carbon nanotubes
Carbon-based catalysts	850 – 970°C	Turbostratic carbon Carbon filaments
Co, Ni, Fe, Pd, Pt, Cr, Ru, Mo, W catalysts	750 – 1050°C	Graphitic carbon Turbostratic carbon
Noncatalytic decomposition	>1200°C	Amorphous carbon

Source: Adapted from Muradov, N.Z., and T.N. Veziroğlu. 2005. *Int. J. Hydrogen Energy* 30: 225–237.

These include activated carbon, carbon black, microcrystalline graphite, and nanostructured carbons [16,17]. However, it should be pointed out that compared to steam reforming of methane, which is a commercialized process, hydrogen production from methane decomposition requires more research on more durable catalysts. A recent review of hydrogen production via hydrocarbon decomposition can be found in Ahmed et al. [18].

9.2.1.1.6 Hydrogen Production from Biomass-Based Volatile Carbohydrates

Corn and switchgrass consist of large amounts of carbohydrates that can be converted into hydrogen. Currently, carbohydrates in corn starch can be used for the production of ethanol, which can be further reformed for hydrogen production. Biomass-based carbohydrates can also be used in conventional steam reforming, partial oxidation, and autothermal processes for the production of hydrogen gas. Dauenhauer et al. [11] investigated the autothermal reforming of three volatile carbohydrates on noble metal catalysts: methanol, ethylene glycol, and glycerol. These three molecules represent the first three carbohydrates of the formula $C_n(H_2O)_nH_2$, and are highly oxygenated, with a hydroxyl group on each carbon atom and an internal carbon-to-oxygen ratio (C/O) equal to 1. Due to high oxygen content, hydrogen production via these three carbohydrates is unique and can be summarized in Table 9.4 [11].

9.2.1.1.7 Hydrogen Production from Hydrogen Sulfide Reforming of Methane

Hydrogen sulfide (H_2S) is a common contaminant in natural gas and must be removed because it deactivates metal-based catalysts. Approximately one-third of U.S. natural gas sources contain high hydrogen sulfide concentrations, and are considered low-quality natural gases. Hydrogen sulfide removal from high sulfur content natural gases is a costly process that prevents its application for hydrogen production. However, thermodynamic

TABLE 9.4

Hydrogen Production Processes from Volatile Carbohydrates

Process	Reaction	Methanol (n = 1)	Ethylene glycol (n = 2)	Glycerol (n = 3)
		$\Delta H°$ (kJ/mol Carbohydrate)		
Decomposition	$C_n(H_2O)_nH_2 \rightarrow nCO + (n+1)H_2$	90	173	245
Complete oxidation	$C_n(H_2O)_nH_2 + (n+0.5)O_2 \rightarrow nCO_2 + (n+1)H_2O$	−676	−1118	−1570
Partial oxidation	$C_n(H_2O)_nH_2 + (n/2)O_2 \rightarrow nCO_2 + (n+1)H_2$	−193	−393	−603
Steam reforming	$C_n(H_2O)_nH_2 + (n)H_2O \rightarrow nCO_2 + (2n+1)H_2$	49	91	123
Autothermal steam reforming	$C_n(H_2O)_nH_2 + (n/2)H_2O + (n/4)O_2 \rightarrow nCO_2 + (3n/2+1)H_2$	−72	−160	−240

Source: Adapted from Dauenhauer, P.J. et al. 2006. *J. Catal.* 244: 238–47.

analysis has shown that during the processes of methane pyrolysis, steam reforming of methane, and autothermal processes at a temperature range of 700 to 1000°C, hydrogen sulfide in natural gas is highly stabilized due to the presence of hydrogen [19,20]. Therefore, hydrogen sulfide can be treated as an inert gas during the processes of hydrogen production. This result indicates, based on a thermodynamic viewpoint, that high hydrogen content natural gas can possibly be treated in two steps: hydrogen production from hydrocarbon processing and hydrogen sulfide reforming of methane. In this approach, no prior hydrogen sulfide separation is needed. Although being thermodynamically favorable, kinetically, hydrogen sulfide can rapidly react with almost all metal-based catalysts before hydrogen is produced, resulting in the deactivation of the catalysts. It is therefore critical that the catalyst deactivation issue in the first step be solved in order to assemble a complete process. As discussed previously, carbon-based catalysts are hydrogen sulfide tolerant if they can overcome deactivation caused by coking. It should be pointed out that handling high sulfur concentration natural gas is an inherent issue and no good solution is available. With continual increases in fossil fuel prices, subquality natural gases may become potential sources for hydrogen production. The other benefit of hydrogen sulfide reforming is that hydrogen sulfide reforming of methane will not produce the greenhouse gas, CO_2. Instead, it produces carbon disulfide (CS_2), which is in liquid form under ambient conditions and is a desirable feedstock for synthesis of sulfuric acid.

The reaction for hydrogen production via the hydrogen sulfide reforming of methane is as follows [21]:

$$H_2S + CH_4 \rightarrow H_2 + CS_2 \qquad \Delta H^\circ_{298\,K} = 232.4 \text{ kJ/mol} \qquad (9.38)$$

Compared to steam reforming of methane ($\Delta H^\circ_{298\,K} = 165.2$ kJ/mol), hydrogen sulfide reforming is a more endothermic process that requires higher thermal energy for hydrogen production. Hydrogen sulfide reforming of methane is accompanied by two side reactions:

$$H_2S \rightarrow 0.5S_2 + H_2 \qquad \Delta H^\circ_{298\,K} = 79.9 \text{ kJ/mol} \qquad (9.39)$$

$$CH_4 \rightarrow C + 2H_2 \qquad \Delta H^\circ_{298\,K} = 74.9 \text{ kJ/mol} \qquad (9.40)$$

Depending on the H_2S/CH_4 ratio, hydrogen sulfide reforming occurs at a temperature higher than 850°C. At these temperatures, the elemental sulfur produced is in the vapor phase and cannot cause deactivation of metal sulfide-based catalysts. However, the formation of solid carbon can be harmful for the catalysts. To avoid coking effect, the molar ratio of H_2S/CH_4 must be greater than 4. As indicated by thermodynamic analyses [21], a higher H_2S/CH_4 ratio can reduce carbon formation to zero at lower temperatures.

In summary, the utilization of low quality natural gases for the production of hydrogen is still in the stage of conceptual design and analysis. The main incentive for the development of this process is the continually rising cost of fossil fuels. More efforts are needed in terms of novel catalyst exploration and process assessments.

9.2.1.2 Hydrogen Production from Biomasses

The majority of hydrogen produced today is derived from fossil fuels. With the increasing demands for energy, the fossil fuel resources have been rapidly depleted in just a few decades. Besides the resource concerns, hydrogen production from fossil fuels generates greenhouse gases. Continuous efforts have been made to explore clean, renewable alternative sources for a sustainable energy future. Biomass is a renewable and abundant energy source [22] since it is a form of solar energy stored in plants. It is well known that biomass is formed by utilizing carbon dioxide from the atmosphere during plants' photosynthesis. Therefore, using biomass for energy production is a carbon neutral process. Biomass resources can be divided into four categories [23]:

1. *Energy crops*: Herbaceous energy crops, woody energy crops, industrial crops, agricultural crops, and aquatic crops
2. *Agricultural residues and waste*: Crop waste and animal waste
3. *Forestry waste and residues*: Mill wood waste, logging residues, and trees and shrub residues
4. *Industrial and municipal waste*: Solid municipal waste, sewage, sludge, and production waste

Energy production processes from biomass can be separated into thermochemical and biological processes. Thermochemical processes include biomass combustion, pyrolysis, liquefaction, and gasification. Direct biomass combustion in air converts the chemical energy of biomass into thermal heat, with an efficiency of 10 to 30%. No hydrogen is produced by this process. Biomass liquefaction is a high pressure process, during which biomass is mixed with water under a pressure about 50 to 200 atm and temperature of 300°C in the absence of air. Solvents or catalysts are needed in the process to extract organic compounds from the biomass. Biomass biological hydrogen production can be classified into five groups [23,24]: (1) direct biophotolysis, (2) indirect biophotolysis, (3) biological water gas shift reaction, (4) photo fermentation, and (5) dark fermentation. All processes are controlled by hydrogen-producing enzymes, such as hydrogenase and nitrogenase. However, as concluded by Levin et al. [24], those technologies are still in their infancy. Existing technologies offer potential for practical hydrogen production, but the slow hydrogen production rates prevent them from becoming competitive commercial systems. Further research and development aimed at increasing rates and yields of hydrogen production are essential.

Based on the discussion above, the near-term biomass application for hydrogen production will most likely be through high temperature thermochemical processes. In line with hydrogen production from natural gas, the utilization of biomass for hydrogen production can be divided into: biomass pyrolysis (biomass decomposition), steaming reforming biomass, biomass partial oxidation, and autothermal reforming of biomass. Similar to the methane thermal decomposition process, biomass pyrolysis converts biomass into liquid oils, solid charcoal, and gaseous products. Additionally, biomass pyrolysis can be classified into fast and slow processes. The slow process is not suitable for hydrogen production because its product is mainly charcoal. Fast pyrolysis, on the other hand, is high temperature processing in which a biomass feedstock is heated rapidly in the absence of air to form vapor, which is subsequently condensed to a dark brown mobile bio-liquid [23] as:

$$\text{Biomass} + \text{heat} \rightarrow H_2 + CO + CH_4 + \text{other products}$$

Methane and other hydrocarbon vapors can then be reformed by steam to produce more hydrogen while carbon monoxide reacts with water vapor to produce carbon dioxide and hydrogen.

Since most biomass compositions are hydrocarbons or oxidative hydrocarbon with a general molecular formula of $C_nH_mO_z$ [25], biomass processing can use similar technologies for the hydrogen production via natural gas. The reactions of steam reforming of biomass, biomass partial oxidation, and biomass autothermal reforming are summarized in Table 9.5 [25].

9.2.1.3 Hydrogen Production from Coal

Coal is the most abundant fossil fuel on Earth and will continue to be an important energy source over the next several hundred years. It is estimated that at present consumption rates coal can last from 216 years to over 500 years [26]. Coal is a complex mixture of organic compounds composed of carbon, hydrogen, oxygen, and smaller amounts of nitrogen and sulfur, as well as moisture and minerals. According to its degree of coalification, coal

TABLE 9.5

Hydrogen Production from Biomass Thermochemical Processes

Process	Reaction
Steam reforming of biomass	$C_nH_mO_z + 2(n - z)H_2O \rightarrow nCO + (n + 0.5m - z)H_2$
Biomass partial oxidation	$C_nH_mO_z + 0.5(n - z)O_2 \rightarrow nCO + 0.5mH_2$
Biomass autothermal reforming	$C_nH_mO_z + yO_2 + 2(n - y - 0.5z)H_2O \rightarrow nCO_2 + 2(n - y - 0.5z + 0.25m)H_2$

can be classified into different ranks: lignite (brown coal), subbituminous coal, bituminous coal, and anthracite, each having a different heating value [27]. Typical compositions of different coal sources are listed in Table 9.6 [27]. Based on Table 9.6, elemental molar percentages have been calculated and listed in Table 9.7.

Compared to production of hydrogen from methane (H/C ratio = 4), the low molar ratio of H/C (= 0.70 to 1.03) shown in Table 9.7 indicates that hydrogen production from coal will generate more CO_2 per mole of hydrogen produced. Based on the low H/C molar ratio and high nitrogen and sulfur contents, it is understandable that the development of hydrogen production technologies from coal should first focus on the reduction of the emissions of NO_x and SO_x, and CO_2.

TABLE 9.6

Proximate and Ultimate Analyses of Coals

Coal	Proximate Analyses (wt%)				Ultimate Analyses (wt%)				
	W	A	VM	FC	C	H	O	N	S
Lignite (Wyoming, U.S.)	15.2	7.0	46.6	31.2	54.1	3.9	19.0	0.8	1.1
Subbituminous (Taiheiyo, Japan)	4.8	7.6	48.8	38.8	67.7	5.8	13.0	1.1	0.2
Bituminous (Datong, China)	4.6	6.4	24.3	64.6	72.3	4.2	11.6	0.9	0.4
Bituminous (Ebenezer, Australia)	17.3	12.9	36.5	48.9	69.7	5.2	9.2	1.3	0.5

Note: W = water; A = ash; VM = volatile matter; and FC = fixed carbon.

Source: Lin, S.Y. 2009. In *Hydrogen fuel: Production, transport, and storage*, 103–125, ed. R.B. Gupta. CRC Press: Boca Raton, FL. With permission.

TABLE 9.7

Molar Percentages of Elements in Coals

Coal	Molar Percentage (%)					
	C	H	O	N	S	H/C
Lignite (Wyoming, U.S.)	4.51	3.90	1.19	0.057	0.034	0.87
Subbituminous (Taiheiyo, Japan)	5.64	5.80	0.81	0.079	0.006	1.03
Bituminous (Datong, China)	6.03	4.20	0.73	0.064	0.013	0.70
Bituminous (Ebenezer, Australia)	5.81	5.20	0.58	0.093	0.016	0.90

Hydrogen production from coal mainly uses coal gasification technology, in which air (or oxygen) and steam are used to react with coal power (CH_xO_y) at temperatures of 1000 to 1500°C to produce a gas mixture containing H_2 and CO. Production of hydrogen from coal consists of three steps: coal gasification, WGS reaction, and CO_2 separation. Coal gasification is a costly process and some innovative approaches have been proposed and developed. This process can be performed in three major types of gasifiers: fixed bed reactors, fluidized bed reactors, and entrained bed rectors. Detailed information can be found in Lin [27]. The overall process can be written as:

$$CH_xO_y + (1-y)H_2O \rightarrow (0.5x + 1 - y)H_2 + CO$$

$$CO + H_2O \rightarrow H_2 + CO_2$$

Overall reaction: $CH_xO_y + (2-y)H_2O \rightarrow (0.5x + 2 - y)H_2 + CO_2$ (9.41)

In detail, coal gasification is performed by the following several reactions [27]:

$$C + O_2 \rightarrow 2CO \qquad\qquad \text{(oxygen gasification)}$$

$$C + H_2O \rightarrow H_2 + CO \qquad\qquad \text{(steam gasification)}$$

$$C + 2H_2 \rightarrow CH_4 \qquad\qquad \text{(hydrogen gasification)}$$

9.2.1.3.1 One Step Hydrogen Production from Coal: HyPr-RING Process

Hydrogen production reaction integrated novel gasification (HyPr-RING) is an integrated process combining hydrogen production, WGS reaction, and carbon dioxide capture reaction in one reactor. The concept is based on the fact that hydrogen production from steam reforming requires a high temperature to ensure a fast reaction rate, but thermodynamically WGS reaction has to be operated at a low temperature. Therefore, two reactors are needed to carry out these two reactions. If two reactions can be operated at one temperature, only one reactor is needed, resulting in a reduction of the time and cost of the process. In order to achieve this strategy, the HyPr-RING method introduces a CO_2 absorption reaction to combine with reforming and WGS reactions. Removing the reaction product, CO_2, from the system manipulates and achieves lower reaction temperatures so that reforming and WGS reactions are possibly operated at an optimized temperature. The reactions of the HyPr-RING process are shown as follows [27]:

$$C + H_2O \rightarrow CO + H_2 \qquad \Delta H°_{298} = 132 \text{ kJ/mol} \qquad (9.42)$$

$$CO + H_2O \rightarrow CO_2 + H_2 \qquad \Delta H°_{298} = -41 \text{ kJ/mol} \qquad (9.43)$$

$$CaO + H_2O \rightarrow Ca(OH)_2 \qquad \Delta H°_{298} = -109 \text{ kJ/mol} \qquad (9.44)$$

$$Ca(OH) + CO_2 \rightarrow CaCO_3 + H_2O \qquad \Delta H°_{298} = -69 \text{ kJ/mol} \qquad (9.45)$$

The ratio of partial pressure, $P_{CO2}P_{H2}/P_{H2O}P_{CO}$, is decreased by the CO_2 absorption. Thus, the equilibrium of the WGS reaction can be operated at a higher temperature with simultaneous CO_2 removal. The overall reaction of the HyPr-RING process can be written as:

$$C + CaO + 2H_2O \rightarrow CaCO_3 + 2H_2 \qquad \Delta H°_{298} = -88 \text{ kJ/mol} \qquad (9.46)$$

Additionally, CaO can react with H_2S and catalyze the decomposition of NH_3 and tar decomposition [27].

$$CaO + H_2S \rightarrow CaS + H_2O \qquad (9.47)$$

$$2NH_3 \rightarrow N_2 + 3H_2 \qquad (9.48)$$

9.2.1.3.2 ZECA Project

The Los Alamos National Laboratory proposed a Zero Emission Coal Alliance (ZECA) process [27] for hydrogen production from coal, in which coal is first gasified via a methanation process without combustion. Like the HyPr-RING process, CaO is also introduced as a CO_2 sorbent. The ZECA method can be expressed as the following steps [27]:

1. $C + 2H_2 \rightarrow CH_4$
2. $CH_4 + 2H_2O \rightarrow CO_2 + 4H_2$; $CO_2 + CaO \rightarrow CaCO_3$
3. $CaCO_3 \rightarrow CaO + CO_2$

The two reactions in step 2 occur simultaneously. With the removal of product CO_2, reaction equilibrium is shifted to favor hydrogen production and decrease the reaction temperature. Concentrated CO_2 released from step 3 can be sequestered without separation from other gases.

9.2.1.3.3 Coal Gasification with CCR Process

The Ohio State University proposed a method to combine coal gasification with a carbonation/calcination reaction (CCR) process to produce hydrogen [27]. CCR is also a sorption-enhanced process. In contrast to the ZECA process, the CCR method uses CaO to promote WGS reaction as follows:

1. $C + H_2O \rightarrow CO + H_2$
2. $CO + H_2O \rightarrow CO_2 + H_2$; $CO_2 + CaO \rightarrow CaCO_3$
3. $CaCO_3 \rightarrow CaO + CO_2$

9.2.1.3.4 Synthesis Gas Chemical Looping (SCL) Processes [28,29]

The SCL process uses specially developed metal oxide composite particles in a cyclic loop consisting of reduction and oxidation steps to convert synthesis gas for the production of hydrogen. The SCL process consists of five major components: an air separation unit, a coal gasifier, a gas cleanup system, a reduction reactor, and an oxidation reactor. After coal gasification, the purified resultant synthesis gas is completely reduced in a reduction reactor by a metal oxide to produce CO_2, H_2O, and low valance metal oxide or metal. In this step, CO is converted to CO_2 and is collected and separated from H_2. The resulting low valance metal oxide or metal particles from the reduction reactor are then introduced into the oxidation reactor where they react with steam to produce high purity hydrogen (>99.7%).

The SCL process based on iron oxide/iron monoxide/iron metal (Fe_2O_3/ FeO/Fe) redox pairs can be written as follows:

Reduction reactions (T = 750 to 900°C, P = 30 atm):

$$Fe_2O_3 + CO \rightarrow 2FeO + CO_2$$

$$FeO + CO \rightarrow Fe + CO_2$$

$$Fe_2O_3 + H_2 \rightarrow 2FeO + H_2O$$

$$FeO + H_2 \rightarrow Fe + H_2O$$

Oxidation reactions (T = 500 to 750°C, P = 30 atm):

$$Fe + H_2O_{(g)} \rightarrow FeO + H_2$$

$$3FeO + H_2O_{(g)} \rightarrow Fe_3O_4 + H_2$$

The magnetite (Fe_3O_4) formed in the oxidation reactor is regenerated to Fe_2O_3 in a unit called the combustion train according to the reaction:

$$4Fe_3O_4 + O_2 \rightarrow 6Fe_2O_3$$

Iron oxide and iron monoxide particles are transferred back to the reduction reactor for the production of CO_2 from synthesis gas to close the loop.

9.2.1.3.5 Coal Direct Chemical Looping (CDCL) Process [28]

A direct chemical looping process uses coal as a feedstock with presence of iron oxide (Fe_2O_3) particles. This process can reduce the consumption of oxygen needed for coal gasification. The CDCL consists of three reactors: coal processing reactor, hydrogen generation reactor, and combustion reactor. A moving bed reactor is served as the fuel reactor operated at 750 to 900°C and 1 to 30 atm. The desirable reaction is:

$$C_{11}H_{10}O(coal) + 26/3Fe_2O_3 \rightarrow 11CO_2 + 5H_2O + 52/3Fe \qquad (9.49)$$

The heat provided for the above reaction is from partial oxidation of coal with a substoichiometric amount of oxygen. Thus, the overall reaction of the Fe_2O_3 reduction and coal partial oxidation can be written as:

$$C_{11}H_{10}O(coal) + 6.44_3 + 3.34O_2 \rightarrow 11CO_2 + 5H_2O + 12.88Fe \qquad (9.50)$$

In a hydrogen production reactor, metal iron produced in the fuel reactor reacts with steam to produce hydrogen and iron oxide that is recycled back to the fuel reactor to close the loop. Similar to the SCL process, some magnetite is formed and can be regenerated by oxidation in a combustion reactor to produce Fe_2O_3 particles.

9.2.1.3.6 Calcium Looping (CLP) Process

The CLP process uses calcium oxide (CaO) as a sorbent in the water gas shift reaction to simultaneously remove CO_2, sulfur, and chloride impurities at a high temperature as:

Water gas shift reaction: $CO + H_2O \rightarrow H_2 + CO_2$
Carbonation: $CaO + CO_2 \rightarrow CaCO_3$
Hydrogen sulfide (H_2S) removal: $CaO + H_2S \rightarrow CaS + H_2O$
Carbonyl sulfide (COS) removal: $CaO + COS \rightarrow CaS + CO_2$
Hydrogen chloride (HCl) removal: $CaO + HCl \rightarrow CaCl_2 + H_2O$
Calcium oxide regeneration (calcination): $CaCO_3 \rightarrow CaO + CO_2$

It should be pointed out that the formation of CaS, $CaCl_2$ mixed with $CaCO_3$, makes the calcination process difficult. Thus, sorbent loss can be predicted.

9.2.1.4 Hydrogen Production from Water

Water is a renewable resource for hydrogen production. When hydrogen chemical energy is converted into electrical energy via a hydrogen fuel cell, water is the only byproduct and it can be recycled forming a closed cycle. Considering the long-term future of hydrogen economics, hydrogen production from water using renewable energy sources, such as solar or electrical energy from wind power, should be the ultimate goal of human energy future. Due to the strong bonding energy between hydrogen and oxygen atoms, water splitting is an energy intensive process that requires innovative technologies to increase the process efficiency and cost. Although currently hydrogen production is mostly from fossil fuel-based feedstocks, a number of water splitting processes for hydrogen production have been under research and development. Based on energy sources (electrical energy, thermal energy, and photonic energy) used for water splitting, these processes include water electrolysis, high temperature electrolysis, high temperature water decomposition, photocatalysis, and thermochemical water-splitting cycles. Despite the differences in energy input or technologies, hydrogen production from water is always based on the following reaction:

$$H_2O \rightarrow 0.5O_2 + H_2 \qquad \Delta H°_{298} = 285.9 \text{ kJ/mol} \qquad (9.51)$$

Direct high temperature water decomposition involves temperatures over 2500°C, requiring a high temperature source and insulation materials. To prevent or reduce the reverse reaction ($0.5O_2 + H_2 \rightarrow H_2O$), a rapid quenching system is needed. Hydrogen production via water electrolysis can produce high purity hydrogen with a minimum potential requirement of 1.23 V. Considering electrode polarization loss, a practical electrolyzer requires a potential over 2.0 V. In addition to the higher electrical energy requirement, precious metals (such as Pt, Ru, and Ir) are needed as electrode materials. The electrode reactions are:

Anode reaction: $2H_2O \rightarrow O_2 + 4H^+ + 4e^-$

Cathode reaction: $4H^+ + 4e^- \rightarrow 2H_2$

Today's electricity grid is not an ideal source to provide the electrical power needed for water electrolysis due to the fact that intensive energy is needed. On the other hand, greenhouse gas emissions may get involved when a fossil fuel is used for electricity generation. Electricity generation using renewable (solar, wind, or hydraulic) or nuclear energy is a possible option for hydrogen production via water electrolysis. However, more research efforts will be needed to reduce the capital cost of electrolyzers and improve energy efficiency of hydrogen production.

High temperature water electrolysis is an integrated thermal and electrolytic process. A high temperature thermal energy is used to reduce the requirement in electrical energy, which is actually the change in Gibbs free energy, $\Delta G = \Delta H - T\Delta S$. Thermodynamic calculations indicate that the electrical energy requirement for water electrolysis decreases as temperature increases. Since thermal energy is cheaper than electricity, high temperature electrolysis is more efficient economically than room temperature electrolysis. Normally, high temperature electrolysis operates between 100 and 850°C. A solid oxide electrolyzer using yttria-stabilized zirconia (YSZ) as an oxygen conductive electrolyte is needed.

Solar energy far exceeds all human energy demands. Natural photosynthesis processes use solar photonic energy to convert carbon dioxide and water into sugar and oxygen, thereby storing solar energy in the form of biomass that can be converted into fossil fuels. Photocatalytic water splitting for the production of hydrogen is regarded as an artificial photosynthesis and is an attractive but a challenging realm in chemistry [30–32].

$$H_2O + h\nu + \text{photocatalyst} \rightarrow H_2 + 0.5O_2$$
$$\Delta G^\circ = 237.1 \text{ kJ/mol} (1.3\text{eV/e}, \lambda_{min} = 1100 \text{ nm})$$

This reaction can be catalyzed by inorganic semiconductor-based photocatalysts that typically consist of a compound semiconductor and a fused metal or metal oxide co-catalyst [33,34]. Water photocatalytic splitting requires a photocatalyst's band gap larger than 2.4 eV. Irradiation of aqueous photocatalyst powder dispersions can lead to the production of hydrogen and oxygen. Because of the contradictory requirements for water splitting and solar light absorption, no idealized photocatalysts have been reported. On one hand, water splitting requires a stable and large band gap photocatalyst, such as TiO_2 (a band gap of 3.1 eV). On the other hand, TiO_2 can only absorb light with a wavelength shorter than 400 nm. In the solar spectrum, less than 5% of photonic energy is in this range. Therefore, 95% of solar spectral energy cannot be utilized. Photocatalysts with low band gaps are normally unstable and cannot be used for water splitting. As yet, photocatalytic splitting of water is still in an early developmental stage and is considered one of the long-term research focuses.

Thermochemical water-splitting cycles (TCWSCs) use connected chemical reactions to form closed processes for the production of hydrogen and oxygen from water dissociation. TCWSCs bypass the separation of hydrogen and oxygen issue in direct water decomposition processes and each step of the cycle operates at relatively moderate temperatures (e.g., ~850°C). Since the concept of the thermochemical cycle was first proposed in the 1970s, approximately 400 cycles have been reported and a few pilot plants have been built and tested. Fundamentally, each TCWSC consists of at least two steps: a hydrogen production step and an oxygen production step. Although some cycles utilize electrical power in the hydrogen step, the majority of energy needed to drive the entire process is low-grade thermal energy. This enables the utilization of thermal heat sources, such as concentrated solar radiation, heat from nuclear power plants, and many other heat sources. Chemical simulation and thermodynamic analyses have indicated that this technology can reach the efficiency of thermal energy to hydrogen chemical energy up to 50% (based on hydrogen's high heating value). Among these cycles, a group of sulfur family cycles seems to be most promising. These include:

Westinghouse hybrid cycle:

$$SO_2(g) + 2H_2O(l) + \Delta E \rightarrow H_2(g) + H_2SO_4(aq) \qquad 25°C \qquad (9.52)$$

$$H_2SO_4(aq) + \Delta H \rightarrow SO_2(g) + 0.5O_2(g) + H_2O(g) \qquad 850°C \qquad (9.53)$$

Sulfur–Iodine cycle:

$$SO_2(g) + 2H_2O(l) + I_2 \rightarrow 2HI(aq) + H_2SO_4(aq) \qquad 25°C \qquad (9.54)$$

$$H_2SO_4(aq) + \Delta H \rightarrow SO_2(g) + 0.5O_2(g) + H_2O(g) \qquad 850°C \qquad (9.55)$$

$$2HI(aq) + \Delta H \rightarrow H_2(g) + I_2(l) \qquad 400°C \qquad (9.56)$$

Sulfur–Bromine hybrid cycle:

$$SO_2(g) + 2H_2O(l) + B_2 \rightarrow 2HB(aq) + H_2SO_4(aq) \qquad 25°C \qquad (9.57)$$

$$H_2SO_4(aq) + \Delta H \rightarrow SO_2(g) + 0.5O_2(g) + H_2O(g) \qquad 850°C \qquad (9.58)$$

$$2HB(aq) + \Delta E \rightarrow H_2(g) + B_2(l) \qquad 400°C \qquad (9.59)$$

Based on the sulfur family cycles, Florida Solar Energy Center (University of Central Florida) has recently developed a new class of purely solar-driven photo/thermal water-splitting cycles in which solar photonic energy is used for the production of hydrogen while solar thermal heat is utilized for oxygen production [34–38]:

Sulfur–Ammonia photo-thermal hybrid cycle:

$$SO_2(g) + H_2O(l) + 2NH_3 \rightarrow (NH_4)_2SO_3(aq) \qquad 25°C \qquad (9.60)$$

$$(NH_4)_2SO_3(aq) + H_2O + hv \rightarrow H_2(g)$$
$$+ (NH_4)_2SO_4 \text{ (photocatalysis)} \qquad 30 \sim 50°C \qquad (9.61)$$

$$(NH_4)_2SO_4(aq) + \Delta H \rightarrow H_2SO_4(g) + 2NH_3(g) + H_2O(g) \qquad 300 \sim 400°C \qquad (9.62)$$

$$H_2SO_4(aq) + \Delta H \rightarrow SO_2(g) + 0.5O_2(g) + H_2O(g) \qquad 850°C \qquad (9.63)$$

Metal Sulfate–Ammonia photo-thermal hybrid cycle:

$$SO_2(g) + H_2O(l) + 2NH_3 \rightarrow (NH_4)_2SO_3(aq) \qquad 25°C \qquad (9.64)$$

$$(NH_4)_2SO_3(aq) + H_2O + hv \rightarrow H_2(g)$$
$$+ (NH_4)_2SO_4 \text{ (photocatalysis)} \qquad 30 \sim 50°C \qquad (9.65)$$

$$(NH_4)_2SO4(aq) + MO(s) + \Delta H \rightarrow MSO_4(g)$$
$$+ 2NH_3(g) + H_2O(g) \qquad 300 \sim 400°C \qquad (9.66)$$

$$MSO_4(aq) + \Delta H \rightarrow SO_2(g) + 0.5O_2(g) + MO(s) \qquad 850°C \qquad (9.67)$$

Where M = Zn, Mg, Mn, Fe etc.

Metal Pyrosulfate–Ammonia photo-thermal hybrid cycle:

$$SO_2(g) + H_2O(l) + 2NH_3 \rightarrow (NH_4)_2SO_3(aq) \qquad 25°C \qquad (9.68)$$

$$(NH_4)_2SO_3(aq) + H_2O + h\nu \rightarrow H_2(g)$$
$$+ (NH_4)_2SO_4 \text{ (photocatalysis)} \qquad 30 \sim 50°C \qquad (9.69)$$

$$(NH_4)_2SO_4(aq) + M_2SO_4(s) + \Delta H \rightarrow M_2S_2O_7(g)$$
$$+ 2NH_3(g) + H_2O(g) \qquad 300 \sim 400°C \qquad (9.70)$$

$$M_2S_2O_7(aq) + \Delta H \rightarrow SO_2(g) + 0.5O_2(g) + M_2SO_4(s) \qquad 850°C \qquad (9.71)$$

Where M = K, Rb, Cs etc.

9.2.2 Overview of Hydrogen Purification Technologies

9.2.2.1 Hydrogen Sulfide Removal Technologies

Hydrogen sulfide in natural gas not only causes severe corrosion to pipelines, but also deactivates catalysts employed for hydrogen production. Hydrogen sulfide is also a well-known toxic air pollutant. Removal of hydrogen sulfide from a feedstock is critical for both hydrogen production processes and hydrogen fuel cell applications. Basically, three technologies are potentially applicable for the separation of hydrogen sulfide from a hydrocarbon feedstock. These include hydrogen sulfide chemical absorption, partial oxidation, and decomposition.

The chemical absorption method is applicable when the concentration of hydrogen sulfide is in the ppm level. Metal oxide and metal hydroxide can be used for this purpose, according to the following reactions:

$$H_2S_{(g)} + CaO_{(s)} \rightarrow H_2O + CaS_{(s)} \qquad (9.72)$$

$$H_2S_{(g)} + ZnO_{(s)} \rightarrow H_2O + ZnS_{(s)} \qquad (9.73)$$

$$H_2S_{(g)} + 2NaOH_{(aq)} \rightarrow 2H_2O + Na_2S_{(aq)} \qquad (9.74)$$

This process is simple and the hydrogen sulfide chemical absorption can be carried out with fast reaction rates under ambient conditions. Additionally, complete hydrogen sulfide removal can be achieved without side reactions. Metal oxide absorption in this method may be applicable for onboard hydrogen sulfide removal. The disadvantage of this method is that a post-treatment is needed for metal sulfide generated in the process.

Hydrogen sulfide partial oxidation is an exothermic process for converting hydrogen sulfide into elemental sulfur and water. A Claus plant (named for London chemist Carl Friedrich Claus) is a typical example of hydrogen sulfide partial oxidation and it is a commercialized technology used in fuel processing industries. In a Claus plant, a portion of hydrogen sulfide is oxidized to produce sulfur dioxide, and the latter is then recombined with the remaining hydrogen sulfide to produce elemental sulfur and water vapor according to:

$$2H_2S + 3O_2 \rightarrow 2SO_2 + 2H_2O \qquad (9.75)$$

$$2H_2S + SO_2 \rightarrow 3S^\circ + 2H_2O \qquad (9.76)$$

Thus, the overall reaction is:

$$2H_2S + O_2 \rightarrow 2H_2O + 2S^\circ \qquad (9.77)$$

It should be noted that partial oxidation of hydrogen sulfide can only recover sulfur and thermal heat value, but hydrogen in the H_2S is wasted as water. A Claus process is a two-step process: separation of hydrogen sulfide and hydrogen sulfide partial oxidation. Prior to hydrogen sulfide partial oxidation, an amine absorption process is needed to separate hydrogen sulfide and carbon dioxide from a hydrocarbon fuel stream. Furthermore, a Claus unit normally generates tail gas, mainly SO_2, and treatment is needed to minimize the environmental impact. In order to more completely remove hydrogen sulfide and overcome the tail gas treatment issue, one-step hydrogen sulfide partial oxidation processes are proposed.

The Fe^{3+}/Fe^{2+} redox system is another well-known process that can be applied for the partial oxidation of hydrogen sulfide without the need of a separating step. The process can be expressed as:

Fe^{3+} reduction:

$$H_2S(g) + \text{Hydrocarbons}(g) + Fe^{3+}(aq) \rightarrow$$
$$Fe^{2+}(aq) + 2H^+(aq) + S^\circ(s) + \text{Hydrocarbons}(g) \qquad (9.78)$$

Fe^{2+} oxidation:

$$Fe^{2+}(aq) + 2H^+(aq) + 0.5O_2(g) \rightarrow Fe^{2+}(aq) + H_2O(l) \qquad (9.79)$$

This process can achieve complete hydrogen sulfide removal and the process can be optimized by adjusting the Fe^{3+} concentration, solution pH, temperature, and contact time. The difficult step is the oxidation of Fe^{2+} because it is a relatively slow reaction. In the case of Fe^{2+} oxidation from air purging, a large reactor is needed to ensure a longer contact time. Fe^{2+} oxidation can also be accomplished using a biomass-based catalyst that can facilitate the oxidation rate. In addition to iron, other metals can also be applied for the partial oxidation of hydrogen sulfide according to:

$$M(s) + xH_2S(g) \rightarrow MS_x(s) + xH_2(g) \qquad (9.80)$$

$$MS_x + xO_2(g) \rightarrow M(s) + xSO_2(g) \tag{9.81}$$

$$\text{Overall reaction: } H_2S(g) + O_2(g) \rightarrow H_2(g) + SO_2(g) \tag{9.82}$$

SO_2 produced by this process can be scrubbed using aqueous NaOH or $(NH_4)OH$ solutions to form Na_2SO_3 or $(NH_4)_2SO_3$ solutions. Aqueous sulfite solutions can be further oxidized via electrolytic or photolytic processes, producing sulfate based fertilizer and hydrogen according to:

$$Na_2SO_3 + H_2O + \Delta E \text{ or } h\nu \rightarrow Na_2SO_4 + H_2 \tag{9.83}$$

$$(NH_4)_2SO_3 + H_2O + \Delta E \text{ or } h\nu \rightarrow (NH_4)_2SO_4 + H_2 \tag{9.84}$$

Unlike partial oxidation, the decomposition of hydrogen sulfide can recover both hydrogen and sulfur as products. The penalty is that the process is highly endothermic and either thermal energy or electrical energy is needed to carry out the decomposition reaction. The overall hydrogen sulfide decomposition can be expressed as:

$$H_2S(g) \rightarrow H_2(g) + S^\circ$$

Direct hydrogen sulfide decomposition requires temperatures greater than 1500°C and a rapid cooling system to avoid the reverse reaction. Hydrogen sulfide decomposition can also be achieved using a high temperature electrolytic process as follows:

$$\text{Cathode: } H_2S + 2e^- \rightarrow S^{2-} + H_2$$
$$\text{Another: } S^{2-} \rightarrow S^\circ + 2e^-$$

Similar to thermochemical water splitting cycles, hydrogen sulfide decomposition can also be achieved via thermochemical cycles. Some metal- or metal oxide-based cycles are listed as follows:

$$M(s) + xH_2S(g) \rightarrow MS_x(s) + xH_2(g) \tag{9.85}$$

$$MS_x(s) \rightarrow M(s) + xS(g) \tag{9.86}$$

Some metals and metal sulfide pairs, such as Fe/FeS and Mo/MoS$_2$, can be used in thermochemical cycles for the decomposition of hydrogen sulfide. Some nonmetal-based thermochemical cycles are listed in the following reactions:

$$\text{CO/COS-based cycle: } \quad H_2S + CO \rightarrow COS + H_2$$
$$COS \rightarrow CO + S^\circ$$

I_2/HI-based cycle: $\quad H_2S + I_2 \rightarrow 2HI + S^\circ$

$\qquad\qquad\qquad\qquad 2HI \rightarrow H_2 + I_2$

H_2SO_4/I_2-based cycle: $\quad H_2S + H_2SO_4 \rightarrow S^\circ + SO_2 + H_2O$

$\qquad\qquad\qquad\qquad 2H_2O + I_2 + SO_2 \rightarrow H_2SO_4 + HI$

$\qquad\qquad\qquad\qquad 2HI \rightarrow H_2 + I_2$

The practicality of these cycles depends on the kinetics of each step and the efficiency of product separation. A side reaction also needs to be taken into account. In the CO/COS cycle, for example, H_2 separation from COS and unreacted CO may require a distillation tower that increases the system capital cost. Additionally, COS decomposition is accompanied by a side reaction: $2COS \rightarrow CO_2 + CS_2$.

An Fe^{3+}/Fe^{2+}-H_2SO_4-based hybrid hydrogen sulfide decomposition process consists of two steps [39]:

Sulfur production step: $H_2S + 2Fe^{3+}(aq) \rightarrow 2Fe^{2+}(aq) + 2H^+(aq) + S^\circ$

Hydrogen production step: $2Fe^{2+}(aq) + 2H^+(aq) + \Delta E \rightarrow H_2(g) + 2Fe^{3+}(aq)$

Electrode reactions for the production of hydrogen from an acidic Fe^{2+} solution are as follows:

Anode reaction: $2Fe^{2+}(aq) \rightarrow 2Fe^{3+}(aq) + 2e^-$ $\quad \Delta E^\circ = 0.77$ V vs. NHE

Cathode reaction: $2H^+(aq) + 2e^- \rightarrow H_2(g)$ $\quad \Delta E^\circ = 0.00$ V vs. NHE

Both $Fe_2(SO_4)_3$/$FeSO_4$ and $FeCl_3$/$FeCl_2$ are applicable in this process except that the $FeCl_3$/$FeCl_2$ system is more acidic than the $Fe_2(SO_4)_3$/$FeSO_4$ redox pair. This method can approach 100% hydrogen sulfide removal and generate high purity hydrogen and elemental sulfur. Detailed hydrogen sulfide removal technologies can be found in Zaman and Chakma [40] and Kohl and Nielsen [41].

9.2.2.2 Carbon Monoxide Removal Technologies

Carbon monoxide is one of the most common impurities in hydrogen fuel streams that can cause significant performance degradation of a hydrogen PEM fuel cell. When carbon-based feedstock is used for the production of hydrogen, the CO impurity in the hydrogen stream is unavoidable. To illustrate this fact, Table 9.8 lists the compositions of hydrogen feed streams produced from carbon-based feedstocks [42].

CO removal technologies can be separated into two categories: hydrogen purification technologies and CO deep removal processes. Four technologies have been most commonly used for separating hydrogen, carbon dioxide, and carbon monoxide from a hydrogen-rich gas mixture [43].

TABLE 9.8

Compositions of Feed Streams Entering Hydrogen Pressure Swing Adsorption (PSA) Purification Units

Component	Steam Reforming	Autothermal Steam Reforming (O_2-based)	Coal Gasification
H_2	94.3%	93.2%	87.8%
CO	0.1%	1.4%	2.6%
CO_2	2.5%	1.7%	3.9%
N_2	0.2%	0.7%	5.0%
Ar	0.0%	0.6%	0.9%
CH_4	2.9%	2.4%	0.01%
Temperature (°C)	33.3	35.0	30.0
Pressure (atm)	26.0	24.7	27.6
Flow rate (Nm^3/h)	17,318	17,631	19,402

Source: Besancon, B.M. et al. 2009. *Int. J. Hydrogen Energy* 34: 2350–60. With permission.

1. Adsorption of a species other than hydrogen to produce pure hydrogen
2. Absorption (physical or chemical) of CO_2 to produce pure CO_2
3. Membrane separation to offer bulk separation of H_2
4. Cryogenic separation to provide multiple pure products, especially for the separation of pure carbon monoxide

9.2.2.2.1 Adsorption-Based Hydrogen Purification

Temperature swing adsorption (TSA) and pressure swing adsorption (PSA) are two types of adsorption-based hydrogen purification technologies. TSA processes are based on the periodic variation of the temperature of an adsorbent bed. The adsorption occurs at low temperature, and the bed regeneration at high temperature. Currently, TSA processes are commonly used for trace impurity removal from air in prepurification unit systems as well as for volatile organic compound abatement from process air streams [44]. The PSA process is used to separate gas species from a mixture of gases under high pressure, according to the species' molecular affinity for an adsorbent material. The adsorbent bed desorption occurs under a low pressure while adsorption is operated under high pressure. Besides, PSA operates at near ambient temperature. The major difference between TSA and PSA is the time required to change the adsorption column from adsorption to desorption or regeneration conditions. Pressure and concentration can be changed much more rapidly than can temperature [45]. Therefore, TSA is limited to situations in which the adsorbate concentration is quite low in the feed stream because, for higher concentrations of adsorbate in a gas mixture, the adsorption time would become quite short compared to the regeneration time.

TABLE 9.9

Relative Strength of Adsorption of
Gas Species

+	+ +	+ + +	+ + + +
He	Ar	CO	C_3H_6
H_2	O_2	CH_4	C_4H_8
	N_2	CO_2	C_5+
		C_2H_6	H_2S
		C_2H_4	NH_3
		C_3H_8	H_2O

Source: Besancon, B.M. et al. 2009. *Int.
J. Hydrogen Energy* 34: 2350–60.
With permission.

PSA is well suited for hydrogen separation from a hydrogen-rich gas mixture because hydrogen is hardly adsorbed on most adsorbents. A PSA process is able to provide H_2 with a purity ranging from 99 to 99.999% [42]. At least three columns are needed in a PSA process and all three columns are operated simultaneously. Each one runs the same cycle, but in different time sequences. Additional adsorbent columns used in a PSA process generate higher hydrogen purity [43,45]. Typical adsorbent materials include silica, alumina, molecular sieves, and activated carbons. The relative strength of adsorption increases from alumina to molecular sieve [42] as:

Alumina < Carbon prefilter < Activated carbon < Molecular sieve

To achieve a higher purification effect, a column can use multiple adsorbent materials placed as individual layers. In the case of a single adsorbent, the relative strength of adsorption of gases is listed in Table 9.9.

A PSA process can easily separate impurities from a hydrogen stream in the rightmost columns with the strongest adsorption strength, especially hydrogen sulfide and ammonia. This is an addition to the desulfurization process for the removal of sulfur species. If multiple packed adsorbent columns are used, a PSA process can also remove components (such as CO, CH_4, etc.) in the + + + column. For retaining inorganic gases, such as Ar, O_2, and N_2, a PSA unit is less effective. In a PSA process, achieving higher hydrogen purity normally leads to more hydrogen loss during the purification process.

9.2.2.2.2 Cryogenic Separation

Conventionally, cryogenic H_2 purification utilizes partial condensation to separate H_2 from impurities with higher boiling points, such as H_2O, CO_2, CO, CH_4, and hydrocarbons. This technology is widely used in refineries for the production of liquid fuels from crude oils. Under a pressure of 12 atm, the cryogenic separation of H_2 from a liquid mixture of CO and CH_4 requires a temperature as low as $-237.0°C$ [46]. Economically, a cryogenic process is not cost effective because of high electrical energy consumption. However, liquid hydrogen is currently the most valuable onboard hydrogen storage technology. Presently, liquid hydrogen is produced by liquefaction of high purity gaseous hydrogen generated by steam reforming of methane followed by PSA purification. Normally, when hydrogen purity is up to 99.95%, hydrogen recovery can reach only 85%, meaning 15% of hydrogen together with CO and CH_4 is burned and wasted. Huang et al [46] have proposed a one-step liquid hydrogen production process that combines cryogenic separation/purification and hydrogen liquefaction. This integrated process can achieve first and second law efficiencies of 85% and 56%, respectively. The hydrogen purity can be as high as 99.9999% and CO_2 captured during this process is 99.54% pure. Since CO and CH_4 separated from liquid hydrogen can be recycled to further produce hydrogen, there are no emission concerns for this process.

9.2.2.2.3 CO Deep Removal Technologies

The outlet hydrogen stream from a PSA unit contains CO concentrations higher than 100 ppm. For PEM fuel cell applications, to reach acceptable CO concentrations ~10 ppm at Pt anodes and <100 ppm at CO-tolerant alloy anodes, a CO deep removal process is essential. Four approaches have been investigated for CO deep removal, including selective diffusion membrane separation, selective CO methanation, preferential CO oxidation, and electrochemical water gas shift reaction [17–49]. These technologies are considered promising for the application of small-scale fuel processors [47].

Hydrogen purification can use various types of membranes, including polymeric constituents to inorganic membranes, such as dense phase metal, metal alloys, and porous ceramic materials [45]. Among these materials, the Pd metal-based membrane is well known for its high hydrogen permeability, good mechanical characteristics, and highly catalytic surface. The process of hydrogen permeation through Pd involves following three steps [47]:

1. Reversible dissociative chemisorption of H_2 on the Pd surface
2. Reversible dissolution of surface atomic hydrogen in the bulk layers of the Pd
3. Diffusion of atomic hydrogen in the membrane

Despite the fact that hydrogen can be separated by diffusion through a membrane, the low hydrogen flux requires a large membrane area. Recently, integrating membrane separation into hydrogen production process has been an attractive topic of research due to the fact that hydrogen removal during the reaction can result in higher equilibrium hydrogen yields.

Deep removal of CO from a hydrogen stream involves the following reactions:

$$CO + 3H_2 \rightarrow CH_4 + H_2O \quad \text{methanation} \quad \Delta H°_{298} = -205.8 \text{ kJ/mol} \quad (9.87)$$

$$CO + H_2O \rightarrow CO_2 + H_2 \quad \text{water gas shift} \quad \Delta H^{\circ}_{298} = -41 \text{ kJ/mol} \quad (9.88)$$

$$CO + 0.5O_2 \rightarrow CO_2 \quad \text{preferential CO oxidation} \quad \Delta H^{\circ}_{298} = -283 \text{ kJ/mol} \quad (9.89)$$

The above reactions are accompanied by side reactions:

$$CO_2 + 4H_2 \rightarrow CH_4 + 2H_2O \quad \text{reverse steam reforming} \quad \Delta H^{\circ}_{298} = -164.6 \text{ kJ/mol} \quad (9.90)$$

$$H_2 + 0.5O_2 \rightarrow H_2O \quad H_2 \text{ oxidation} \quad \Delta H^{\circ}_{298} = -241.8 \text{ kJ/mol} \quad (9.91)$$

Differing from a high CO concentration conversion, CO deep removal is difficult because the ppm level of CO is thermodynamically stable in an H_2 stream at ambient conditions. In order to overcome this obstacle, to remove extremely low concentration CO from an H_2 stream requires a two-step process. The first step is CO adsorption onto a catalyst surface to increase its local concentration. This step can be accomplished because CO has a higher catalytic adsorption capability than H_2. The second step is the thermochemical conversion of adsorbed CO to CO_2 or CH_4. The overall process can be expressed as:

$H_2 + CO$ (ppm) + M (metal based catalyst) $\rightarrow H_2 + M - CO$ (preferential adsorption)

$M - CO + 0.5O_2 \rightarrow CO_2 + M$ (CO preferential oxidation)

$M - CO + 3H_2 \rightarrow CH_4 + H_2O + M$ (CO methanation)

$M - CO + H_2O + \text{electricity} \rightarrow H_2O + CO_2 + M$ (electrochemical WGS reaction at room temperature process)

The electrode reactions of electrochemical WGS are:

Cathode reaction: $H_2O + CO \rightarrow CO_2 + 2H^+ + 2e^-$

Anode reaction: $2H^+ + 2e^- \rightarrow H_2$

CO methanation is an effective means of removing trace CO in a hydrogen steam. The process is technically simple because it does not require the introduction of any gases, such as air, as is required in CO preferential oxidation. The process is normally operated at temperatures from 180 to 280°C [47]. However, CO methanation is not suitable for an onboard application as compared to preferential oxidation of CO, which requires temperatures of only 80 to 180°C. In the CO preferential oxidation process, since air must be mixed with the hydrogen stream, its onboard application raises concerns with hydrogen safety issues. The unique feature of the CO electrochemical WGS reaction is that it can operate under ambient conditions. No heating elements are needed and the CO removal rate is rapid in comparison with both CO methanation and preferential oxidation.

9.2.2.3 Carbon Dioxide Removal Technologies

A PSA-based hydrogen purification unit can separate CO_2 from hydrogen. In a PSA unit, carbon-based species, CO_2, CO, and CH_4, are accumulated. The tail gas from a PSA can be burned to recover heat. If CO_2 capture is desired to reduce greenhouse gas emission, a CO_2 absorption-based system would be attractive since it produces pure hydrogen as well as pure carbon dioxide. There are basically two types of CO_2 removal processes: CO_2 separation based on a low temperature liquid absorbent solution, and CO_2 separation based on solid sorbents at a temperature of about 600°C [43].

CO_2 absorption by amines is the most common process. All amines are relatively weak bases that can react with CO_2, H_2S, and COS, and separate them from H_2 or hydrocarbon streams. CO_2 is typically absorbed by amines at 25 to 80°C under pressure ~70 atm. CO_2 removal from an H_2 stream by monoethanolamine is as follows:

CO_2 separation and H_2 collection:

$$(H_2 + CO_2) + 2 \ HOC_2H_4NH_2 \rightarrow H_2 + HOC_2H_4NHCOO^- + NH_3C_2H_4OH$$

Absorbent regeneration and CO_2 collection:

$$HOC_2H_4NHCOO^- + NH_3C_2H_4OH \rightarrow 2HOC_2H_4NH_2 + CO_2$$
$$HOC_2H_4NHCOO^- + NH_3C_2H_4OH \rightarrow 2HOC_2H_4NH_2 + CO_2$$

Potassium carbonate can also be used for CO_2 removal and the reactions can be written as follows:

CO_2 separation and H_2 collection:

$$(H_2 + CO_2) + K_2CO_3 + H_2O \rightarrow H_2 + 2KHCO_3$$

Absorbent regeneration and CO_2 collection:

$$2KHCO_3 + \Delta H \rightarrow K_2CO_3 + H_2O + CO_2$$

CO_2 absorption and potassium carbonate regeneration occur at about the same temperature (107°C). Note that CO_2 separation requires a higher pressure, but the regeneration process is operated under a lower pressure.

In addition to CO_2 liquid absorption processes, CO_2 can also be separated by solid sorbents, typically at high temperatures. Integrating high temperature CO_2 sorption into hydrogen production from a carbonaceous feedstock has become a novel and attractive research focus, entitled Sorption-Enhanced Hydrogen Production (SEHP). SEHP combines hydrocarbon reforming, WGS, and CO_2 separation reactions to produce H_2 in a single step. The simultaneous reactions occur over a mixture of a reforming catalyst and CO_2 sorbent [50,51]. Considering a conventional hydrogen production via steam reforming of methane as an example, the SEHP process can be expressed as:

$$CH_4(g) + 2H_2O(g) + Sorbent(s) \rightarrow 4H_2(g) + Sorbent - CO_2(s)$$

The sorbent may truly react with CO_2 to form a solid carbonate or may be physically or chemically adsorbed on the surface of the sorbent. The potential advantages of SEHP include the following [50]:

1. Replacement of the high-temperature, high-alloy steels required in the reforming reactor with less expensive materials
2. Simplification (or elimination in some cases) of the hydrogen purification section due to higher H_2 purity and lower concentrations of CO and CO_2
3. Elimination of the WGS reactors
4. Reduction or possible elimination of carbon deposition in the reformer
5. Decreased size of heat exchanging equipment because of the reduction in heat loss

Table 9.10 shows a comparison of hydrogen results from an SEHP process and conventional steam reforming of methane (SRM) [52].

To date, a number of candidate CO_2 sorbents have been investigated, including Ca-based oxides (CaO), potassium-based hydrotalcites, lithium silicates ($LiSiO_4$), and sodium zirconate (Na_2ZrO_3) [53]. The reversible carbonation/decarbonation reactions are:

TABLE 9.10

Comparison of H_2 Production from SRM and SEHP at 80% Methane to Hydrogen Conversion

Process	Catalyst/ Sorbent	Reaction	Reaction Temp.	H_2 Purity	CO_2 Conc.	CH_4 Conc.
SRM	Ni-based	$CH_4 + H_2O = CO + 3H_2$ $CO + H_2O = CO_2 + H_2$	>650°C	~75 mol%	~20 mol%	
SEHP	Ni-based catalyst + CaO	$CH_4 + 2H_2O + CaO = CaCO_3 + 4H_2$	450°C	>95 mol%	~50 ppm	<5 mol%

$$CaO + CO_2 = CaCO_3 \qquad\qquad T = 600°C$$
$$Li_4SiO_4 + CO_2 = Li_2CO_3 + Li_2SiO_3 \qquad T = 100 \text{ to } 650°C$$
$$CaO - MgO + CO_2 = CaCO_3 + MgO \quad T > 250°C$$
$$Na_2ZrO_3 + CO_2 = Na_2CO_3 + ZrO_2 \qquad T = 100 \text{ to } 785°C$$

9.2.3 Summary

Carbon based feedstocks are currently major sources for hydrogen production and steam reforming of natural gas is a commercialized process. Sorbent-enhanced steam reforming of methane is a recently developed process that integrates the reforming reaction and CO_2 removal in one reaction that is possible to reduce the reaction temperatures, increase process efficiency, and produce higher purity hydrogen for the application of PEM fuel cells. This technology has also found application in coal gasification and biomass processing. In addition to the natural gas based hydrogen production, biomass can potentially be a promising feedstock for a carbon neutral hydrogen production. Due to its availability, hydrogen from coal gasification may represent a new research direction. The major impurities associated with hydrogen production from carbonaceous materials are hydrogen sulfide, carbon monoxide, and carbon dioxide. Hydrogen sulfide removal is critical for hydrogen production from carbon based feedstocks. Due to its reactivity, a number of hydrogen sulfide removal technologies have been proposed and some are available for in a large scale. Basically, those technologies can be classified into hydrogen sulfide adsorption, decomposition, and partial oxidation. An Fe^{3+}/Fe^{2+}-based redox system can completely remove hydrogen sulfide from a hydrocarbon stream and provides a sulfur-free feedstock for hydrogen production. For the application, a PEM fuel cell carbon monoxide removal is an important step. The technologies used for CO removal can be separated into two types: onsite removal and deep removal. A well-optimized pressure swing adsorption process currently can remove most impurities in hydrogen stream and produce hydrogen with purity up to 99.9999%. This process can also remove carbon dioxide, ammonia, and unreacted hydrocarbons. Depending on application environments, CO deep removal can be grouped into onground and onboard processes. Onground CO removal can use higher temperature and large reaction systems and, therefore, achieve a higher quantity of CO removal. Due to the limits of temperature and reactor volume, onboard CO removal requires innovative approaches. Electrochemical WGS CO removal may be applicable for a PEM fuel cell-powered vehicle because of its rapid kinetic rate and lower temperature. Removing carbon dioxide from a hydrogen fuel stream mainly relies on onproduction site treatment. CO_2 absorption and sorbent-enhanced processes can both remove CO_2 to a very low level.

9.3 System Level Mitigation Strategies

9.3.1 Fuel Cell System and Source of Contaminants

The core of a fuel cell system is the fuel cell stack, which includes individual cells, bipolar plates, cooling plates, end (pressure) plates, current collection plates, gaskets, manifolds, and conduits for fuel, oxidant, and coolant. Auxiliary components are necessary to support the operation of a fuel cell stack. These extra components are collectively called balance of plant (BOP). The BOP components for a fuel cell system depend on the fuel cell type (e.g., PEM fuel cell, solid oxide fuel cell, molten carbonate fuel cell, phosphoric acid fuel cell, etc.), the type of fuel used, and the application requirements. Essential components in the BOP commonly seen in most fuel cell systems include a subsystem for moving air through the stack, a subsystem for moving fuel through the stack, a subsystem for water and thermal management, a subsystem for conditioning the electrical power output of the stack, and a power plant monitoring and control unit. If hydrocarbon fuel is to be used, a reformer is typically required to convert the fuel into a hydrogen-rich reformate gas. For PEM fuel cells, humidifiers (water management subsystem) are used to control the water contents of the air and fuel going into the stack.

The air moving subsystem, fuel moving subsystem, the fuel reformer (if used), and cooling subsystem communicate fluid media directly with the stack and, hence, may introduce and/or spread contaminates in the system. The materials used in the cell/stack and the BOP components can be the major source of contaminants if not carefully selected and processed. Trace anion (e.g., F^-, Cl^-) and cation impurities may reside in gas diffusion media, the gasket, the membrane, and the electrocatalyst; they can leach out and contaminate other parts of the fuel cell system. Contaminants can be released from the gradual corrosion of metallic components and/or the decomposition of gaskets and electrolyte membranes. Corrosion of metallic components (metal bipolar plate, end plate, tubing, regulators, etc.) can release metallic ions into the stack. Certain metallic ions (Cu^{++}, Fe^{++}) are known to be detrimental to an electrolyte membrane. The polymeric components can gradually decompose (hydrolyze) and form small organic molecules, which can be detrimental to electrocatalysts or the ionomer membrane.

The contaminants in a fuel cell system, if not effectively mitigated, can result in fuel cell performance degradation and premature failure of the fuel cell system. The adverse effect of certain contaminants (e.g., CO, H_2S, NH_4) on fuel cell performance can be detected quickly. These effects tend to be somewhat reversible. The effects of other contaminants are not easily detected, and tend to be irreversible. For example, certain trace metallic ions (Fe^{++}, Cu^{++}) can catalyze the decomposition of polymer electrolyte membranes without affecting ionic conductivity or the gas crossover rate of the membrane until a mechanical breach formed in the membrane [54]. Development of materials (e.g., catalysts) that have better tolerance to the impurities is one way to reduce the impact of contaminants. Section 9.4 of this chapter discusses in detail materials and cell-level approaches for contaminant mitigation.

At a system level, there are several basic and effective strategies for reducing or eliminating the adverse effects of the contaminants: (1) proper selection and screening of materials used in a fuel cell power plant, (2) prevention of contaminants (from air or fuel supplies) from entering the fuel cell system, (3) continuous elimination of contaminants from fluids circulating in the fuel cell power plant, and (4) periodic decontamination operation to flush contaminants out of the fuel cell system. Most current commercial or developmental fuel cell systems have one or multiple contamination mitigation strategies built in. Implementation of contamination mitigation strategies may add to the system complexity, cost, or negatively impact system performance; trade-offs between fuel cell system performance and contamination mitigation methods are sometimes necessary.

9.3.2 Materials Selection and Screening

Proper selection and screening of materials to be used in a fuel cell system is critical to prevent contaminants from being generated within the fuel cell. Material selection is highly dependent on the type of fuel cells, the type of fuel, and operating conditions and environments. For low temperature fuel cells, the corrosion resistance of metallic components that are in contact with warm, moist environments needs to be carefully considered. Materials such as stainless steels are preferred over regular steel, copper, or aluminum. For metallic components (e.g., bipolar plate) that come in direct contact with the membrane electrode assembly, usually a protective coating is necessary for adequate corrosion protection. Electroplated gold coating is the most commonly used protective coating, although gold coating drastically increases the cost of the bipolar plate. As such, very thin layers of gold deposited by a physical vapor deposition method, and other compound coatings (e.g., nitride), have been investigated [55]. Polymers (PTFE, HDPE, PVDF, EDPM, silicone) are used in various places in a fuel cell stack, such as tubing, pipe, o-ring, gaskets, sealant, etc. Because most polymers contain additives, residual solvent, and catalysts, it is important to understand the nature of these additives and the effect of these additives on the fuel cell performance.

For PEM fuel cells, grade 316 stainless steel, graphite, Teflon®, and gold are commonly regarded as acceptable materials; copper, iron, and aluminum are usually not regarded as suitable materials. Caution should be exercised whenever a new material is to be used in a fuel cell system. Careless insertion of new materials into a fuel cell system may later cause problems that can be extremely hard to trace down. For materials with uncertain constituents, screening tests are necessary to ensure no undesirable contaminants will be released from them. A simple screening test can be carried out by placing the materials (e.g., bipolar plate materials or gas diffusion media) in deionized water or dilute acid for a period of time at a controlled temperature, then analyzing the solution, e.g., pH meter, conductivity probe, ion-chromatography, atomic adsorption spectroscopy, or inductively coupled plasma mass spectrometry (ICP-MS), to determine the level and type of species released from the material under investigation. The measured value can be compared to a predetermined threshold value to define the acceptability of the material.

9.3.3 Prevention of Contaminants from Entering a Fuel Cell System

Contaminants may enter a fuel cell stack through the air inlet, the fuel inlet, or the gaskets/sealants. "Barrier" elements at the air or fuel inlet can be set up to reduce the amount of incoming contaminants. Stack sealant and gaskets are barriers for inboard, offboard, and cross-leaking. Maintaining the integrity of sealants is also very important for preventing the ingress of unwanted substances into the fuel cell stack, or the transfer of contaminant materials from one compartment to another compartment in a fuel cell system.

Most terrestrial fuel cell systems use ambient air as the oxidant in the cathode. Depending on the location and environments, ambient air contains various amounts of dust, smoke, salt particles (deicing compound), aerosols, pollen, spores, mold, bacteria, virus, SO_x, NO_x, volatile organic compounds, etc. Filtration is the most commonly used method for reducing airborne contaminants entrained with intake air. Air filter assembly also prevents insects, birds, and rodents from entering and clogging the air intake conduit. Air filtration is widely used for the air intake of internal combustion engines, air compressors, air conditioning units for buildings, etc. Similar filtration materials, devices, and configurations can be applied to the fuel cell systems. Commonly used filtering media include foam, nonwoven fiber mats, pleated paper, spun fiberglass filter, and cotton filters. Pleated paper filters are typically used for automobile engine air filtration. Air filtration is a highly effective method to reduce airborne particulate matters; in fact, high efficiency particulate air (HEPA) filters, commonly used in industrial and commercial environments, can remove 99.97% or more of airborne particles $0.3\,\mu$ in diameter.

To remove gaseous impurities from the air, it is necessary to use "chemical" filters or scrubbers, which are typically composed of high surface area active materials, such as activated carbons, ion-exchange resin beds, molecular sieves, etc. The impurity gases passing through the filters are eliminated by selective adsorption (chemisorption or physicorption) on the surface of the active filtering media. The Donaldson Company, Inc. (Minneapolis) obtained numerous patents [56] on the air filter assembly for fuel cell applications. Stenersen et al. with the Donaldson Company disclosed a filter assembly (U.S. Patent 6,783,881) for intake air of fuel cell. The assembly includes a particulate filter portion, a chemical filter portion, and a noise suppression element. The noise suppression element is arranged so that the noise emanating from compressor impellers or blades is significantly attenuated. Both particulate filter and chemical filter need to be replaced or regenerated to avoid excessive pressure drop or the loss of the gas absorbing capacity. Schroeter [57] with Daimler AG disclosed an air filter assembly for a fuel cell. The patent disclosed an arrangement and method for regeneration of the particles and the chemical filter by monitoring the pressure drop and impurity gas concentration downstream.

However, the filters and scrubbers tend to add to the parasitic loss of the system due to increased air flow resistance. The trade-off needs to be considered for choosing proper air side filter assembly. Kennedy et al. [58] applied a programmable algorithm to an optimal design of cathode air filter by balancing compressor power requirement and efficiency loss from catalyst poisoning. They found that, due to their high contact efficiency, microfibrous materials provide the lowest pressure drop for cases requiring high contaminant removal and packed beds have the lowest pressure drop for applications requiring a high capacity. However, the bed depth of a packed bed increases substantially when meeting high removal requirements. The most favorable solution is provided through the use of an optimized composite bed for cases requiring both high contact efficiency and high capacity. Filters must be optimized for each contaminant type and concentration to achieve the required operating life of the fuel cell and to minimize pressure drop, which can consume as much as 1% of total output power.

Fuel side contaminants depend on the source of fuel. For fuel cell systems that draw high purity hydrogen from a storage system, fuel contamination is generally not a big issue. The burden of cleaning up hydrogen is shifted to the hydrogen production process. Section 9.2 of this chapter reviewed the contaminant mitigation techniques for hydrogen production processes. For fuel cell systems that draw fuel from a reformer, there is typically a residual amount of contaminants (ammonia, CO, CO_2, etc.) contained in the reformate stream. The details of the contaminants and their impact on the fuel cell performance have been described in the earlier chapters of this book. Certain contaminants tend to get trapped in the humidifier; these dissolved contaminants subsequently have to be removed from the water. Certain species, such as CO, that enter the fuel cell anode side can be actively removed by chemical and electrochemical methods, e.g., oxygen bleeding and potential cycling.

Polymer-based adhesives and sealants are used in PEM stacks in numerous places to separate fuel, oxidant, and coolant from offboard and inboard leakage (intermixing). The performance and the stability of sealants are critical in ensuring the long-term durability of PEM stacks. Several observed failure mechanisms of PEM stacks are related to seals or edge seal structures. In PEM stacks, sealant materials are exposed to a hot, humid, and acidic environment. As a result, contact pressure, which is necessary to maintain the gas-tightness of the seal, may relax due to the aging of sealant materials in such environments. Thermal cycling and freeze-thaw cycling also adversely affect the sealing pressure, causing gradual loss of hermetic sealing. Air that leaks in through the seal chemically interacts with hydrogen and Pt-based catalysts. Such interaction generates hydrogen peroxide, which in turn attacks the catalysts and the electrolyte membrane and causes gradual loss of electrode catalytic activities and the degradation of the membrane electrolyte. In the long term, chemical interaction between the polymer electrolyte and the sealant material may degrade the ionic conductivity of the electrolyte. For example, silicone-based gasket material can decompose via hydrolysis to produce siloxane compounds, which accelerate the degradation of the electrolyte membrane; fluorides and sulfate ions resulting from decomposing electrolyte membrane, in turn, accelerate platinum dissolution and cell performance reduction. Improvement of chemical and mechanical stability of sealant materials is necessary for improved long-term reliability of the fuel cell system.

9.3.4 Continuous Elimination of Contaminants within a Fuel Cell System

Water is produced as a byproduct in the electrochemical reaction of the hydrogen air fuel cells. For PEM fuel cells, water is needed in the humidifier. For a fuel cell system operating on natural gas, water is also needed in the WGS reactions. Water can also be used as the coolant in the fuel cell thermal management subsystem. As such, a water management subsystem is usually present in a fuel cell system to collect, store, and re-distribute water. Ideally, a fuel cell system should be water neutral, meaning no extra makeup water is necessary during normal operation of the fuel cell system. In such a close-circulated water system, contaminants tend to accumulate in the deionized water reservoir in the humidification and/or cooling subsystem. These include metal ions, ammonia, carbonates (dissolved carbon dioxide), organic materials, silica, iron oxide, trace metal ions, etc. St-Pierre et al. [59] studied the relationship between water management, contamination, and lifetime degradation of PEM fuel cell stack. They concluded that contaminants resulted in increased activation, ionic conduction, and mass transport losses after long-term operation of PEM fuel cells. Huang et al. [60] studied the effect of water management schemes on the membrane durability in PEM fuel cells. The authors reported the degradation behavior of polymer–electrolyte membrane fuel cells tested under two types of water management schemes: the solid plate (SP) cells that manage water via flow and dew point control of the reactant gases, and water transport plate (WTP) cells that actively remove liquid through microporous bipolar plates. The WTP experiments were conducted both with and without an ion filter in the water loop. It was found that the cell voltage decay rate was much lower when an ion filter was used.

It is advantageous to continuously clean up the contaminants from the reservoirs. Almost all fuel cell systems utilize ion filters or a dematerializing bed to clean up these contaminants in the water loop. Typical ion filters contain mixed resin beds that can clean up both cation and anion species. In addition to ion filters, Fisher et al. [61] with NuCellSys GmbH (Nabern, Germany) disclosed a filter to remove silica from water in the fuel cell water management subsystem. It is reported that silica, if not removed, may form insoluble metal silicates that may reduce the lifetime of the fuel cell system. Horinouchi et al. [62] patented a water treatment apparatus for a fuel cell system. The apparatus contains a decarbonation column that removes carbon dioxide gas in the exhaust gas condensates, a membrane filter that removes iron oxide from the recovered water, an adsorption device to remove iron ion, and an electrode-ionizer to remove trace ions

Another purpose of removing ionic species from the water or coolant loop is to maintain the desirable low ionic conductivity of these fluids to avoid shunt current, which is an internal short current (ionic) and also accelerates local corrosion. Typically, a bypass filtration loop through a deionizing (demineralizing) filter is used to continuously clean up a portion of the liquid. Replacement or regeneration of the ion filter or other filters is necessary to ensure the effectiveness of those filters.

For fuel cell systems operating on reformates, high levels of ammonia and CO_2 may be present in the fuel stream. Ammonia is known to be detrimental to PEM fuel cell catalysts. It is desirable to reduce the concentration of ammonia in the fuel stream to below 2 ppm. However, ion filters, if used alone, will be quickly saturated due to the high concentration of ammonia in the water. A water "contact cooler" has been used to transfer the ammonia into a reservoir of water, which needs to be dumped or cleaned. To achieve water neutral operation, Bonville et al with International Fuel Cells Corporation (Japan) [63] patented a method to remove the ammonia in the contaminated stack water via a steam stripping process, which removes ammonia around 400 ppm (and CO_2 as well) from contaminated water to about 30 ppm. An ammonia-laden steam is discharged to the environment. A demineralization bed is then used to further reduce ammonia concentration in the water to an acceptable low level.

9.3.5 Periodic Decontamination Procedure to Flush Out Contaminants from the Fuel Cell System

The "barrier-setting" approach for contaminant mitigation cannot be 100% effective, and contaminants will inevitably be generated within a fuel cell system due to material corrosion and degradation, so a small amount of contaminants will eventually accumulate inside a fuel cell system over time. Platinum catalysts and the ionomer membrane are two susceptible locations where gaseous and ionic contaminants tend to accumulate. The contaminants contribute to the ionic transport resistance across the membrane and reduce the rates of charge transfer reactions occurring at the anode and the cathode sides of the cell, among adverse others. Certain operations on the fuel cell stack can result in the adsorbed impurities on the catalyst surface to be oxidized and removed. A well-known method to recover from CO poisoning of an electrocatalyst is by periodic air bleeding. Discussion of this recovery method is given in section 9.4. It is frequently observed that the startup and shutdown process itself has the effect of restoring the open circuit voltage (OCV) of PEM fuel cells; the detailed mechanisms for recovery are still not clearly understood. Repeated potential cycling is another effective technique demonstrated for eliminating certain adsorbate on the platinum surface. It is desirable to periodically remove these contaminants by so-called decontamination procedures, which have been sought after by many fuel cell developers. The findings appear mostly in the patent literature.

Oko et al. [64] with Plug Power LLC (Latham, New York) disclosed methods and kits for removing contaminants from fuel cells. The procedure involves passing a wash solution through the fuel cell and analyzing the contaminants in the washed out solution, then selecting a removal substance targeted to the discovered contaminants and flushing the fuel cell with the removal substance. It is suggested that metallic contaminants (e.g., iron, magnesium, calcium, aluminum, cobalt, nickel ions) can be flushed out of the cell with an acidic solution or a chelating agent, sulfur-based contaminants (e.g., SO_x) can be removed by an alkaline solution, and organic contaminants can be removed by flushing the cells with a strong oxidant. It is also stated that many compounds can be removed to some degree by flushing the cells with deionized water. Performance data of a cell before and after flushing were not given, and the effectiveness of the flushing method was not clearly demonstrated.

Reiser et al. [65] disclosed a method for recovering fuel cell performance by cyclic oxidant starvation. The procedure involves shutting down or reducing the oxidant supply to a fuel cell power plant and drawing current from the fuel cell until the cell voltage drops to less than 0.1 volts, and repeating the procedure. Experimental data were presented demonstrating that fuel cell performance loss over long-term operation can be improved (rejuvenated) considerably: cell voltage improved by 20 to 50 mV across a wide range of current density; the improvement at high current density was found to be more significant. It was also demonstrated that voltage cycling was more effective than low voltage hold (at H_2 potential) on performance recovery. The experimental data also showed that the voltage decay rates reduced after the rejuvenation procedure. The method essentially cycled the potential of the platinum in the cathode catalyst from OCV to a low value close to hydrogen potential. It is suspected that the cyclic oxygen starvation not only caused cathode catalyst reduction, but might also help in removing contaminant species adsorbed on the platinum surface.

Patterson et al. [66] with UTC Fuel Cells, LLC (Irvine, California), disclosed a decontamination procedure for a fuel cell power plant. The procedure involves exposing the electrodes to flowing oxygen; to heated flowing oxygen; to a number of start–stop cycling; or to controlled potential. The procedure can be performed on one stack in a fuel cell system with two or multiple stacks without shutting down the entire power plant. Unlike the decontamination procedure disclosed by Plug Power [64], the procedure does not require any specialized cleaning liquid, or liquid pumping and valve system. The decontamination procedure can be carried out on a predetermined fixed schedule for each stack, or it can be executed when the stack power output decreases to a threshold level, e.g., 20% under a given operating condition. The procedure has been demonstrated on a hydrogen sulfide (10 ppm exposure on anode for 2 h) contaminated cell that has shown significant performance degradation. After exposing the anode and the cathode of the contaminated cell with an oxidant and applying a DC voltage of 0.4 V for about 5 min, the performance of the degraded cell was found to be fully recovered.

9.4 Onboard Fuel Cell Impurity Mitigation Strategies

9.4.1 Introduction

Proton exchange membrane fuel cells (PEMFCs) are promising energy sources for automobile, stationary, and portable power applications due to their many advantages, such as zero emissions and high efficiencies. However, the efficiency and long-term operation of PEMFCs are limited by contamination [67]. Impurities causing contamination can be introduced through the supplied gases (anode fuel and cathode oxidant), the fuel cell hardware (bipolar plates, end plates, connectors, and sealing gaskets), the supplemental accessories (coolants, saturators, tubes, gas compressors), and the fabrication residues or degradation products of the membrane electrode assemblies (MEAs), including gas diffusion layers (GDLs), catalyst layers, ionomers, and electrolyte membranes. Methods to mitigate contamination and to restore fuel cells that have been damaged by contamination are currently being developed to minimize or eliminate the effects.

Common impurities include those formed in reformate fuel streams (carbon oxides (CO_x), methane (CH_4), hydrogen sulfide (H_2S), and ammonia (NH_3)), pollutants found in air (sulfur oxides (SO_x), nitrogen oxides (NO_x), ammonia (NH_3), and organic compounds (propane, benzene, toluene)), catalyst fines carryover, rust from piping, salts, and dust. Impurities may adsorb onto the Pt surface (CO_x, H_2S, SO_x, and Cl^-), carbon support (H_2S and SO_x), or gas diffusion layers (salts, dust, and organic compounds), and adsorb into the ionomer (silica, cations (M^+), NH_4^+), or simply plug up the flow passages.

Contamination effects of impurities on a PEMFC may be classified into three categories: (1) kinetic effects, caused by adsorbing onto the catalyst surface and poisoning active sites on both the anode and cathode catalyst layers; (2) mass transfer effects, due to changes in the structure, pore size, pore size distribution, and hydrophobicity/hydrophilicity of the catalyst layers or gas diffusion layers; and (3) conductivity effects, caused by adsorbing into the membrane and ionomer in the catalyst layer, increasing proton resistance.

This section focuses on the contamination mitigation strategies on the fuel stream and the oxidant gas (air). Mitigation strategies on other impurities (including silica, organic compounds, cations, and solid particles) will not be discussed.

9.4.2 Fuel Side Contamination Mitigation

As mentioned in section 9.2, hydrogen is mainly produced through the reformation of hydrocarbons and/or oxygenated hydrocarbons [68–70], although it is also produced by electrolysis, partial oxidation of small organic molecules [71], and hydrolysis of sodium borohydride [72]. Production of hydrogen through reformation unavoidably results in impurities, such as CO_x [73,74] and H_2S [75], which can cause significant poisoning of the anode catalysts for the hydrogen oxidation reaction (HOR) in fuel cell systems. Ammonia (NH_3) can also be found in the reformate gas due to its use as a tracer gas in natural gas distribution systems. Another issue is the heavy hydrocarbons that are formed during fuel cell startup.

9.4.2.1 CO Mitigation

The major contaminants in PEMFC anodes are CO and CO_2, which are produced by reforming or partial oxidation of hydrocarbons and, thus, enter cells through the anode fuel stream. CO poisons the anode catalyst even at very low concentrations (10 ppm). CO poisoning is caused by preferential adsorption of CO onto the catalysts, forming a CO monolayer that blocks active sites for HOR, therefore, reducing cell voltage. CO is more strongly bonded to Pt than hydrogen and has a greater potential required for oxidation than hydrogen [76]. At low CO concentrations (<100 ppm), the observed PEMFC performance degradation can be explained by bridge-site adsorption of CO, followed by electrochemical oxidation of CO to CO_2. Any CO_2 impurity in the fuel adsorbs weakly on the Pt surface. At higher CO_2 concentrations, the decrease in cell potential due to dilution of the H_2 also contributes to the performance degradation [77].

The general approaches to mitigate CO poisoning are to either generate an oxidative environment so that the adsorbed CO can be removed by oxidation to CO_2 or to minimize CO adsorption. There are several methods available to eliminate CO poisoning in PEMFCs, such as introducing an anode oxidant bleed, developing CO tolerant catalysts, and optimizing fuel cell operating conditions [78].

9.4.2.1.1 Oxidant Bleeding

Injection of air or oxygen, called *oxidant bleeding*, into the fuel stream ahead of the fuel inlet of the fuel cell is a method that has been employed to oxidize the

CO from the catalyst surface and to restore the cell performance [77,79–86]. Oxidant bleeding is a promising method in reducing CO contamination due to its simplicity, effectiveness, and economic value. Blending low levels of an oxidant into the anode fuel stream reduces the levels of CO in the fuel by the WGS reaction mechanism and selective oxidation of CO.

Gottesfeld and Passford [79] reported that the addition of 2 to 5% O_2 to a H_2 fuel stream resulted in restoration of CO-free cell performance for a fuel cell fed with CO at levels as high as 500 ppm. Murthy et al. [80] also found that a 5% air bleed resulted in almost a complete recovery in cell performance for a CO level of 500 ppm at 70°C and atmospheric pressure. However, 5 and 10% air bleed were not sufficient to improve the cell performance in the presence of 3000 ppm CO, and even 15% air bleed resulted in only partial recovery with 3000 ppm CO. As opposed to uninterrupted oxidant bleeding, Chen et al. [81] found that dosing air for 10 s intervals of 10 s yielded similar results (i.e., complete recovery) to that of continuous air bleeding for CO concentrations of 20 ppm and 52.7 ppm in the anode fuel stream.

Although air and oxygen bleeding is effective, only one out of every 400 O_2 molecules oxidizes an adsorbed CO molecule [82] and the remaining oxygen chemically combusts with the hydrogen. The combustion reaction lowers the fuel efficiency, and might also accelerate the sintering of catalyst particles, resulting in a performance decline with time [79,83–85]. As an alternative to using air and oxygen bleeding, adding hydrogen peroxide (H_2O_2) to the fuel stream also minimizes CO contamination [86]. H_2O_2 evolves oxygen under operating conditions of the fuel cell according to reaction (9.92).

$$H_2O_2 \rightarrow H_2O + \tfrac{1}{2} O_2 \qquad\qquad (9.92)$$

Using H_2O_2 offers several advantages over oxygen and air. For example, it is possible to store liquid hydrogen peroxide separately and add it to the humidification water. This indirect oxygen bleeding prevents the safety issues that arise when handling H_2/O_2 gas mixtures in the presence of highly active metal catalysts for the nonelectrochemical recombination of H_2 and O_2. The addition of 5% H_2O_2 to an anode humidifier restored cell performance to a CO-free hydrogen level when 100 ppm CO in a H_2 feed was used [87]. Bellows et al. [82] demonstrated that even 0.75% H_2O_2 in the anode humidifier eliminated 96 ppm CO in an impure H_2 feed.

Although interesting in a laboratory setting, the addition of oxidant, particularly H_2O_2, to the fuel stream can result in degradation and stability issues. Inaba et al. [88] reported that the presence of H_2O_2 in a cell, either added to the fuel or formed through the reaction of oxygen and hydrogen, affected the long-term durability of the fuel cell. Introduction of a small amount of O_2 into the H_2 atmosphere at the anode produced a large amount of H_2O_2 according to reaction (9.93) to reaction (9.95). H_2O_2 may cause serious membrane degradation in long-term operation with air bleeding.

$$H_2 + Pt \rightarrow Pt\text{-}H \text{ (at anode)} \tag{9.93}$$

$$Pt\text{-}H + O_2 \text{ (added to fuel stream or diffused through PEM to anode)} \rightarrow$$
$$\cdot OOH \tag{9.94}$$

$$\cdot OOH + Pt\text{-}H \rightarrow H_2O_2 \tag{9.95}$$

Inaba et al. [88] carried out a long-term test (4600 h) of a fuel cell with 5% air bleeding to investigate the effects of air bleeding on membrane degradation in PEMFCs. The rate of membrane degradation was negligible (1.3 × 10^{-10} mol cm^{-2}h^{-1}) up to 2000 h, but increased after 2000 h. The CO tolerance of the anode gradually dropped, indicating that the anode catalyst deteriorated during the test. It was concluded that air bleeding degraded the anode catalyst, which enhanced H_2O_2 formation upon air bleeding, resulting in increased membrane degradation rate after 2000 h.

In summary, although it has been shown to be quite effective in eliminating CO poisoning from PEMFCs, oxidant bleeding of the fuel streams has several drawbacks. Adding oxygen to the fuel stream allows it to chemically combust with the hydrogen, which lowers the fuel efficiency. Addition of both air and H_2O_2 to the fuel stream has been shown to accelerate MEA degradation, resulting in a performance decline with time. For air bleeding to be efficient, precise system control is needed because CO output from reformer systems is not constant and depends on operating conditions.

9.4.2.1.2 CO-Tolerant Catalysts

Although platinum is a very efficient catalyst for hydrogen oxidation, it is readily poisoned by low levels of carbon monoxide (ppm levels), leading to a rapid degradation of activity. Researchers have attempted several approaches to reduce the problems associated with CO contamination of the anode catalyst, including the utilization of one or more other metals in Pt-based anode catalysts [89–113], using two or three layer anode configurations [114–118], and improving catalyst preparation methods [119–121].

The most widely investigated method of making CO-tolerant catalysts has been alloying or co-depositing Pt with other metals. Combining Pt with other metals enables the catalysts to form oxygenated species (metal-OH) at potentials lower than that with pure Pt. The use of the other metal(s) mitigates CO contamination through a "bifunctional mechanism" [110], where adsorbed CO species are oxidized by OH species generated on the other metal surface atoms or by electronic effects where the presence of the other metal involves a change in the electronic density state of platinum, leading to the weakening of the CO-Pt bond, or a mixing of both effects. The metal-OH species is a source of oxygen, which allows the oxidation of adsorbed CO to CO_2 and frees Pt sites, where the adsorption and oxidation of hydrogen can take place.

PtRu alloy catalysts, which have been commercially used in the PEMFC industry, have been the most widely investigated and have been demonstrated to be highly successful CO-tolerant catalysts, but they are not stable. The use of PtRu alloys as anode catalysts allows CO to be oxidized at more negative potentials compared to pure Pt [89–90]. Watanabe et al. [96] found that alloying with Co or Ru modified the electronic structure of Pt atoms, which resulted in weakened Pt-CO interactions. CO adsorption became weaker in the order Pt>Pt-Co>Pt-Ru [91]. Using PtRu catalysts in combination with oxidant bleeding also yielded promising results. Haug et al. [92–93] placed a layer of Ru on a Pt anode and this anode structure increased the CO tolerance when oxygen was added to the fuel stream. Gubler et al. [94] reported that if PtRu was used as anode catalyst, 1% of oxygen was sufficient to achieve full performance recovery when H_2 + 100 ppm CO was used as fuel, whereas 2% O_2 was required using pure Pt in the cell hardware.

Luna et al. [95] demonstrated that the catalytic activity of PtRu/C is intimately related to the preparative process. These researchers obtained PtRu/C catalysts by thermally treating the sample at 300°C in H_2 atmosphere. With this catalyst and using H_2 + 100 ppm CO, they obtained cell performances similar to that of the commercial E-Tek (BASF Fuel Cell, Inc., New Jersey) PtRu/C samples with pure hydrogen.

Because ionic ruthenium has stability issues, in that it diffuses into the polymeric membrane and exchanges onto the sulfonic acid, increasing the ohmic resistance, Pozio et al. [97] developed an alternate method to prepare PtRu/C catalysts. These researchers combined carbon black powder, $Pt(NH_3)_4Cl_2$, $RuNO(NO_3)_x(OH)_y$, and NaOH, and then added the reducing agent sodium borohydride, resulting in a catalyst that, when tested in a fuel cell, yielded CO tolerance and electrochemical performance in hydrogen similar to that obtained when fuel cell testing the commercial E-Tek catalyst. The new catalyst was more stable in ethanol and methanol even if heated or ultrasonicated [97].

Other Pt-based alloy catalysts have also displayed high tolerance to CO poisoning, such as:

- Binary (PtM where M = Mo, Nb, Ta, Sn, Co, Ni, Fe, Cr, Ti, Mn, V, Zr, Pd, Os, Rh) [98–107]
- Ternary (PtRuM where M = Mo, Nb, Ta, Sn, Co, Ni, Fe, Cr, Ti, Mn, V, Zr, Pd, Os, Rh) [99–106]
- Quaternary ($PtRuM_1M_2$ where M = Ru, Mo, Nb) [99–106]
- Pt-based metal oxide catalysts ($PtWO_x$, $PtRuSnO_x$, $PtRuWO_x$) [107]
- Pt-based composite supported ($PtRu-H_xWO_3/C$, $PtRu-H_xMoO_3/C$) [108]
- Pt-based organic metal complexes [109]

For example, Garcia et al. [110] achieved high performance with H_2/O_2 PEMFCs using anode catalysts PdPt/C and PdPtRu/C with 0.4 mg metal cm^{-2}

loading. When using these catalysts and feeding the anode with H_2 + 100 ppm CO, the polarization curves showed higher tolerance to CO compared to when Pd/C and Pt/C were used as anode catalysts. No CO_2 formation was seen at the PEMFC anode outlet for the PdPt/C catalysts, indicating that CO tolerance improved due to increased free sites for H_2 oxidation, likely due to a weaker Pt-CO interaction compared to pure Pt. CO_2 was present in the anode outlet for PdPtRu/C, indicating that this catalyst oxidized the CO [110]. For PtCu anode catalysts, although the current densities were much smaller, the rate of CO consumption was comparable to that on Pt anodes [111].

Despite having sufficient activity for HOR and high CO tolerance, platinum-based alloys (such as PtRu) have limited long-term stability, especially under nonoptimal operating conditions [112]. Christian et al. have overcome this problem through the use of tungsten-based catalysts, which had catalytic activity toward HOR approaching 25% of that with pure platinum as well as high tolerance to both CO and H_2S [113].

Other anode configurations with two to three layer electrodes were also explored to improve CO tolerance [83,94,114–118]. Most of these electrodes contained an inner layer of Pt and an outer layer of PtRu. In this bilayer anode, the region in which the oxygen with CO reaction occurred was moved to the gas phase catalyst layer, thereby reducing the poisoning of the catalyst and enhancing the tendency toward O_2 bleeding in the cell hardware [83,94,115]. The outer layer acted as a CO barrier. This anode catalyst layer structure exhibited superior performance and CO tolerance, which was attributed to the filtering effect of the anode catalyst layer.

Janssen et al. [116] achieved improved CO and CO_2 tolerance utilizing an alternative bilayer anode, containing PtMo on the inner layer and PtRu on the outer layer. In this alternative bilayer anode structure, the inner layer enabled CO oxidation at low potentials and had a lower rate of H_2 oxidation, while the outer layer enabled faster H_2 oxidation and had limited CO adsorption. Materials like PtRu do not catalyze the oxidation of CO at low potentials and are relatively inactive for the reverse water gas shift (WGS) reaction. The PtMo/PtRu bilayer structure showed improved CO tolerance when operated with H_2 + CO [116]. In another alternative bilayer structure, Shi et al. [117] placed Pt or Au particles in the diffusion layer, allowing the chemical oxidation of CO to occur at the Pt or Au particles before it reached the catalyst layer when trace amounts of oxygen were injected into the anode. This structure also avoided the disadvantageous heating effects to the catalyst and membrane caused by the oxygen injection. All MEAs composed of Pt or Au diffusion layers performed better than the traditional MEAs when 100 ppm CO + H_2 and 2% air were fed. Without oxygen injections, when 100 ppm CO + H_2 was fed, MEAs composed of Au-refined diffusion layers had slightly better performance than traditional MEAs because Au particles in the diffusion layer have activity in the WGS reaction at low temperatures [117]. Uribe et al. also formed a different bilayer structure [118], placing a composite film onto the GDL that was able to catalyze the oxidation of CO with O_2 to form CO_2.

Additionally, improvements in catalyst preparation methods have resulted in better control of particle-size distribution and the chemical composition of catalysts. CO tolerance was enhanced by Pt-based catalysts prepared using high energy ball milling [119], sulfiding (synthesized using $Na_2S_2O_3$) [120], and combustion synthesis [121].

In summary, PtRu catalysts have been the most widely investigated and have provided the most effective results in terms of improving CO tolerance, with cell performance comparable to that of using pure Pt. PtRu catalysts are often used in combination with oxidant bleeding. However, these PtRu catalysts can suffer from lack of stability, which can be overcome through alternative preparation methods or alternative catalyst configurations. Layered electrode configurations are promising in that they can completely shield the Pt catalyst layer from the CO contamination.

9.4.2.1.3 Optimizing Operating Conditions

Fuel cell operating conditions, such as flow rate [122], temperature, relative humidity, and pressure, greatly affect CO contamination. An increase in flow rate increases CO contamination. CO adsorption is also strongly favored at lower temperatures due to a high negative value of standard entropy and standard free energy for CO adsorption. CO adsorption is increased at lower relative humidity due to lower H_2O and OH coverage and at higher anode pressure due to higher partial pressure of CO and lower partial pressure of water. Higher cathode pressure results in decreased CO adsorption due to increased CO oxidation [122,123].

With consideration of the effects of operating conditions on CO contamination, Kawatsu [124] developed a method that employed the adjustment of temperature and pressure to maintain the partial pressure of hydrogen within a predetermined range, thus maintaining the cell performance when a fuel stream with harmful levels of CO was used.

CO poisoning is highly temperature dependant and, as a result, temperature has been the most widely investigated operating condition with regard to CO poisoning. High-temperature operation has a benefit of reducing the effect of CO contamination, in addition to other benefits, such as faster kinetics of HOR and ORR and improved water management [125,126]. For example, when operating at high temperatures, CO tolerance was greatly improved from 10 to 20 ppm at 80°C to 1000 ppm at 130°C, and 30,000 ppm at 200°C [125].

Murthy et al. [73] investigated the effects of both temperature and pressure on CO tolerance in a PEMFC. For 500 ppm CO + H_2 mixtures without air bleeding, the fuel cell operated at 202 kPa and 0.6 V exhibited a steady-state current density of 1.0 A cm^{-2} at 90°C, but only 0.4 A cm^{-2} at 70°C. At 101 kPa and 70°C, exposure to 500 ppm CO + H_2 mixtures required 5% air bleeding to obtain a current density of 1.0 A cm^{-2} at 0.4 V. Transient experiments with these CO levels indicated that there was up to a four-fold decrease in the poisoning rate at 202 kPa versus 101 kPa. At 202 kPa, increasing the cell temperature from 70 to 90°C resulted in approximately a 14-fold decrease in the poisoning rate for 3000 ppm CO + H_2 mixtures and approximately four times decrease for 10,000 ppm CO + H_2 mixtures [73].

CO contamination can be greatly diminished through careful control of fuel cell operating conditions, such as temperature, pressure, flow rate, and relative humidity. However, complications are associated with the adjustment of any of these parameters to achieve improved CO tolerance. For example, increasing the temperature can dehydrate the PEM, resulting in loss of conductivity. A key issue with high temperature operation is to develop alternative PEMs that can be operated at temperatures higher than 100°C. Adjustments of operating conditions are seemingly straightforward changes to make to fuel cells, but become complicated to carry out when ensuring that the cell performance will not diminish as a result of the harsher operating conditions.

9.4.2.1.4 *Other CO Contamination Mitigation Methods*

Current Cycling—This is a method that has been used to restore the catalyst from CO poisoning. Carette at al. cycled the current and raised the anode overpotential to enhance the electrochemical oxidation of CO [85]. Lakshmanan et al. [127] utilized pulsing techniques for the electrochemical oxidation of CO, using a PEMFC as a flow reactor for continuous preferential oxidation of CO over H_2 from 1% CO in H_2 under pulse-potential control. By varying the pulse profile, the CO and H_2 oxidation currents were varied independently. Zhang et al. [128] proposed an electrochemical preferential oxidation process for selectively removing CO from H_2-rich gas by utilizing the anode potential oscillations in a device with a structure similar to that of a PEMFC. Current cycling allowed almost complete recovery of the catalyst from CO contamination. Although current cycling is effective in removing CO from the catalysts, it can lead to the dissolution of Pt particles promoted by the change of potential from low to high [129].

Fuel Starvation—Starvation of fuel can improve performance of a PEMFC exposed to CO [130]. Periodic fuel starvation causes the anode potential to increase and allows the oxidation and removal of the catalyst poisons from the anode catalysts, improving fuel cell performance. The preferred method has successive localized portions of the fuel cell anode momentarily and periodically fuel starved, while the remainder of the fuel cell anode remains electrochemically active and saturated with fuel so the fuel cell can continually generate power. However, when the cell is deprived of fuel, cell voltages can become negative as the anode is elevated to positive potentials and the carbon is consumed (carbon corrosion) instead of the absent fuel [131]. When this happens, the anodic current is generally provided by carbon corrosion to form carbon dioxide, resulting in permanent damage to the anode catalyst layer.

9.4.2.2 *H_2S Mitigation*

Similar to CO, hydrogen sulfide is usually present as an impurity in the fuel stream of a fuel cell and strongly adsorbs on the Pt catalyst. Under the same conditions, the H_2S poisoning rate is much higher than CO because H_2S adsorbs more strongly than CO. The rate of H_2S poisoning increases with increases in the H_2S concentration and H_2 flow rate. The poisoning rate of H_2S decreases when cell temperature increases and the level of anode humidification does not greatly affect the poisoning rate [134]. There are not as many successful methods for eliminating H_2S contaminants from fuel cells as there are for CO contaminants.

Although the performance of the CO-poisoned cell can be recovered by introducing neat H_2, when neat H_2 is introduced to an H_2S-poisoned cell, the recovery is only partial. Shi et al. [132] tested the contamination effect of multiple contaminants (25 ppm CO + 25 ppm H_2S) in the fuel stream. It was found that the fuel cell performance deteriorated more quickly than 50 ppm CO alone, but more slowly than 50 ppm H_2S alone, and the performance could only be partially recovered by reintroducing neat H_2.

Even though high temperature operation is a promising measure of mitigating CO poisoning, it only slows down H_2S poisoning. Mohtadi et al. [133] reported that sulfur adsorbed more strongly at a cell operating temperature of 50°C than at 90°C, but sulfur eventually saturated the Pt at 50, 70, and 9°C after prolonged exposure to 5 ppm H_2S in H_2.

While using novel catalysts has led to some success in the mitigation of CO poisoning, the presence of the Ru in a Ru-Pt alloy catalyst did not provide tolerance against H_2S poisoning of the anode catalyst [75]. After H_2S poisoning, the Ru-Pt alloy catalyst demonstrated lower recovery in neat H_2 compared to the Pt anode, which was attributed to the lower Pt loading of the Ru-Pt alloy catalyst.

Current cycling has been the most effective measure to recover the fuel cell performance from H_2S poisoning. The adsorption of H_2S on the anode is dissociative and this dissociation can produce adsorbed sulfur. The dissociation potentials of H_2S are ~0.4 V at 90°C, 0.5 V at 60°C, and 0.6 V at 30°C. The adsorbed sulfur can be oxidized at higher potential [134]. Cyclic voltammetry (CV) spectra, obtained by Mohtadi et al., revealed the presence of two sulfur species at 70°C and 95% recovery of the original performance was obtained after a CV treatment beyond the oxidation potential of sulfur [75]. Similarly, Shi et al. recovered the performance of the H_2S poisoned Pt anode by applying a high voltage pulse (1.5 V for 20 s) followed by a low voltage pulse (0.2 V for 20 s) in a single cell. During 10 poisoning-recovery cycles, the ohmic resistance and electrochemical surface area did not change significantly [135].

9.4.3 Air Side Contamination Mitigation

The effects of air impurities, such as SO_2, NO_2, H_2S, and Cl^-, on the performance of PEMFCs have not been extensively studied in the literature, although it is believed that continual exposure to low concentration of these impurities is detrimental to cell performance. Contamination mitigation on the air side gases will mainly focus on sulfur-containing compounds, nitrogen oxides (NO_x), and chloride ion (Cl^-). These impurities can lead to serious poisoning effects on the cathode electrocatalysts for ORR.

9.4.3.1 Cl^- Mitigation

Fuel cell catalysts are often synthesized from halide containing precursors, such as Cl^-, which are not always completely removed after synthesis. Chloride is also a likely contaminant brought into the fuel cell through the humidifier. Schmidt et al. reported that the presence of Cl^- affected the ORRs on both the Pt(111) and Pt(100) planes, resulting in reduced ORR activity and increased H_2O_2 production [136].

Matsuoka et al. reported that chloride caused significant degradation of the cell performance, oxygen gain, and a 30% loss of ECA. It promoted the dissolution of Pt in the inlet side of the MEA and Pt particles, whose sizes were 3 to 200 nm, and formed a Pt band in the membrane. The partial loss of Pt in the inlet-side of the MEA led to increases in local current density at the middle and outlet sides of the MEA, which could have caused the observed degradation of cell voltage and oxygen gains [137].

In an attempt to recover a cell poisoned by Cl^-, Zhao et al. reported that flushing the electrode with hot water for 20 h did not remove the adsorbed Cl^- [137], indicating that Cl^- is difficult to remove once it is in a fuel cell. Therefore, Cl^- should be avoided in catalyst synthesis and highly pure water should be used for the humidifier.

9.4.3.2 Sulfur Compounds Mitigation

Sulfur compounds, such as SO_2, H_2S, and COS, significantly degrade the performance of cathode Pt catalysts when they are present in the air stream of PEMFCs. Sulfur compounds deactivate catalysts by adsorbing onto active sites, forming strong chemical bonds between the sulfur atoms and the active metal catalysts [139]. Loss of performance due to the presence of sulfur compounds on the air side can be attributed to kinetic losses, for example, due to the adsorption of sulfate generated by oxidation of sulfur species [140]. Two sulfur species are formed on the Pt cathode after exposure to either SO_2 or H_2S, identified as strongly and weakly adsorbed sulfur on the Pt cathode [141].

Sulfur can be removed from the cathode catalyst through electrochemical cycling, flushing with water, and exposure to pure air. Sulfur species adsorbed on the Pt can be electrochemically oxidized to water-soluble sulfate by a six electron process at potentials greater than 0.8 V versus RHE [142,143]:

$$Pt\text{-}S^\circ + 4H_2O \rightarrow SO_4^{2-} + 8H^+ + 6e^- + Pt \tag{9.96}$$

The resulting sulfate desorbs from the catalyst surface, allowing recovery of the catalyst performance. Gould et al. [144] examined the deactivation behavior of a commercial Pt/C catalyst when the cathode was exposed to three different sulfur species: SO_2, H_2S, and COS in air at a constant voltage of 0.6 V. The loss of cell performance over time was the same for all three species. The drop in cell performance over time leveled off as exposure time increased and the surface became saturated when 45% of the Pt surface was covered by adsorbed sulfur. CV was a successful means of removing sulfur and restoring 92% of the power density. Using successive polarization curves led to an 85% recovery of power density [144]. Similarly, Jing et al. reported that the cell voltage dropped by 240 mV after exposure to 1 ppm SO_2 over 100 h and 84% of the cell performance was recovered after CV scanning [145]. In an alternative recovery route, Nagahara et al. achieved almost complete performance recovery from sulfur compound poisoning by using high relative humidity (RH) to flush the sulfate from the catalyst layer [146].

9.4.3.3 NO₂ Mitigation

NO_2 in the air stream degrades the performance of the fuel cell. The poisoning effects of NO_2 are mainly due to the overlapping of the ORR, NO, and HNO_2 oxidation reactions, and increased cathodic impedance [141,147].

NO_2 can be eliminated from the cell by applying clean air for 12 to 24 h [141,147]. Another method that allows a fuel cell to recover from NO_2 poisoning is CV scanning. Using an air mixture with impurities similar to ambient impurity levels, i.e., 0.8 ppm NO_2 + 0.2 ppm NO + 1 ppm SO_2, Jing et al. discovered that the cell voltage decreased by 160 mV and the voltage recovered by 94% after CV scanning [148].

9.4.4 Ammonia Mitigation

Ammonia (NH_3) or ammonium (NH_4^+) can exist in both the fuel and air streams. The diffusion of ammonium is fast, therefore, the ammonium entering the fuel cell from either side can quickly diffuse to the other side causing the contamination effect on both sides. For instance, for a typical membrane with a thickness of 10 to 100 μm, the estimated characteristic time constant for diffusion is 1 to 100 sec [149]. Ammonia may affect the PEMFC performance in different ways: (1) by the reduction of the ionic conductivity of the membrane, which in its ammonium form is a factor of 4 lower than in the protonated form [149–151]; (2) by poisoning the cathode catalyst [151]; and (3) by poisoning the anode catalyst [149]. Recently, fuel cell tests have shown that the reduced membrane conductivity is not the major reason for performance losses induced by ammonia [149,150]. The effect of ammonia on the HOR was found to be minor at current densities below 0.5 A cm^{-2}, but would increase with increasing current densities. The current density did not exceed 1 A cm^{-2} in the presence of ammonia [149].

Several different efforts to recover cells exposed to NH_3 have been unsuccessful. Although it seems that the replacement of H^+ by NH_4^+ ions within the

PEM and catalyst layers is a primary reason for cell performance drop, the addition of acid more than 10 times the original amount of base (NH_3) into the anode did not affect the recovery rate. Bleeding 6% air in the fuel stream containing 30 ppm NH_3 also had no effect on the recovery rate [153]. There was no difference in NH_4^+ poisoning using either a Pt or a Pt-Ru catalyst [149].

The exposure time of the fuel cell to ammonia affects the fuel cell's ability to recover. With short times of exposure (1 to 3 h), the original performance can be fully recovered. Longer exposure times (15 h) decrease the cell performance to unusable levels and the cell does not recover even after several days of operation on pure H_2 [148,152]. Uribe et al. found that the cell resistance started to increase only when the cell had been exposed to ammonia for more than 1 h, and the cell resistance more than doubled when the cell was exposed to 30 ppm ammonia for 15 h. Exposure to 30 ppm NH_3 for 17 h was not fully recoverable within 4 days of operation on pure hydrogen [153].

The only method that has served to mitigate contamination by ammonia is by passing the fuel stream through a tube containing a H^+ exchange resin, which prevented ammonia contamination of the fuel cell [153].

9.4.5 Summary and Conclusion

The most straightforward method of eliminating harmful effects of contamination on both the air and fuel sides of PEMFCs is to remove the impurities from the gas streams and component materials before they enter the cell. This could be accomplished through the development of appropriate filtration devices in the air stream and the pretreatment of the reformate in the fuel stream. However, with this type of contamination mitigation strategy it is difficult to remove the contaminants to values less than 10 to 50 ppm, which are still harmful levels. Therefore, it is necessary to develop contamination mitigation strategies within the fuel cell.

Great measures have been developed for CO contamination mitigation in the fuel stream. Oxidant bleeding, CO tolerant catalysts, and optimizing fuel cell operating conditions have all been found to eliminate the harmful effects of CO. Cycling the current and fuel starvation allow fuel cells to recover from CO contamination. A combination of these contamination mitigation strategies would result in the elimination of most CO poisoning effects.

Other fuel cell contaminants have not been as well studied as CO and it is often more difficult to eliminate their effects on a fuel cell. H_2S contamination is somewhat reversible when the contaminant is removed from the fuel stream and the performance can be restored to 95% of the original value through current cycling. Similarly, NO_2 poisoning is reversible by removing the contaminant from the air stream and current cycling restores the performance to 99% of the original value. However, when using current cycling as a recovery strategy, it is important to remember that platinum dissolves rapidly when the fuel cell experiences transitions from low to high potentials. There are many methods of restoring a PEMFC suffering from sulfur contamination from the air side, such as by flushing with water, current cycling, or using pure air. With NH_3 exposures of less than 3 h, complete recovery occurs by applying pure gases. However, the only successful method that has been found to remove ammonia poisoning after exposure for longer than 3 h is by passing the gas stream through a H^+ exchange column. The most difficult contaminant to remove seems to be Cl^-, therefore Cl^- should be avoided in catalyst synthesis and highly pure water should be used for the humidifier. However, completely preventing PEMFCs from contacting Cl^- is difficult due to the inevitable presence of Cl^- in the atmosphere in coastal regions and cold areas where road deicers are often used.

Future work in contamination mitigation in PEMFCs should include improving gas stream filtering methodologies as well as developing more contamination tolerant fuel cell components, such as the catalyst, catalyst layers, and membranes.

References

1. Chemical composition of natural gas: http://www.uniongas.com/aboutus/aboutng/composition.asp (accessed May 19, 2009).
2. U.S. Department of Energy, Energy Efficiency and Renewable Energy, Hydrogen, Fuel Cells and Infrastructure Technologies Program–Hydrogen Production: http://www1.eere.energy.gov/hydrogenandfuelcells/production (accessed May 19, 2009).
3. Ogden, J.M. 2001. Review of small stationary reformers for hydrogen production, Princeton University Center for Energy and Environmental Studies, a report for the International Energy Agency: http://www.afdc.energy.gov/afdc/pdfs/31948.pdf (accessed May 19, 2009).
4. York, A.P.E., T. Xiao, and M.L.H. Green. 2003. Brief overview of the partial oxidation of methane to synthesis gas. *Top. Catal.* 22: 3–4.
5. Ogden, T., S. Kreutz, Kartha, and L. Iwan. 1996. Assessment of technologies for producing hydrogen from natural gas at small scale. *Princeton University Center for Energy and Environmental Studies Draft Report*, Nov. 26, Princeton, NJ.
6. Lee, S.H.D., D. Applegate, S. Ahmed, and S. Galderone. 2005. Hydrogen from natural gas: Part I, Autothermal reforming in an integrated fuel processor. *Int. J. Hydrogen Energy* 30: 829–842.
7. Bharadwai, S.S. and L.D. Schmidt. 1995. Catalytic partial oxidation of natural gas to syngas. *Fuel Process. Technol.*, 42: 109–27.
8. Heinzel, A., B. Vogel, and P. Hubner. 2002. Reforming of natural gas–Hydrogen generation form small scale stationary fuel cell systems. *J. Power Sources* 105: 202–7.
9. Dybkjaer, I. 1995. Tubular reforming and autothermal reforming of natural gas: An overview of available processes. *Fuel Process. Technol.* 42: 85–107.
10. Freni, S., G. Galogero, and S. Cavallaro. 2000. Hydrogen production from methane through catalytic partial oxidation reactions. *J. Power Sources* 87: 28–38.

11. Dauenhauer, P.J., J.R. Salge, and L.D. Schmidt. 2006. Renewable hydrogen by autothermal reforming of volatile carbohydrates. *J. Catal.* 244: 238–47.
12. Steinberg, M. 1999. Fossil fuel decarbonization technology for mitigating global warming. *Int. J. Hydrogen Energy* 24: 771.
13. Muradov, N.Z. and T.N. Veziroğlu. 2005. From hydrocarbon to hydrogen-carbon to hydrogen economy. *Int. J. Hydrogen Energy* 30: 225–237.
14. Aiello, R., J. Fiscus, H. Loye, and M. Amiridis. 2000. Production via direct cracking of methane over Ni/SiO₂: Catalyst deactivation and regeneration. *Appl. Catal., A* 15: 1528–34.
15. Shah, N., D. Panjala, G. Huffman. 2001. Hydrogen production by catalytic decomposition of methane. *Energy Fuels* 15: 1528–34.
16. Muradov, N. 1998. CO_2-free production of hydrogen by catalytic pyrolysis of hydrocarbon fuel. *Energy Fuels* 12: 41.
17. Muradov, N. 2001. Catalysis of methane decomposition over elemental carbon. *Catal. Commun.* 2: 89.
18. Ahmed, S., A. Aitani, F. Rahman, A. Al-Dawood, and F. A.-Muhaish. 2009. Decomposition of hydrocarbons to hydrogen and carbon. *Appl. Catal., A* 359: 1–24.
19. Huang, C. and A. T-Raissi. 2007. Thermodynamic analyses of hydrogen production from sub-quality natural gas, Part I: Pyrolysis and autothermal pyrolysis. *J. Power Sources* 163: 645–52.
20. Huang, C. and A. T-Raissi. 2007. Thermodynamic analyses of hydrogen production from sub-quality natural gas, Part II: Steam reforming and autothermal steam reforming. *J. Power Sources* 163: 637–44.
21. Huang, C. and A. T-Raissi. 2008. Liquid hydrogen production via hydrogen sulfide methane reforming. *J. Power Sources* 175: 464–72.
22. Demirbaş, A. 2001. Biomass resource facilities and biomass conversion processing for fuels and chemicals. *Energy Convers. Manage.* 42: 1357–78.
23. Ni, M., D.Y.C. Leung, M.K.H. Leung, and K. Sumathy. 2006. An overview of hydrogen production from biomass. *Fuel Process. Technol.* 87: 461–72.
24. Levin, D.B., L. Pitt, and M. Love. 2004. Biohydrogen production: Prospects and limitations to practical application. *Int. J. Hydrogen Energy* 19: 173.
25. Xuan, J., M.K.H. Leung, D.Y.C. Leung, and M. Ni. 2009. A review of biomass-derived fuel processors for fuel cell systems. *Renewable Sustainable Energy Rev.* 13: 1301–13.
26. Shoko, E., B. McLellan, A.L. Dicks, and J.C. Diniz da Costa. 2006. Hydrogen from coal: Production and utilization technologies. *Int. J. Coal Geo.*, 65: 213–22.
27. Lin, S.Y. 2009. Hydrogen production from coal. In *Hydrogen fuel: Production, transport, and storage,* 103–125, ed. R.B. Gupta. CRC Press: Boca Raton, FL.
28. Fan, L., F. Li, and S. Ramkumar. 2008. Utilization of chemical looping strategy in coal gasification processes. *Particuology* 6: 131–141.
29. Gnanapragasam, N.V., B.V. Reddy, and M.A. Rosen. 2009. Hydrogen production from coal using direct chemical looping and syngas chemical looping combustion systems: Assessment of system operation and resource requirements. *Int. J. Hydrogen Energy* 34: 2606–2615.
30. Kudo, A. and Y. Miseki. 2009. Heterogeneous photocatalyst material for water splitting. *Chem. Soc. Rev.* 38: 253–78.
31. Osterloh, F.E. 2008. Inorganic materials as catalysts for photochemical splitting of water. *Chem. Mater.* 20(1): 35–54.

32. Dhere, N.G. and R.S. Sennur. 2009. Use of solar energy to produce hydrogen. In *Hydrogen fuel: Production, transport, and storage*, 227–282, ed. R.B. Gupta, CRC Press: Boca Raton, FL.

33. Compton, O.C. and F.E. Osterloh. 2009. Niobate nanosheets as catalysts for photochemical water splitting into hydrogen and hydrogen peroxide. *J. Phys. Chem. C* 113: 479–485.

34. Yao, W., C. Huang, N. Muradov, and A. T-Raissi. 2009. Cr_2O_3 loading effects on the photocatalytic activity of palladium/cadmium sulfide photocatalyst. Paper presented at the *Proceedings of Energy Sustainability 2009*, San Francisco, CA, July 19–23.

35. Huang, C. and A. T-Raissi. 2004. A new solar thermochemical water splitting cycle from hydrogen production. Paper presented at *The 15th World Hydrogen Energy Conference*, Yokohama, Japan, June 27–July 2.

36. T-Raissi, A., N. Muradov, C. Huang, and O. Adebiyi, 2007. Hydrogen from solar via light-assisted high-temperature water splitting cycles, *J. Sol. Energy Eng.* 129: 184–89.

37. Huang, C., A. T-Raissi, and N. Muradov. 2007. A Thermochemical Cycle for Production of Hydrogen and/or Oxygen via Water Splitting Processes. (PCT Patent, Pub. No.: WO2007002614 (A2), January 4, 2007).

38. Huang, C., A. T-Raissi, L. Mao, S. Fenton, and N. Muradov. 2008. A new family of solar metal sulfate–ammonia based thermochemical water splitting cycles for H_2 production. Paper presented at the *Proceedings of Solar 2008*, San Diego, CA, May 3–8.

39. Mizuta, S., W. Kondo, and K. Fuji, 1991. Hydrogen production from hydrogen sulfide by the Fe-Cl hybrid process. *Ind. Eng. Chem. Res.* 30: 1601–1608.

40. Zaman, Z. and A. Chakma. 1995. Production of hydrogen and sulfur from hydrogen sulfide. *Fuel Process* 41: 159–198.

41. Kohl, A. and R. Nielsen. *Gas purification.* 5th ed. Gulf Publishing Co.: Houston, TX, 670–730.

42. Besancon, B.M., V. Hasanov, R. Imbault-Lastapits, R. Benesch, M. Barrio, and M.J. Molnvik. 2009. Hydrogen quality from decarbonized fossil fuels to fuel cells. *Int. J. Hydrogen Energy* 34: 2350–60.

43. Damle, A. Hydrogen separation and purification. *Hydrogen fuel: Production, transport, and storage*, 283–324, ed. R.B. Gupta, CRC Press: Boca Raton, FL.

44. Bonjour, J., J.B. Chalfen, and F. Meunier. 2002. Temperature swing adsorption process with indirect cooling and heating. *Ind. Eng. Chem. Res.* 41: 5802–11.

45. Humphrey, J.L. and G.E. Keller II. 1997. *Separation process technology*, McGraw-Hill: New York, 153–223.

46. Huang, C. and A. T-Raissi. 2007. Analyses of one-step hydrogen production from methane and landfill gas. *J. Power Sources* 173: 950–58.

47. Park, E.D., D. Lee, and H.C. Lee. 2009. Recent progress in selective CO removal in a H_2-rich stream. *Catal. Today* 139: 280–90.

48. Trimm, D.L. 2005. Review – Minimization of carbon monoxide in a hydrogen stream for fuel cell application. *Appl. Catal., A* 296: 1–11.

49. Huang, C., R. Jiang, M. Elbaccouch, N. Muradov, and J.M. Fenton. 2007. On-board removal of CO and other impurities in hydrogen for PEM fuel cell application. *J. Power Sources* 162: 563–71.

50. Harrison, D.P. 2008. Sorption-enhanced hydrogen production: A review. *Ind. Eng. Chem. Res.* 47: 6486–501.

51. Barelli, L., G. Bidini, F. Gallorini, and S. Servili. 2008. Hydrogen production through sorption-enhanced steam methane reforming and membrane technology: A review. *Energy* 33: 554–70.

52. Hufton, J.R., S. Mayorga, and S. Sircar. 1999. Sorption-enhanced reaction process for hydrogen production. *Separations AIChE Journal* 45(2): 248–56.

53. Bretado, M.E., M.D. Delgado, D. Vigil, V.H.C. Martinez, and A.L. Ortiz. 2008. Thermodynamic analysis for the production of hydrogen through sorption enhanced water gas shift (SEWGS). *Int. J. Chem. Reactor Eng.* 6: A51.

54. Pozio, Z., R.F. Silva, M. De Francesco, and L. Giorgi. 2003. Nafion degradation in PEFCs from end plate iron contamination. *Electrochim. Acta* 48(11): 1543–49.

55. Yoon, W., X. Huang, P. Fazzino, K. Reifsnider, M. Akkaoui. 2008. Evaluation of coated metallic bipolar plates for polymer electrolyte membrane fuel cells. *J. Power Sources* 179(1): 265–273.

56. U.S. Patents 7,416,580; 7,306,658; 7,270,692; 7,138,008; 7,101,419; 7,090,712; 7,002,158; 7,008,465; 6,997,977; 6,984,465; 6,974,490; 6,951,697; 6,875,249; 6,797,027; 6,783,882; 6,783,881; 6,780,534; 6,673,136; 6,645,271; 6,638,339; 6,4321,77; 6,368,368; 6,123,751.

57. Schroeter, D. 2008. Method and arrangement for purifying gases fed to a fuel cell by removing operational unfavorable constituents. U.S. Patent 7,449,046 B2.

58. Kennedy, D.M., D.R. Cahela, W.H. Zhu, K.C. Westrom, R.M. Nelms, and B.J. Tatarchuk. 2007. Fuel cell cathode air filters: Methodologies for design and optimization. *J. Power Sources* 168(2): 391–399.

59. St Pierre, J., D.P. Wilkinson, S. Knights, and M.L. Bos. 2000. Relationships between water management, contamination and lifetime degradation in PEFC. *J. New Mater. Electrochem. Syst.* 3: 99–106.

60. Huang, X., X. Wang, J. Preston, L. Bonville, R. Kunz, M. Perry, and D. Condit. 2008. Effect of water management schemes on the membrane durability in PEMFCs. *ECS Transactions* 16(2): 1697–1703.

61. Fischer, R. and M.L. Bos. 2007. Water filter for an electrochemical fuel cell system. U.S. Patent 7250230 B2.

62. Horinouchi, H., K. Nishizaki, S. Shigiaki, F. Azakami, and T. Deguchi. 1999. Water treatment apparatus for a fuel cell system. U.S. Patent 5,980,716.

63. Bonville, L.J., A.P. Grasso, and R.A. Sederquist. 1989. Removal of ammonia and carbon dioxide from fuel cell stack water system by steam stripping. U.S. Patent 4,816,040.

64. Oko, U.M. and N. Childs. 2002. Methods and kits for decontaminating fuel cells. U.S. Patent 6,358,639 B2.

65. Reiser, C.A. and R.J. Balliet. 2005. Fuel cell performance recovery by cyclic oxidant starvation. U.S. Patent 6,841,278 B2.

66. Patterson, Jr, T.W., M.L. Perry, T. Skiba, P. Yu, T.D. Jarvi, J.A. Leistra, H. Chizawa, and T. Aoki. 2008. Decontamination procedure for a fuel cell power plant. U.S. Patent, 7,442,453 B1.

67. Cheng, X., Z. Shi, N. Glass, L. Zhang, J. Zhang, D. Song, Z.-S. Liu, H. Wang, and J. Shen. 2007. A review of PEM hydrogen fuel cell contamination: Impacts, mechanisms, and mitigation. *J. Power Sources* 165: 739–756.

68. Dicks, A.L. 1996. Hydrogen generation from natural gas for the fuel cell systems of tomorrow. *J. Power Sources* 61: 113–124.

69. Hohlein, B., M. Boe, J. Bogild-Hansen, P. Brockerhoff, G. Colsman, B. Emonts, R. Menzer, and E. Riedel. 1996. Hydrogen from methanol for fuel cells in mobile systems: Development of a compact reformer. *J. Power Sources* 61: 143–147.

70. Schmidt, V.M., P. Brockerhoff, B. Hohlein, R. Menzer, and U. Stimming. 1994. Utilization of methanol for polymer electrolyte fuel cells in mobile systems. *J. Power Sources* 49: 299–313.

71. Parsons, R. and T. VanderNoot. 1988. The oxidation of small organic molecules: A survey of recent fuel cell related research. *J. Electroanal. Chem.* 257: 9–45.

72. Wee, J.H., K.Y. Lee, and S.H. Kim. 2006. Sodium borohydride as the hydrogen supplier for proton exchange membrane fuel cell systems. *Fuel Process. Technol.* 87: 811–819.

73. Murthy, M., M. Esayian, W.-K. Lee, and J.W. Van Zee. 2003. The effect of temperature and pressure on the performance of a PEMFC exposed to transient CO concentrations. *J. Electrochem. Soc.* 150:A29–A34.

74. Giorgi, L., A. Posio, C. Bracchini, R. Giorgi, and S. Turtu. 2001. H_2 and H_2/CO oxidation mechanism on Pt/C, Ru/C and Pt–Ru/C electrocatalysts. *J. Appl. Electrochem.* 31: 325–334.

75. Mohtadi, R., W.-K. Lee, S. Cowan, M. Murthy, J.W. Van Zee. 2003. Effects of hydrogen sulfide on the performance of a PEMFC. *Electrochem. Solid State Lett.* 6(12): A272–A274.

76. Vogel, W., J. Lundquist, P. Ross, and P. Stonehart. 1975. Reaction pathways and poisons II: The rate controlling step for electrochemical oxidation of hydrogen on Pt in acid and poisoning of the reaction by CO. *Electrochim. Acta* 20: 79–93.

77. Ahluwalia, R.K. and X. Wang. 2008. Effect of CO and CO_2 impurities on performance of direct hydrogen polymer electrolyte fuel cells. *J. Power Sources* 180: 122–131.

78. Zhang, J.L., Z. Xie, J.J. Zhang, Y.H. Tang, C.J. Song, T. Navessin, Z.Q. Shi, D.T. Song, H.J. Wang, D.P. Wilkinson, Z.S. Liu, and S. Holdcroft. 2006. High temperature PEM fuel cells. *J. Power Sources* 160(2): 872–891.

79. Gottesfeld, S. and J. Pafford. 1988. A new approach to the problem of carbon monoxide poisoning in fuel cells operating at low temperatures. *J. Electrochem. Soc.* 135: 2651–2652.

80. Murthy, M., M. Esayian, A. Hobson, S. MacKenzie, W.-K. Lee, and J.W. Van Zee. 2001. Performance of a polymer electrolyte membrane fuel cell exposed to transient CO concentrations. *J. Electrochem. Soc.* 148(10): A1141–A1147.

81. Chen, C.H., C.C. Hung, H.H. Lin, and Y.Y. Yan. 2008. Improvement of CO tolerance of proton exchange membrane fuel cell by an air bleeding technique. *J. Fuel Cell Sci. Tech.* 5: 014501–1.

82. Bellows, R.J., E. Marucchi-Soos, and R.P. Reynolds. 1998. The mechanism of CO mitigation in proton exchange membrane fuel cells using dilute H_2O_2 in the anode humidifier. *Electrochem. Solid State Lett.* 1: 69–70.

83. Wan, C.H., Q.H. Zhuang, C.H. Lin, M.T. Lin, and C. Shih. 2006. Novel composite anode with CO "filter" layers for PEFC. *J. Power Sources* 162: 41–50.

84. Thomason, A.H., T.R. Lalk, and A.J. Appleby. 2004. Effect of current pulsing and "self-oxidation" on the CO tolerance of a PEM fuel cell. *J. Power Sources* 135: 204–211.

85. Carrette, L.P.L., K.A. Friedrich, M. Huber, and U. Stimming. 2001. Improvement of CO tolerance of proton exchange membrane (PEM) fuel cells by a pulsing technique. *Phys. Chem. Chem. Phys.* 3: 320–324.
86. Divisek, J., H.F. Oetjen, V. Peinecke, V.M. Schmidt, and U. Stimming. 1998. Components for PEM fuel cell systems using hydrogen and CO containing fuels. *Electrochim. Acta* 43: 3811–3815.
87. Schmidt, V.M., H.F. Oetjen, and J. Divisek. 1997. Performance improvement of a PEMFC using fuels with CO by addition of oxygen-evolving compounds. *J. Electrochem. Soc.* 144(9): L237–L238.
88. Inaba, M., M. Sugishita, J. Wada, K. Matsuzawa, H. Yamada, and A. Tasaka. 2008. Impacts of air bleeding on membrane degradation in polymer electrolyte fuel cells. *J. Power Sources* 178: 699–708.
89. Ianniello, R., V.M. Schmidt, U. Stimming, J. Stumper, and A. Wallau. 1994. CO adsorption and oxidation on Pt and Pt-Ru alloys: Dependence on substrate composition. *Electrochim. Acta* 39: 1863–1869.
90. Oetjen, H.F., V.M. Schmidt, U. Stimming, and F. Trila. 1996. Performance data of a proton exchange membrane fuel cell using H_2/CO as fuel gas. *J. Electrochem. Soc.* 143: 3838–3842.
91. Wakisaka, M., S. Mitsui, Y. Hirose, K. Kawashima, H. Uchida, and M. Watanabe. 2006. Electronic structures of Pt-Co and Pt-Ru alloys for CO-tolerant anode catalysts in polymer electrolyte fuel cells studied by EC-XPS. *J. Phys. Chem. B* 110: 23489–23496.
92. Haug, A.T., R.E. White, J.W. Weidner, and W. Huang. 2002. Development of a novel CO tolerant proton exchange membrane fuel cell anode. *J. Electrochem. Soc.* 149: A862–A867.
93. Haug, A.T., R.E. White, J.W. Weidner, W. Huang, S. Shi, N. Rana, S. Grunow, T.C. Stoner, and A.E. Kaloyeros. 2002. Using sputter deposition to increase CO tolerance in a proton-exchange membrane fuel cell. *J. Electrochem. Soc.* 149: A868–A872.
94. Gubler, L., G.G. Scherer, and A. Wokaun. 2001. Effects of cell and electrode design on the CO tolerance of polymer electrolyte fuel cells. *Phys. Chem. Chem. Phys.* 3: 325–329.
95. Luna, A.M.C., G.A. Camara, V.A. Paganin, E.A. Ticianelli, and E.R. Gonzalez. 2000. Effect of thermal treatment on the performance of CO-tolerant anodes for polymer electrolyte fuel cells. *Electrochem. Commun.* 2: 222–225.
96. Watanabe, M., M. Uchida, and S. Motoo. 1987. Preparation of highly dispersed Pt^+ Ru alloy clusters and the activity for the electrooxidation of methanol. *J. Electroanal. Chem.* 229: 395–406.
97. Pozio, A., R.F. Silva, M. DeFrancesco, F. Cardellini, and L. Giorgi. 2002. A novel route to prepare stable Pt-Ru/C electrocatalysts for polymer electrolyte fuel cell. *Electrochim. Acta* 48: 255–262.
98. Adcock, P.A., S.V. Pacheco, K.M. Norman, and F.A. Uribe. 2005. Transition metal oxides as reconfigured fuel cell anode catalysts for improved CO tolerance: Polarization data. *J. Electrochem. Soc.* 152: A459–A466.
99. Götz, M. and H. Wendt. 1998. Binary and ternary anode catalyst formulations including the elements W, Sn and Mo for PEMFCs operated on methanol or reformate gas. *Electrochim. Acta* 43: 3637–3644.

100. Papageorgopoulos, D.C., M. Keijzer, J.B.J. Veldhuis, and F.A. de Bruijn. 2002. CO tolerance of Pd-rich platinum palladium carbon-supported electrocatalysts. *J. Electrochem. Soc.* 149: A1400–A1404.

101. García, G., J.A. Silva-Chong, O. Guillén-Villafuerte, J.L. Rodríguez, E.R. González, and E. Pastor. 2006. CO tolerant catalysts for PEM fuel cells: Spectroelectrochemical studies. *Catal. Today* 116: 415–421.

102. Mukerjee, S., R.C. Urian, S.J. Lee, E.A. Ticianelli, and J. McBreen. 2004. Electrocatalysis of CO tolerance by carbon-supported PtMo electrocatalysts in PEMFCs. *J. Electrochem. Soc.* 151: A1094–A1103.

103. Santiago, E.I., M.S. Batista, E.M. Assaf, and E.A. Ticianelli. 2004. Mechanism of CO tolerance on molybdenum-based electrocatalysts for PEMFC. *J. Electrochem. Soc.* 151: A944–A949.

104. Liang, Y., H. Zhang, H. Zhong, X. Zhu, Z. Tian, D. Xu, and B. Yi. 2006. Preparation and characterization of carbon-supported PtRuIr catalyst with excellent CO-tolerant performance for proton-exchange membrane fuel cells. *J. Catal.* 238: 468–476.

105. Papageorgopoulos, D.C., M. Keijzer, and F.A. de Bruijn. 2002. The inclusion of Mo, Nb and Ta in Pt and PtRu carbon supported electrocatalysts in the quest for improved CO tolerant PEMFC anodes. *Electrochim. Acta* 48: 197204.

106. Gustavo, L., S. Pereira, F.R. dos Santos, M.E. Pereira, V.A. Paganin, and E.A. Ticianelli. 2006. CO tolerance effects of tungsten-based PEMFC anodes. *Electrochim. Acta* 51: 4061–4066.

107. Yu, P., M. Pemberton, and P. Plasse. 2005. PtCo/C cathode catalyst for improved durability in PEMFCs. *J. Power Sources* 144: 11–20.

108. Hou, Z., B. Yi, H. Yu, Z. Lin, and H. Zhang. 2003. CO tolerance electrocatalyst of PtRu-H$_x$MeO$_3$/C (Me = W, Mo) made by composite support method. *J. Power Sources* 123: 116–125.

109. Yano, H., C. Ono, H. Shiroishi, and T. Okada. 2005. New CO tolerant electrocatalysts exceeding Pt–Ru for the anode of fuel cells. *Chem. Commun.* 1212–1214.

110. Garcia, A.C., V.A. Paganin, and E.A. Ticianelli. 2008. CO tolerance of PdPt/C and PdPtRu/C anodes for PEMFC. *Electrochim. Acta* 53: 4309–4315.

111. Sapountzi, F.M., M.N. Tsampas, and C.G. Vayenas. 2007. Methanol reformate treatment in a PEM fuel cell-reactor. *Catal. Today* 127: 295–303.

112. Ferreira, P.J., G.J. la O', Y. Shao-Horn, D. Morgan, R. Makharia, S. Kocha, and H.A. Gasteiger. 2005. Instability of Pt/C electrocatalysts in proton exchange membrane fuel cells. *J. Electrochem. Soc.* 152: A2256–A2271.

113. Christian, J.B., S.P.E. Smith, M.S. Whittingham, and H.D. Abruna. 2007. Tungsten based electrocatalyst for fuel cell applications. *Electrochem. Commun.* 9: 2128–2132.

114. Antolini, E. 2004. Review in applied electrochemistry. Number 54, Recent Developments in Polymer Electrolyte Fuel Cell Electrodes. *J. Appl. Electrochem.* 34: 563–576.

115. Yu, H., Z. Hou, B. Yi, and Z. Lin. 2002. Composite anode for CO tolerance proton exchange membrane fuel cells. *J. Power Sources* 105: 52–57.

116. Janssen, G.J.M., M.P. de Heer, and D.C. Papageorgopoulos. 2004. Bilayer anodes for improved reformate tolerance of PEM fuel cells. *Fuel Cells* 4: 169–174.

117. Shi, W., M. Hou, Z. Shao, J. Hu, Z. Hou, P. Ming, and B. Yi. 2007. A novel proton exchange membrane fuel cell anode for enhancing CO tolerance. *J. Power Sources* 174: 164–169.

118. Uribe, F.A., J.A. Valerio, F.H. Garzon, and T.A. Zawodzinski. 2004. PEMFC reconfigured anodes for enhancing CO tolerance with air bleed. *Electrochem. Solid State Lett.* 7: A376–A379.
119. Denis, M.C., G. Lalande, D. Guay, J.P. Dodelet, and R. Schulz. 1999. High energy ball-milled Pt and Pt–Ru catalysts for polymer electrolyte fuel cells and their tolerance to CO. *J. Appl. Electrochem.* 29: 951–960.
120. Venkataraman, R., H.R. Kunz, and J.M. Fenton. 2004. CO-tolerant, sulfided platinum catalysts for PEMFCs. *J. Electrochem. Soc.* 151: A710–A715.
121. Chinarro, E., B. Moreno, G.C. Mather, and J.R. Jurado. 2004. Combustion synthesis as a preparation method for PEMFC electrocatalysts. *J. New Mater. Electrochem. Syst.* 7: 109–115.
122. Zhang, J., T. Thampan, and R. Datta. 2002. Influence of anode flow rate and cathode oxygen pressure on CO poisoning of proton exchange membrane fuel cells. *J. Electrochem. Soc.* 149: A765–A772.
123. Springer, T.E., T. Rockward, T.A. Zawodzinski, and S. Gottesfeld. 2001. Model for polymer electrolyte fuel cell operation on reformate feed: Effects of CO, H_2 dilution, and high fuel utilization. *J. Electrochem. Soc.* 148: A11–A23.
124. Kawatsu, S. 1999. Fuel-cells generator system and method of generating electricity from fuel cells. U.S. Patent 5,925,476.
125. Li, Q.F., R. He, J.-A. Gao, J.O. Jensen, and N.J. Bjerrum. 2003. The CO poisoning effect in PEMFCs operational at temperatures up to 200°C. *J. Electrochem. Soc.* 150: A1599–A1605.
126. Gasteiger, H.A., J.E. Panels, and S.G. Yan. 2004. Dependence of PEM fuel cell performance on catalyst loading. *J. Power Sources* 127: 162–171.
127. Lakshmanan, B., W. Huang, and W.J. Weidner. 2002. Electrochemical filtering of CO from fuel-cell reformate. *Electrochem. Solid State Lett.* 5: A267–A270.
128. Zhang, J. and R. Datta. 2005. Electrochemical preferential oxidation of CO in reformate. *J. Electrochem. Soc.* 152: A1180–A1187.
129. Darling, R.M. and J.P. Meyers. 2005. Mathematical model of platinum movement in PEM fuel cells. *J. Electrochem. Soc.* 152: A242–A247.
130. Wilkinson, D.P., C.Y. Chow, D.E. Allan, E.P. Johannes, J.A. Roberts, J. St-Pierre, C.J. Langley, and J.K. Chan. 2000. Method and apparatus for operating an electrochemical fuel cell with periodic fuel starvation at the anode. U.S. Pat. 6,096,448.
131. Knights, S.D., K.M. Colbow, J. St-Pierre, and D.P. Wilkinson. 2004. Aging mechanisms and lifetime of PEFC and DMFC. *J. Power Sources* 127: 127–134.
132. Shi, W., B. Yi, M. Hou, and Z. Shao. 2007. The effect of H_2S and CO mixtures on PEMFC performance. *Int. J. Hydrogen Energy* 32: 4412–4417.
133. Mohtadi, R., W.-K. Lee, and J.W. Van Zee. 2005. The effect of temperature on the adsorption rate of hydrogen sulfide on Pt anodes in a PEMFC. *Appl. Catal., B* 56: 37–42.
134. Shi, W., B. Yi, M. Hou, F. Jing, H. Yu, and P. Ming. 2007. The influence of hydrogen sulfide on proton exchange membrane fuel cell anodes. *J. Power Sources* 164: 272–277.
135. Shi, W., B. Yi, M. Hou, F. Jing, and P. Ming. 2007. Hydrogen sulfide poisoning and recovery of PEMFC Pt-anodes. *J. Power Sources* 165: 814–818.
136. Schmidt, T.J., U.A. Paulus, H.A. Gasteiger, and R.J. Behm. 2001. The oxygen reduction reaction on a Pt/carbon fuel cell catalyst in the presence of chloride anions. *J. Electroanal. Chem.* 508: 41–47.

137. Matsuoka, K., S. Sakamoto, K. Nakato, A. Hamada, and Y. Itoh. 2008. Degradation of polymer electrolyte fuel cells under the existence of anion species. *J. Power Sources* 179: 560–565.

138. Zhao, X., G. Sun, L. Jiang, W. Chen, S. Tang, B. Zhou, and Q. Xin. 2005. Effects of chloride anion as a potential fuel impurity on DMFC performance. *Electrochem. Solid- State Lett.* 8(3): A149–A151.

139. Maxted, E.B. 1951. *Adv. Catal.* 3: 129.

140. Nagahara, Y., S. Sugawara, and K. Shinohara. 2008. The impact of air contaminants on PEMFC performance and durability. *J. Power Sources* 182: 422–428.

141. Mohtadi, R., W.-K. Lee, and J.W. Van Zee. 2004. Assessing durability of cathodes exposed to common air impurities. *J. Power Sources* 138: 216–225.

142. Loucka, T.J. 1971. Adsorption and oxidation of sulphur and of sulphur dioxide at the platinum electrode. *Electroanal. Chem.* 31: 319–332.

143. Spotnitz, R.M., J.A. Colucci, and S.H. Langer. 1983. The activated electro-oxidation of sulphur dioxide on smooth platinum. *Electrochim. Acta* 28: 1053–1062.

144. Gould, B.D., O.A. Baturina, and K.E. Swider-Lyons. 2009. Deactivation of Pt/VC proton exchange membrane fuel cell cathodes by SO2, H2S and COS. *J. Power Sources* 188: 89–95.

145. Jing, F., M. Hou, W. Shi, J. Fu, H. Yu, P. Ming, and B. Yi. 2007. The effect of ambient contamination on PEMFC performance. *J. Power Sources* 166: 172–176.

146. Nagahara, Y., S. Sugawara, and K. Shinohara. 2008. The impact of air contaminants on PEMFC performance and durability. *J. Power Sources* 182: 422–428.

147. Yang, D., J. Ma, L. Xu, M. Wu, and H. Wang. 2006. *Electrochim. Acta* 51: 4039–4044.

148. Jing, F., M. Hou, W. Shi, J. Fu, H. Yu, P. Ming, and B. Yi. 2007. The effect of ambient contamination on PEMFC performance. *J. Power Sources* 166: 172–176.

149. Halseid, R., P.J.S. Vie, and R. Tunold. 2006. Effect of ammonia on the performance of polymer electrolyte membrane fuel cells. *J. Power Sources* 154: 343–350.

150. Soto, H.J., W.K. Lee, J.W. VanZee, and M. Murthy. 2003. Effect of transient ammonia concentrations on PEMFC performance. *Electrochem. Solid State Lett.* 6: A133–A135.

151. Halseid, R., P.J.S. Vie, and R. Tunoid. 2004. Influence of ammonium on conductivity and water content of Nafion 117 membranes, *J. Electrochem. Soc.* 151: A381–A388.

152. Halseid, R., T. Bystron, and R. Tunoid. 2006. Oxygen reduction on platinum in aqueous sulphuric acid in the presence of ammonium. *Electrochim. Acta* 51: 2737–2742.

153. Uribe, F.A., S. Gottesfeld, and T.A. Zawodzinski. 2002. Effect of ammonia as potential fuel impurity on proton exchange membrane fuel cell performance. *J. Electrochem. Soc.* 149(3): A293–A296.

Index

A

Accelerated stress test (AST), 30
AC impedance, *see* Alternating current impedance
Activation polarization, 23, 214
AFC, *see* Alkaline fuel cell
Air/O_2 bleed, 133, 143
Alcohol fuels, 211
Alkaline fuel cell (AFC), 12, 135
Allis-Chalmers Manufacturing Company, 8
Alternating current (AC) impedance, 28
  calculations, 321
  measurement, 97, 104, 327
  spectroscopy, 317
  testing, 28
Ambient Air Quality Standard, 98
Ammonia (NH_3)
  adsorption, 66
  anode contamination, 64–66, 275–276
  cathode contamination, 72–73
  cell voltage degradation, 167
  contamination, 167–169, 275–276
  deleterious effects on PEMFC, 140
  diffusion of through GDL, 169
  effect of on HOR, 390
  effect of on Pt, 276
  fuel cell performance, 40
  membrane contamination, 167–169
  mitigation, 390–391
  oxidized, 294
  reformate gas, 381
Anode
  adsorption potential, 60
  current densities, critical, 242
  degradation, 38–39
  kinetics, 215
  overpotentials, 322, 330
  polarization, 26, 139, 257
Anode contaminants, 41–42, 54–66
  CO, 54–56, 56–62
  CO_2, 54–56, 62
  CO_x, 54–62

H_2S, 62–64
NH_3, 64–66
air/O_2 bleed, 133, 143
Anode contamination, 86, 115–150,
    see also Anode contamination modeling
anode polarization, 139
autothermal reforming, 119
back-donation, 125
bifunctional mechanism, 132, 142
carbon (coke) formation, 119
coal gasification, 118, 120
CO oxidation, 119
CO poisoning model, 123
CO_x, 123, 124–136
  addition of ruthenium as second metal, 132–133
  combined impacts of CO and CO_2, 135–136
  deleterious effect of CO, 127–134
  impact of anode CO on cathode, 133–134
  impact of CO_2, 134–135
  preferential adsorption of CO on platinum, 125–127
  reduction of CO_2 to CO, 127
  severe poisoning impact of trace amounts of CO, 127–129
  temperature effect, 130–131
crossover process, 133
H_2S, 124, 136–139
  adsorption behavior of H_2S, 136–137
  cyclic voltammetry study, 138–139
  impact of H_2S, 137–138
  open circuit voltage operation, 139
hydrogen economy, 116, 121
hydrogen production processes, 117–121
  autothermal reforming, 119
  carbon (coke) formation, 119
  CO oxidation, 119
  electrolysis, 120–121
  partial oxidation, 119

401